OXFORD STATISTICAL SCIENCE SERIES

SERIES EDITORS

OXFORD STATISTICAL SCIENCE SERIES

1. A. C. Atkinson: *Plots, transformations, and regression*
2. M. Stone: *Coordinate-free multivariable statistics*
3. W. J. Krzanowski: *Principles of multivariate analysis: a user's perspective*
4. M. Aitkin, D. Anderson, B. Francis, and J. Hinde: *Statistical modelling in GLIM*
5. Peter J. Diggle: *Time series: a biostatistical introduction*
6. Howell Tong: *Non-linear time series: a dynamical system approach*
7. V. P. Godambe: *Estimating functions*
8. A. C. Atkinson and A. N. Donev: *Optimum experimental designs*
9. U. N. Bhat and I. V. Basawa: *Queuing and related models*
10. J. K. Lindsey: *Models for repeated measurements*
11. N. T. Longford: *Random coefficient models*
12. P. J. Brown: *Measurement, regression, and calibration*
13. Peter J. Diggle, Kung-Yee Liang, and Scott L. Zeger: *Analysis of longitudinal data*
14. J. I. Ansell and M. J. Phillips: *Practical methods for reliability data analysis*
15. J. K. Lindsey: *Modelling frequency and count data*
16. J. L. Jensen: *Saddlepoint approximations*
17. Steffen L. Lauritzen: *Graphical models*
18. A. W. Bowman and A. Azzalini: *Applied smoothing methods for data analysis*
19. J. K. Lindsey: *Models for repeated measurements* second edition
20. Michael Evans and Tim Swartz: *Approximating integrals via Monte Carlo and deterministic methods*
21. D. F. Andrews and J. E. Stafford: *Symbolic computation for statistical inference*
22. T. A. Severini: *Likelihood methods in statistics*
23. W. J. Krzanowski: *Principles of multivariate analysis: a user's persepective* updated edition
24. J. Durbin and S. J. Koopman: *Time series analysis by state space models*
25. Peter J. Diggle, Patrick Heagerty, Kung-Yee Liang, and Scott L. Zeger: *Analysis of longitudinal data*second edition
26. J. K. Lindsey: *Nonlinear models in medical statistics*
27. Peter J. Green, Nils L. Hjort & Sylvia Richardson: *Highly Structured Stochastic Systems (forthcoming—2003)*
28. Margaret S. Pepe: *Statistical Evaluation of Medical Tests (forthcoming—2003)*
29. Christopher G. Small, Jinfang Wang: *Numerical methods for nonlinear estimating equations*

Numerical Methods for Nonlinear Estimating Equations

CHRISTOPHER G. SMALL
University of Waterloo

JINFANG WANG
Obihiro University of Agriculture and Veterinary Medicine

CLARENDON PRESS • OXFORD
2003

OXFORD
UNIVERSITY PRESS
Great Clarendon Street, Oxford OX2 6DP
Oxford University Press is a department of the University of Oxford.
It furthers the University's objective of excellence in research, scholarship,
and education by publishing worldwide in
Oxford New York
Auckland Bangkok Buenos Aires Cape Town Chennai
Dar es Salaam Delhi Hong Kong Istanbul Karachi Kolkata
Kuala Lumpur Madrid Melbourne Mexico City Mumbai Nairobi
São Paulo Shanghai Taipei Tokyo Toronto
Oxford is a registered trade mark of Oxford University Press
in the UK and in certain other countries
Published in the United States
by Oxford University Press Inc., New York

First published 2003

A catalogue record for this title is available from the British Library
Library of Congress Cataloging in Publication Data
(Data available)
ISBN 0 19 850688 0
10 9 8 7 6 5 4 3 2 1
Typeset by Newgen Imaging Systems (P) Ltd., Chennai, India
Printed in Great Britain
on acid-free paper by
Biddles Ltd., Guildford & King's Lynn

Preface

This book has grown from the authors' investigations into the problem of finding consistent and efficient solutions to estimating equations which admit multiple solutions. While there have been studies of multiple roots in likelihood estimation, most notably by Barnett (1966), we felt that there was a need to incorporate more recent research by studying the problem in other contexts such as semiparametrics and the construction of artificial likelihoods.

While the book began as the study of multiple roots in estimating functions, it soon expanded to become a study of nonlinearity in estimating equations and iterative methods for the solution of these equations. Nonlinearity appears in many different ways within a model, and in differing degrees of severity as an obstacle to statistical analysis. Some simple forms of nonlinearity can be removed by a reparametrisation of the model. This type of nonlinearity is usually no major obstacle to the construction of point estimators and the study of their properties. More entrenched forms of nonlinearity cannot be removed by reparametrising the model, and may force the researcher to use more intensive numerical methods to construct estimators. The use of root search algorithms or one-step estimators is standard for these sorts of models. Perhaps the most severe type of nonlinearity in an estimating equation is that which affects the monotonicity of the estimating function, making the function redescending, with the possibility of multiple roots. When estimates are constructed by maximising an objective function, such as a likelihood or quasi-likelihood, an analogous problem occurs when the objective function is not concave or not strongly concave. Hill climbing algorithms which start at points of nonconcavity often have very poor convergence properties.

The approach of the book is that the properties of estimators should not be separated from the properties of the estimating functions or objective functions which are used to construct them. In particular, by ignoring nonlinear characteristics of the estimating function, we can overlook important information about model interpretations and model fit. Part of the shift in viewpoint will be to see a point estimator as a dynamic object rather than a static one. That is, we shall adopt the philosophy that a point estimator is the result of a process such as an iterated function system, rather than simply a function on the data taking values in the parameter space. Looked at in this way, estimating equations lead to iterative methods within the parameter space and estimates become fixed points of these iterations. Of course, we must avoid excessive pedantry in the reinterpretation of the classical problem of point estimation. Nevertheless, there are certain advantages to this change of perspective within the theory of estimation. To separate estimators from the iterative methods which produce them can lead to difficulties. We may not know, for example, how delicately the final step of the iteration is dependent upon the starting point. Nor can we judge, on the basis of a reported statistic, whether the researcher has terminated the algorithm with sufficient care. It is not sufficient to assume we are close to a fixed point of an iteration simply because two successive values in an iteration are close.

Acknowledgements

The expectations under which this book project began were vague at best. Along the way, the project's goals were clarified through the advice of colleagues. We are grateful to numerous people for conversations that they may well have forgotten. Thanks go to (in alphabetical order) some anonymous referees for the journal *Statistical Science*, Shinto Eguchi, Omar Foda, V. P. Godambe, Chris Heyde, Don McLeish, Ilya Molchanov, Masaaki Taguri, and the members of the Waterloo Nonlinear Estimating Equations Seminar. We are also grateful to many colleagues, who when told that we were writing a book on estimating equations, did not reject the project out of hand.

Some of the figures in this book were produced by Zejiang Yang during the period when he was a Ph.D. student under one author (CGS). The material discussed in Section 5.11 is work he did for his Ph.D. thesis.

The book uses the MAPLE symbolic programming language, MATLAB and Mathematica. MAPLE is a product of Waterloo Maple, Inc., MATLAB is a product of MathWorks, Inc. and Mathematica is a product of Wolfram Research Inc.

Part of the work in this book was done during the second author's sabbatical leave to the University of Waterloo under the support of the Japanese Ministry of Education, Science, Sports and Culture. Both authors thank the Ministry for partial support under Grand-in-Aid 12780181 during the project of writing this book.

Contents

1
Introduction

1.1 Background to the problem

It is only in simple models, such as the linear model for regression or the exponential family model for parametric inference, that ideal point estimators can be written explicitly in closed form as a function of the sample. In many other cases, the construction of a point estimator is more computationally intensive, and involves an iteration to search for a solution to one or more estimating equations of the form $g(\theta) = 0$, where $g(\theta)$ is a function of the data and the parameter. The construction of maximum likelihood estimators (MLEs), where

$$g(\theta) = \frac{\partial}{\partial \theta} \log L(\theta)$$

and L is the likelihood function, is the best-known example of this. In many models, the likelihood equations cannot be solved explicitly, and the investigator must resort to some numerical method to construct a point estimate.

A basic numerical problem arises when the score function $g(\theta)$ or the likelihood $L(\theta)$ cannot be represented in simple form. This can occur when the distribution of the data has no closed form representation of the density function. For example, most symmetric stable distributions have densities which can only be represented by integrals or infinite series. This makes the estimation of a location, scale or shape parameter of a symmetric stable distribution difficult unless the distribution is a special case such as the normal or Cauchy. To overcome this obstacle, we can simply tackle the difficulty head on, by attempting to evaluate the infinite series or numerical integral. Alternatively, we can attempt to perform a likelihood analysis or to construct estimating equations in a transformed setting such as that provided by a Fourier transform.

Even when the likelihood has a simple form, there may be other obstacles to the construction of a point estimate when maximising the likelihood or solving the likelihood equations. After the initial enthusiasm for the method of maximum likelihood proposed by Fisher (1922, 1925), questions arose as to the existence and uniqueness of a root of the likelihood equation and whether that root corresponded to a maximum of the likelihood function.

For example, Huzurbazar (1948) noted that proofs of the consistency and asymptotic efficiency of the MLE were more properly proofs of the existence of a consistent and asymptotically efficient root of the likelihood equations. Could such a root also be a local minimum of the likelihood function? In regular models, the answer turned out to be no. Huzurbazar (1948) showed that a consistent root of the likelihood equation is asymptotically unique and corresponds to a local maximum of the likelihood function. More precisely, if $\hat{\theta}_1$ and $\hat{\theta}_2$ are two consistent roots of the likelihood equation for a sample of size n, then

$$P(\hat{\theta}_1 = \hat{\theta}_2) \to 1$$

as $n \to \infty$, provided that the standard regularity due to H. Cramér is satisfied. It should be noted that this result of Huzurbazar does not rule out the existence of extraneous inconsistent solutions of the likelihood equation.

Simple as this result of Huzurbazar would appear to be, the statement that consistent roots are asymptotically unique contains mathematically problematic features. This is because a root $\hat{\theta}$ is not a real number, but rather a measurable function

$$\hat{\theta} = \hat{\theta}(y_1, \ldots, y_n)$$

satisfying

$$\left(\frac{\partial \log L}{\partial \theta}\right)_{\theta=\hat{\theta}} = 0.$$

So if the estimating equation has multiple roots, then it has infinitely many roots if these are understood to be infinitely many distinct measurable functions. In particular, it will have infinitely many distinct consistent roots, understood in the same sense.

Huzurbazar (1948) proved that all such distinct consistent roots are asymptotically equivalent. However, the precise mathematical statement of this needs to be given with some care. Perlman (1983) provided a more precise formulation of Huzurbazar's result, by showing that for sufficiently small $\delta > 0$, with probability one there exists exactly one solution to the likelihood equation in the interval $[\theta_0 - \delta, \theta_0 + \delta]$ for all but finitely many n, where θ_0 is the true value of the parameter. Once again, some regularity is required to achieve this result. This regularity is usually satisfied by standard models.

One of the simplest ways to solve an equation of the form $g(\hat{\theta}) = 0$ is to write it as

$$\hat{\theta} = \hat{\theta} + g(\hat{\theta}),$$

or more generally as

$$\hat{\theta} = \hat{\theta} + a(\hat{\theta})g(\hat{\theta}),$$

where $a(\theta) \neq 0$. Such an identity suggests the iteration

$$\theta^{(j+1)} = \theta^{(j)} + a(\theta^{(j)})g(\theta^{(j)}),$$

which is known as *simple iterative substitution*. Unfortunately, this method usually converges rather slowly, or may not converge at all. Attempting to choose $a(\theta)$ to optimise the rate of convergence leads to *Newton–Raphson* iteration, where

$$\theta^{(j+1)} = \theta^{(j)} - \left\{\dot{g}(\theta^{(j)})\right\}^{-1} g(\theta^{(j)}) \tag{1.1}$$

where $\dot{g}$ represents the derivative of g with respect to θ. The Newton–Raphson formula also appears in the classical derivation of the first-order asymptotics for the maximum likelihood estimate, and is a basic tool for relating the efficiency of more general estimating equations to the asymptotics of their consistent roots. Nevertheless, there are other choices for $a(\theta)$ which are also well motivated for estimating functions, and will be developed in this book. For example in likelihood estimation, using

$$\theta^{(j+1)} = \theta^{(j)} - \left(\frac{\partial \log L}{\partial \theta}\right)_{\theta^{(j)}} \Big/ \left\{E\left(\frac{\partial^2 \log L}{\partial \theta^2}\right)\right\}_{\theta^{(j)}},$$

known as Fisher's scoring of parameters, is often superior to Newton–Raphson when the sample size is large and the estimating function is not well approximated by its tangent line at $\theta^{(j)}$. See Fisher (1925). In some cases, Fisher scoring is also easier to compute than Newton–Raphson. There is also an argument to be made for choosing $a(\theta)$ so that the iterative steps are smaller than the steps of Newton–Raphson at an early stage of the iteration. This follows from the fact that Newton–Raphson is motivated by local linearity around $\hat{\theta}$, and may be unstable when started far from the root.

For many estimating functions with multiple roots, Newton–Raphson can be made to converge to any root provided the algorithm is started within an appropriate domain. So it is only iterations which start close to the true value of the parameter which have a reasonable probability of producing a consistent root. Most available iterative methods for root-finding all share this property with Newton–Raphson. The numerical problems of finding all the roots of the likelihood equation were considered by Barnett (1966). In view of the advances in computation that have been made since the 1960s, some of the comments in Barnett's paper are now dated. However, many of the basic insights into the advantages and disadvantages of iterative searches for the roots of estimating equations remain as true today as they were then. Barnett considered five methods for iterating toward a root of the likelihood equation, and performed a simulation study of the likelihood equation for the Cauchy location model. Barnett (1966) considered Newton–Raphson iteration and compared it to Fisher's scoring of parameters and the fixed-derivative Newton method as well as the method of false positions using the Cauchy location model as a test case. The method of false positions is of particular importance for locating a root in a single-parameter model, or in a model with several parameters in which all but one of the parameters can be solved for explicitly. While the procedure can be generalised into higher dimensions, the advantage of using the method of false positions is that for a continuous real-valued estimating function, the procedure is guaranteed to converge to a root—a guarantee provided by the intermediate value theorem.

A survey of the modern theory of efficient likelihood estimation for likelihoods with multiple local maxima can be found in Lehmann (1983, pp. 420–427). In particular, if $\hat{\theta}^{(1)}$ is a $\sqrt{n}$-consistent estimator for θ, then the *one-step estimator*, $\hat{\theta}^{(2)}$, found by applying Newton–Raphson iteration (1.1) to $\hat{\theta}^{(1)}$, will be asymptotically efficient. While this would seem to provide a satisfactory theory for root approximation and selection, the method leaves unanswered the problem of how to appropriately select a $\sqrt{n}$-consistent estimator to use as a starting point. Moreover, asymptotic considerations can only take us so far: there are infinitely many asymptotically efficient estimators which can be constructed even in the most regular of models. How to choose among them remains a problem.

Recently, attention has turned to the problem of multiple roots which arise with more general estimating equations. An example is the *quasi-score*. Suppose that Y is a random column vector of dimension n. A standard semiparametric model for Y assumes a known form for the mean vector and the covariance matrix

$$\mu(\theta) = E_\theta(Y), \qquad \Sigma(\theta) = \mathrm{Cov}_\theta(Y)$$

as functions of an unknown k-dimensional parameter θ. The quasi-score function is

$$g(\theta) = \dot{\mu}^{\mathrm{t}}(\theta)\Sigma^{-1}(\theta)\{Y - \mu(\theta)\}, \tag{1.2}$$

where once again the dot operator denotes (vector) differentiation with respect to θ, and $(\cdot)^{\mathrm{t}}$ denotes the transpose operation. This, and other such estimating functions share certain properties with the score function: they are *unbiased*

$$E_\theta\{g(\theta)\} = 0 \tag{1.3}$$

and *information unbiased*

$$-E_\theta\{\dot{g}(\theta)\} = E_\theta\left\{g(\theta)g^{\mathrm{t}}(\theta)\right\} \tag{1.4}$$

for all θ. However, while these are standard properties of estimating functions, neither property is essential for studying the phenomenon of multiple roots. Under mild regularity, an estimating function will have a consistent root (see Crowder, 1986). In addition, under reasonable regularity, any consistent root of g is unique with probability tending to one; see Tzavelas (1998) for a proof of uniqueness for quasi-score functions. Thus in these, and many other cases, our central interest is how the consistent root of the estimating function can be determined.

While many of the multiple root issues for estimating functions are identical to multiple root problems for the likelihood equations, the major difference is that estimating functions cannot typically be represented as the derivative of an objective function, such as the log-likelihood. This means that it is not possible to distinguish between the roots of a general estimating function as a likelihood does among its relative maxima. In likelihood estimation, there is a solution to the likelihood equations which is distinguished from the others by being a global maximum of the likelihood. Moreover, extraneous roots of the likelihood equation that correspond to local minima can usually be dismissed. However, for a quasi-score estimating equation, there is no comparable procedure in general. An exception to this is the case where we have a quasi-score function for a real-valued parameter. If θ is a real-valued parameter, it is possible to artificially construct an objective function, say $\lambda(\theta)$, whose derivative is $g(\theta)$, by integrating the estimating function:

$$\lambda(\theta) = \int_{\theta_0}^{\theta} g(\eta)\, d\eta. \tag{1.5}$$

Here θ_0 is arbitrary and may be chosen for computational convenience. If g is the score function, then, up to an additive constant, λ is the log-likelihood. On the other hand, if g is the quasi-score function, then λ, as defined by (1.5), is the *quasi-likelihood*. See McCullagh and Nelder (1989). However, for estimating functions other than the score function, $\lambda(\theta)$ is not justified by the usual theoretical considerations which justify likelihoods directly as measures of agreement between parameters and data. Moreover, when θ is a vector-valued parameter, the estimating function g also becomes vector-valued, and the line integral

$$\lambda(\theta) = \int_{\theta_0}^{\theta} g^{t}(\eta)\, d\eta \tag{1.6}$$

where $\eta = (\eta_1, \ldots, \eta_k)^t$ is typically *path-dependent*. Therefore it is not well defined. If g is the score vector, this ambiguity is avoided, because the vector field defined by g on Θ is conservative, being the gradient vector field of the log-likelihood. If $g(\theta)$ is a conservative vector field, then (1.6) is path-independent. However, this is not the case for general estimating functions such as quasi-score functions and others.

1.2 Organisation of the book

This book is organised as follows. Chapter 2 gives a survey of the basic concepts of estimating functions, which are used in subsequent chapters. In Section 2.1, the concept of unbiasedness for estimating functions is introduced as a generalisation of the concept of an unbiased estimator. Godambe efficiency, also known as the Godambe optimality criterion, is introduced in Section 2.2 by generalising the concept of minimum variance unbiased estimation. Under standard regularity conditions in a parametric model, within the class of estimating functions which are unbiased and information unbiased, the score function is characterised (Section 2.3) as the estimating function with maximal Godambe efficiency. Extensions to the multiparameter case are given in Section 2.4, and the connection to the Riesz representation theorem is described briefly in Section 2.5. The rest of the chapter in Section 2.6 is given over to a number of examples from semiparametric models (Section 2.6.1), martingale estimating functions for stochastic processes (Section 2.6.2), empirical characteristic function methods (Section 2.6.3) and quadrat sampling (Section 2.6.4). In some of these examples, the estimating equations can have more than one solution. However, the detailed discussion of these problems is postponed until Chapter 4.

Chapter 3 surveys a variety of root-finding and hill-climbing algorithms that are useful for solving estimating equations or maximising artificial likelihoods. While Newton–Raphson forms the cornerstone of many numerical methods for estimating functions, the chapter starts with an even more basic technique known as iterative substitution. The linear convergence of iterative substitution is illustrated in Section 3.1. Other methods, including Newton–Raphson, are motivated as attempts to improve the rate of convergence of iterative substitution. However, there are many possible ways to modify the Newton–Raphson algorithm, include the popular class of quasi-Newton algorithms (Section 3.6). The method of false positions and Muller's method are introduced in Sections 3.3 and 3.4, respectively. The contractive mapping theorem, which provides general conditions for the convergence of a multiparameter algorithm, is stated and proved in Section 3.5. Unlike many of the algorithms discussed in the chapter, the EM-algorithm is quintessentially statistical. Its popularity is due in part to its reliability and its applicability to data sets which can be represented as marginalised versions of some full and more tractable data set. The EM-algorithm is described in generality and illustrated with examples in Section 3.7. Among root-finding algorithms, the linear convergence of iterative substitution and the EM-algorithm is considered fairly slow. In Section 3.8, Aitken's method for accelerating linear convergence is developed, along with a refinement known as Steffensen's method. Accelerated algorithms of this kind are sometimes not as reliable as their original slow versions, but can converge at a rate that approximates Newton–Raphson in speed. Bernoulli's method and the quotient-difference algorithm (Section 3.9) are of secondary interest in the solution of polynomial estimating equations, but are included because they show the close relationship between solving polynomial equations and solving linear recursions. Often, statisticians fail to use additional tools that are available for analysing polynomials for those special cases where the estimating functions are polynomial in nature. Much of the rest of the chapter is concerned with special methods that can be implemented in this case. Sections 3.10 and 3.11 discuss Sturm's method and the QR-algorithm for finding and studing the roots of polynomials. In Section 3.12, the Nelder–Mead algorithm is described. The advantage of the Nelder–Mead algorithm is its flexible character in maximising or minimising objective functions: the algorithm can converge to local maxima even when the objective function is not differentiable. Finally, in Section 3.13, we consider the problem of matrix inversion which arises in the solution to quasi-likelihood equations. The method of Jacobi iteration for approximate inversion of matrices is described and applied to a problem in quadrat sampling from Chapter 2.

Chapter 4 is perhaps the most eclectic chapter in the book. Its basic purpose is to study the broad class of models which have difficult problems of non-linearity and multiple roots. We take the theory of estimating functions out for a test drive in some rough terrain to assess its performance capabilities. Mixture models are studied in

Section 4.2, and the relationship between non-identifiability and multiple roots is examined. A model for bivariate normal paired data with standardised marginals is presented in Section 4.3, where it is shown that the unusual restriction on the marginals leads to a cubic estimating equation for the correlation coefficient with as many as three roots. In Section 4.4, the Cauchy distribution and the symmetric stable laws are discussed in detail. The problem of evaluating the symmetric stable densities is used as a starting point for another numerical technique: accelerating the convergence of an infinite series using the Euler transformation. A variety of estimating functions for the location, scale and exponent parameters are also described. The chapter takes a philosophical turn in Section 4.5, where examples with inconsistent global maximum likelihood estimates are studied. A natural question to ask is whether the likelihood function can be used to order the roots of the likelihood equation from the most to the least plausible. The existence of such inconsistent MLE's raises difficult questions about the widely accepted principle that the likelihood function gives relative 'plausibility' weights to the points in the parameter space. In Sections 4.6, 4.7 and 4.8, three models, the stratified normal with common mean, regression with measurement error, and weighted likelihood, are considered and found to have multiple roots in their estimating equations.

The task of detecting multiple roots is also important. As Section 4.9 asks, how can we be sure that an estimating equation does not have multiple solutions or that an artificial likelihood does not have several local maxima? The Tobit model provides an instructive example. When we find that we have more than one root, how do we know that we have found all of them (Section 4.10)? We can also treat multiple roots as a point process (Section 4.11) and try to assess their frequency throughout the parameter space. We can also try to modify the estimating function or the artificial likelihood so as to remove unstable extraneous roots by smoothing (Section 4.12).

In Chapter 5, we turn our attention to the specific problem of selecting a root of an estimating equation with more than one root. In Section 5.2, the problem of solving an estimating equation is discussed and distinguished from the problem of estimating the interest parameter from the viewpoint of constructing an iterative algorithm. A class of irregular estimating functions is introduced in Section 5.3, where we see that multiple roots arise due to the irregularity of the estimating functions. It is seen that one may have a unique solution to an estimating equation if the irregularity of the estimating function is removed. The most common method for efficient estimation—iterating using one step of the Newton–Raphson from a consistent estimator—is described in Section 5.4. The concept of an efficient likelihood estimator is defined and generalised to a Godambe efficient estimator for a general estimating function. Modifications to the basic Newton–Raphson iterative procedure are suggested in Sections 5.5 and 5.6 that are particularly applicable to estimating functions. In Section 5.5, a modified Newton–Raphson algorithm is presented and motivated by consideration of a related dynamical system, and in Section 5.6, the information identity is used to motivate a another modification of the basic procedure. Continuing in a vein of modifying standard numerical algorithms for statistical purposes, Section 5.7 proposes a modified Muller's method for root-finding.

One of the most direct ways to assess a root-selection method is to see whether its asymptotic properties are correct. This is the approach of Section 5.8, following a suggestion of Heyde and Morton (1998). Other methods, using weighted least squares or the information identity, are discussed in Section 5.9. The bootstrap quadratic likelihood is introduced in Section 5.10. Following this definition, it is natural to pick the root of an estimating equation whose bootstrap distribution is closest to χ^2 in distribution, as dictated by the standard asymptotics at the true value of the parameter. Another information theoretic approach for location models is discussed in Section 5.11 based upon the shifted information function.

Arguably, it is not too radical to suggest that a model is a hypothetical construct whose utility to the researcher is at least partly determined by its mathematical tractability. There are many cases of models with multiple roots in their likelihood equations which can be extended to larger models where the root is unique. This is the strategy proposed in Section 5.12. Section 5.13 considers the much neglected problem of estimating equations which have no solutions at all.

Finally, Section 5.14 concerns the construction of confidence intervals using unbiased estimating functions. Section 5.14.2 shows that the computational cost can be quite substantial for setting second-order accurate confidence intervals using traditional bootstrap methods even though the original estimating equation admits a unique solution. The computational problem can be avoided using the method of the estimating function bootstrap introduced in Section 5.14.3. Confidence intervals using the latter method are second-order accurate, and, in addition, have the nice property of being transformation-respecting, a property enjoyed by more computational methods such as the $\mathrm{BC_a}$ method.

When an estimating equation has more than one root, the selection of a point estimator can be problematic. The construction of test statistics can also be difficult as well. For example, a Wald test based on the root of an estimating equation may behave poorly if the root is badly chosen. For parametric inference one may resort to the likelihood function to order the roots according to their relative plausibiltiy as estimates. However, in general there does not exist any objective function such that a given estimating function can be regarded as the gradient vector of the scalar objective function. This is because an estimating function is not generally obtained by minimising a 'risk' function, or by maximising a 'plausibility weighting' function such as a likelihood. Chapter 6 is devoted to the problem of building an artificial objective function for use in conjunction with an estimating function. Whenever possible the problem of selecting an appropriate root of an estimating equation using the various artificial likelihoods is discussed throughout this chapter.

The first approach discussed in Section 6.2 is by projecting the likelihood ratio onto a possibly largest subspace so that the projection can be computed using only the information of mean and variance functions. The derived artificial likelihood function may be regarded as an objective function for the quasi-score based on the same semiparametric information. The approach of Section 6.2 is generalised in Section 6.3 where we consider the projection of the log likelihood onto a subspace spanned by some functionally independent elementary estimating functions. Since the projection onto this space can not be computed in general using only the first few moment specifications, so an approximate form of the log likelihood function is used in the projection instead. This approach results in an approximate objective function for the optimally combined estimating function using those elementary estimating functions. The artificial projected likelihoods of Sections 6.2 and 6.3, as well as other alternative approaches in later sections, share many properties in common with a genuine likelihood functions. The null distribution of an approximate likelihood ratio statistic, for instance, follows a χ^2 distribution with appropriate degrees of freedom.

The idea of integrating an estimating function in the parameter space is briefly discussed in Section 6.4. An objective function defined in this fashion inevitably depends in general on the path one chooses. This approach, however, may be of some value when only local paths are of interest or when the estimating function is nearly conservative.

In Section 6.5, an entirely different approach is taken to constructing an artificial likelihood. Since the integrated quasi-score function is path-dependent, we examine the non-conservative vector field defined by the quasi-score. A conservative component of this vector field is extracted using a generalised version of the Helmholtz decomposition. It is this component which can be integrated in a path-independent manner to yield a quasi-likelihood. This technique is not without its problems, however, because the Helmholtz decomposition is not unique. Since a non-conservative component of the quasi-score is to be discarded, the natural way to perform the Helmoltz decomposition is to make sure that the non-conservative component is as statistically uninformative as possible. Such a decomposition is possible when the estimating function is linear. This motivates the discussion of quadratic local likelihoods in Sections 6.5.2–6.5.5. The generalised method of moments and quadratic inference functions are described in Section 6.6. Our approach to quadratic inference functions follows Qu *et al.* (2000). This method provides another way of building an artificial objective function for inference.

In Chapter 7, we explore the relationship between the numerical methods described in earlier chapters and the theory of dynamical systems. Our attention will turn from the original estimating function to a dynamical estimating system, whose domains of attraction and repulsion can be studied in relation to the estimation problem. In Section 7.1, we define a dynamical estimating system using an estimating function. Basic terminologies on dynamical systems are established in this section. In Section 7.2, we define the concepts of stability of a fixed point of a dynamical system. These concepts are studied via linear dynamical systems. Stability of roots to an estimating equation can be studied using the theories given in Section 7.2 through the linearisation method. In Section 7.3, a different method, the Liapunov's method, is introduced and applied to the theory of dynamical estimating systems. In particular, we prove that a consistent root of an estimating equation, under mild conditions, is an asymptotically stable fixed point of the associated dynamical estimating system. In Section 7.4 we give derivations of and formal proofs for the properties of the modified Newton's methods introduced in Sections 5.5 and 5.6. For this purpose, the Newton–Raphson method is reexamined in detail from the perspective of the theory of dynamical systems.

While parameter estimation is usually performed in parameter spaces with real-valued parameters, the domains of attraction of dynamical systems are often more clearly understood in the complex plane. While recognising the spurious nature of imaginary parameter values, Section 7.5 explores the Julia sets and domains of attraction of estimating functions, taking the estimation of the correlation coefficient for bivariate normal data as an example.

Chapter 8, the last chapter of the book, demonstrates that the numerical methods of earlier chapters are not constrained by statistical philosophy. We shall develop the theory of Bayesian estimating functions and show that this theory has Bayesian analogues for many of the concepts introduced in earlier chapters. While point estimation is often considered of secondary importance to Bayesians, the Bayesian estimating function methodology does have important applications in actuarial science, as explained in Section 8.5.

2
Estimating functions

2.1 Basic definitions

In this chapter, we shall develop some of the basic theory of estimating functions using an information criterion that is essentially frequentist in nature. Methods such as least squares, maximum likelihood and minimum chi-square estimation have the common property that they involve the construction and solution to an *estimating equation* $g(\hat{\theta}, y) = 0$, where y is the observed data whose distribution is governed by the value of θ, an unknown parameter.

In what follows, the parameter space Θ will be a k-dimensional subset of the Euclidean space $\mathbb{R}^k$. While it will not be essential to the development of the theory of estimating functions, we will usually find it convenient to assume that Θ is an open subset of $\mathbb{R}^k$. Let us also suppose that Y is a random quantity whose distribution is governed by the parameter $\theta \in \Theta$. Let Y be an element of some set $\mathcal{Y}$, which is called the sample space. We shall usually suppose that $\mathcal{Y}$ is a subset of $\mathbb{R}^n$ for some positive integer n. However, this need not always be the case. The quantity Y can also represent the realised value of some continuous time stochastic process. For such an application, the sample space $\mathcal{Y}$ will be a space of paths.

By an *estimating function*, we shall mean a function

$$g : \Theta \times \mathcal{Y} \to \mathbb{R}^k$$

with finite second moment, in the sense that $E_\theta \, g^2(\theta, Y) < \infty$ for all $\theta \in \Theta$. By constructing an appropriate estimating function, it is our intention to find an estimator $\hat{\theta} = \hat{\theta}(Y)$ which is a solution to the estimating equation $g(\hat{\theta}, Y) = 0$.

In order to ensure that the solution $\hat{\theta}$ to this equation is a reasonable estimator, it is usual to suppose that the estimating function is *unbiased* in the sense that

$$E_\theta \{g(\theta, Y)\} = 0 \tag{2.1}$$

for all $\theta \in \Theta$. The concept of an unbiased estimating function can be traced far back in early statistical work, but was formalised in Kendall (1951) as a concept distinct from that of the unbiasedness of an estimator. In fact, the requirement that an estimating function g be unbiased allows us to generalise the idea of the unbiasedness of an estimator. For example, suppose that $s(Y)$ is an estimator for θ. Defining $g(\theta, Y) = s(Y) - \theta$, we can see that the estimating function g is unbiased if and only if the estimator $s(Y)$ is unbiased. While every unbiased estimator with finite variance can be represented as the root of an unbiased estimating function, it is not the case that the root of a general unbiased estimating function is itself unbiased. The reason for this is that most estimating functions do not separate as a difference of the form $s(Y) - \theta$.

2.2 Godambe efficiency: one-parameter models

There are many possible estimating functions that could be used to estimate a parameter for any particular model. How can we choose among them? In the theory of best unbiased estimation, we restrict to the class of all unbiased estimators of a parameter and choose that estimator from the class which has minimum variance. So it is natural to try to do much the same thing for the class of unbiased estimating functions. Suppose that θ is a real parameter, and that $g(\theta, Y)$ is a real-valued function of θ and Y. By analogy with best unbiased estimation, we could seek that function g which minimises

$$\mathrm{Var}_\theta \{g(\theta, Y)\} = E_\theta \{g(\theta, Y)\}^2$$

in the class of unbiased estimating functions. However, this approach is not fruitful: if $g(\theta, Y)$ is unbiased, and $c(\theta)$ is a real-valued function which does not depend on Y, then the product $g_1(\theta, Y) = c(\theta)\, g(\theta)$ is also an unbiased estimating function. So the function $g \equiv 0$ is the unbiased estimating function with minimum variance in the sense described above. Unfortunately, such a function has no utility for statistical inference.

A better approach is to try to minimise the variance within a class of estimating functions which have been *standardised* so as to prevent the zero function or other inappropriate multiples from being considered. One such standardisation is through the expected derivative of the estimating function. For example, suppose we define $g_s(\theta, Y) = g(\theta, Y)/E_\theta \{\dot{g}(\theta, Y)\}$. (Of course, we have to assume that g is differentiable with respect to θ for this to be meaningful.) This standardisation leads to the following definition.

Definition 2.1 *Suppose that $E_\theta \{g_s(\theta, Y)\}^2$ is minimised among the class of all unbiased standardised estimating functions g_s by $g_s = g_s^*$, for all $\theta \in \Theta$. Then we shall say that g_s^* is a* Godambe efficient *estimating function.*

This definition can be modified to avoid direct reference to standardised estimating functions as follows.

Definition 2.2 *Let g be an unbiased estimating function that is differentiable with respect to θ. We define the* Godambe efficiency *of g by*

$$\mathrm{eff}_\theta(g) = \frac{[E_\theta \{\dot{g}(\theta, Y)\}]^2}{E_\theta \left[\{g(\theta, Y)\}^2\right]}. \tag{2.2}$$

An estimating function g^* is said to be Godambe efficient *if* $\text{eff}_\theta(g^*) \geq \text{eff}_\theta(g)$ *for all differentiable* g *and for all* θ.

The concept of Godambe efficiency was first proposed in Godambe (1960) and Durbin (1960). It is the extensive promotion and elaboration of the consequences of this definition by V. P. Godambe that leads us attach his name to the concept in Definitions 2.1 and 2.2.

Henceforth, we shall simplify the notation $g(\theta, Y)$ by denoting it as $g(\theta)$ with the data Y suppressed. Let us consider the relationship between Godambe efficient estimating functions and the theory of best unbiased estimation. Suppose $s(Y)$ is an unbiased estimator for θ, so that $g(\theta) = s(Y) - \theta$ is an unbiased estimating function. Then $\dot{g}(\theta) = -1$ and

$$\text{eff}_\theta(g) = \frac{1}{\text{Var}_\theta\{s(Y)\}}.$$

So if $g^*(\theta) = s^*(Y) - \theta$ is Godambe efficient, it will follow that $\text{Var}_\theta\{s^*(Y)\} \leq \text{Var}_\theta\{s(Y)\}$ for any unbiased estimator $s(Y)$. We can conclude that $s^*(Y)$ is the best unbiased estimator in the classical sense.

2.3 The score function: one-parameter models

Unlike best unbiased estimators, which do not exist in many models, it is always possible to find a Godambe efficient unbiased estimating function, provided the model possesses sufficient regularity. For example, suppose that $\mathcal{Y} = \mathbb{R}^n$, and that Y has a continuous distribution with joint probability density function (PDF) $f(y;\theta)$, where θ lies in some subset of $\mathbb{R}$. The *log-likelihood function* for θ is defined to be $\ell(\theta) = \log f(y;\theta)$ and the *score function* is then defined as $u(\theta) = \dot{\ell}(\theta)$.

A maximum likelihood estimate (MLE) $\hat{\theta}$ is defined to be a value of θ at which the log-likelihood attains its global maximum. This value need not exist, and even when it does exist, need not be unique. However, when it exists and is unique, we usually speak of the maximum likelihood estimate or the MLE for short. Provided the MLE is to be found in the interior of the parameter space, and that the log-likelihood is a differentiable function of θ, then the MLE will be a solution of the estimating equation $u(\hat{\theta}) = 0$. The function $u(\theta)$ is typically unbiased, because

$$\begin{aligned}
E_\theta\{u(\theta)\} &= \int_{\mathcal{Y}} \left[\frac{\partial}{\partial\theta}\{\log f(y;\theta)\}\right] f(y;\theta)\,dy \\
&= \int_{\mathcal{Y}} \frac{\partial}{\partial\theta}\{f(y;\theta)\}\,dy \\
&= \frac{\partial}{\partial\theta}\left\{\int_{\mathcal{Y}} f(y;\theta)\,dy\right\} \\
&= \frac{\partial}{\partial\theta}(1) = 0.
\end{aligned}$$

For $u(\theta)$ to be unbiased, we can see that the model should be sufficiently regular as to permit the interchange of a derivative and integral as seen above.

The score function $u(\theta)$ typically has finite second moment with $I(\theta) = E_\theta\{u^2(\theta)\}$. The function $I(\theta)$ is called the *Fisher information* or the *expected information*. The Fisher information exists in many models even when the MLE $\hat{\theta}$ has infinite second moment.

Henceforth, let us presume that the model is sufficiently regular as to allow us to interchange derivatives and integrals as needed in the calculations which follow. With this regularity in hand, let us find the correlation between a given unbiased estimating function $g(\theta)$ and the score function $u(\theta)$. First, we have

$$\begin{aligned} E_\theta\{u(\theta)\,g(\theta)\} &= \int_{\mathcal{Y}} g(\theta, y)\,\frac{\partial}{\partial\theta}\{f(y;\theta)\}\,\mathrm{d}y \\ &= \int_{\mathcal{Y}} \frac{\partial}{\partial\theta}\{g(\theta, y)\,f(y;\theta)\}\,\mathrm{d}y - \int_{\mathcal{Y}} \left\{\frac{\partial}{\partial\theta} g(\theta, y)\right\} f(y;\theta)\,\mathrm{d}y \\ &= \frac{\partial}{\partial\theta}\int_{\mathcal{Y}} g(\theta, y)\,f(y;\theta)\,\mathrm{d}y - \int_{\mathcal{Y}} \left\{\frac{\partial}{\partial\theta} g(\theta, y)\right\} f(y;\theta)\,\mathrm{d}y \\ &= \frac{\partial}{\partial\theta}(0) - \int_{\mathcal{Y}} \left\{\frac{\partial}{\partial\theta} g(\theta, y)\right\} f(y;\theta)\,\mathrm{d}y \\ &= -E_\theta\{\dot{g}(\theta)\}. \end{aligned}$$

So the correlation between g and u is

$$\{\mathrm{Corr}_\theta(g, u)\}^2 = \frac{\mathrm{eff}_\theta(g)}{I(\theta)}. \tag{2.3}$$

It follows from equation (2.3) that an unbiased estimating function $g(\theta)$ has maximum Godambe efficiency when the correlation coefficient is ± 1. This will hold if there exist functions $a(\theta)$ and $b(\theta)$ which do not depend upon y such that

$$g(\theta) = a(\theta)\,u(\theta) + b(\theta).$$

In this representation, $b(\theta)$ must vanish because both $u(\theta)$ and $g(\theta)$ are unbiased. Thus $g(\theta)$ must be a multiple of $u(\theta)$.

It is immediate from this result that any stationary point (local maximum or minimum, or point of inflection) of the likelihood will be a root of an estimating function which is Godambe efficient. If the function $a(\theta)$ is nowhere vanishing, the converse will also hold: any root of $g(\theta)$ will be a stationary point of the likelihood. From the perspective of estimation theory, there is no value in choosing a function $a(\theta)$ which does have roots in Θ. Any such root would be functionally independent of the data, and would therefore have no value for inference. So, in practice, we can choose $a(\theta) = 1$. By maximising the efficiency criterion, we will estimate θ using the MLE $\hat{\theta}$, if this is uniquely defined.

Among the class of non-zero unbiased estimating functions $g(\theta)$ for which $\mathrm{eff}_\theta(g)$ is maximum, the score function is uniquely determined by the property that it is *information unbiased.* An unbiased estimating function $g(\theta)$ for a one-parameter model is said to be information unbiased if

$$-E_\theta\{\dot{g}(\theta)\} = E_\theta\{g^2(\theta)\}. \tag{2.4}$$

For if $g(\theta) = a(\theta)\,u(\theta)$, then equation (2.4) implies that $-a(\theta)E_\theta\{\dot{u}(\theta)\} = a^2(\theta)\,I(\theta)$. This requires that $a(\theta) = 0$ or $a(\theta) = 1$ for all θ.

However, unlike the method of maximum likelihood, the Godambe efficiency criterion provides us with no easy way to choose among the roots of $g(\theta)$ if there are more than one.

2.4 Godambe efficiency: multiparameter models

The concept of a Godambe efficient estimating function can be generalised to vector-valued estimating functions for multiparameter models. If

$$g(\theta) = (g_1(\theta), \ldots, g_k(\theta))^{\mathrm{t}}$$

then we let $E_\theta\{\dot{g}(\theta)\}$ be the $k \times k$ matrix whose (j,k)-th element is $E_\theta\,\partial g_j(\theta)/\partial\theta_k$.

Definition 2.3 *The* Godambe efficiency *for $g(\theta)$ is defined to be the $k \times k$ symmetric matrix*

$$\mathrm{eff}_\theta(g) = \{E_\theta\,\dot{g}(\theta)\}^{\mathrm{t}}\left[E_\theta\{g(\theta)\,g^{\mathrm{t}}(\theta)\}\right]^{-1}\{E_\theta\,\dot{g}(\theta)\}. \tag{2.5}$$

We can also interpret this definition of efficiency as the inverse of the covariance matrix for a standardised estimating function. We can write

$$\{\mathrm{eff}_\theta(g)\}^{-1} = E_\theta\left(\left[\{E_\theta\,\dot{g}(\theta)\}^{-1}g(\theta)\right]\left[\{E_\theta\,\dot{g}(\theta)\}^{-1}g(\theta)\right]^{\mathrm{t}}\right). \tag{2.6}$$

This can be written as a covariance matrix, namely

$$E_\theta\left\{g_s(\theta)\,g_s(\theta)^{\mathrm{t}}\right\}$$

of the standardised function

$$g_s(\theta) = \{E_\theta\,\dot{g}(\theta)\}^{-1}g(\theta). \tag{2.7}$$

So an equivalent formulation of Definition 2.3 is the following.

Definition 2.4 *The* Godambe efficiency *for $g(\theta)$ can also be defined to be the inverse of the $k \times k$ symmetric matrix*

$$E_\theta\left\{g_s(\theta)\,g_s(\theta)^{\mathrm{t}}\right\}$$

where $g_s(\theta)$ is the standardised estimating function defined by (2.7) above.

In the case of a one-parameter model, as discussed in the previous section, we sought to minimise the variance of a standardised estimating function. So it is natural that for the multi-parameter case we should seek to minimise the size of the covariance matrix of $g_s(\theta)$, in some sense. Alternatively, we can try to maximise the matrix $\text{eff}_\theta(g)$ defined in (2.5). To maximise or minimise such matrices requires a way of partially ordering the set of all $k \times k$ non-negative definite symmetric matrices. A linear ordering is out of the question, as it is for most objects in dimensions greater than one.

One way to partially order matrices is to use the *Loewner ordering*. Let A and B be non-negative definite symmetric matrices of dimension $k \times k$. We say that A dominates B in the sense of Loewner ordering if the matrix $A - B$ is non-negative definite. It is convenient to write $A \geq B$ when A dominates B in this sense. Loewner ordering is a partial ordering of the class of non-negative definite symmetric matrices in the sense that it is antisymmetric, reflexive and transitive. Additionally, the inversion $A \to A^{-1}$ is antitonic in the sense that $A \geq B$ if and only if $B^{-1} \geq A^{-1}$, assuming that the matrices are of full rank. To prove this last statement, note that if $A \geq B$, then

$$(A - B)\, B^{-1}\, (A - B) + (A - B) = A\, B^{-1} A - A$$

is non-negative definite. Thus $A\, B^{-1}\, A \geq A$. Multiplying by A^{-1} on both the right and the left gives the desired result. A consequence of this result is that maximising $\text{eff}_\theta(G)$ with respect to Loewner ordering is equivalent to minimising $\{\text{eff}_\theta(g)\}^{-1}$ with respect to this ordering.

This leads us to the following definition.

Definition 2.5 *An unbiased estimating function $g^*(\theta)$ shall be said to be* Godambe efficient *if, for any other unbiased estimating function $g(\theta)$, the matrix*

$$\text{eff}_\theta(g^*) - \text{eff}_\theta(g)$$

is non-negative definite for all $\theta \in \Theta$.

An equivalent formulation of this is the following.

Definition 2.6 *If a standardised multiparameter estimating function $g_s^*(\theta)$ has the property that the matrix*

$$E_\theta\, \left\{g_s(\theta)\, g_s(\theta)^{\mathrm{t}}\right\} - E_\theta\, \left\{g_s^*(\theta)\, g_s^*(\theta)^{\mathrm{t}}\right\}$$

is non-negative definite for every θ and for every standardised estimating function $g_s(\theta)$ then $g_s^*(\theta)$ is Godambe efficient.

With this concept of efficiency, Kale (1962) was able to show that the multiparameter score function

$$u(\theta) = \left(\frac{\partial}{\partial \theta_1}\ell(\theta),\ \frac{\partial}{\partial \theta_2}\ell(\theta), \ldots,\ \frac{\partial}{\partial \theta_k}\ell(\theta)\right)^{\mathrm{t}}$$

has maximum Godambe efficiency for sufficiently regular models. Additional work on multiparameter efficiency can be found in Ferreira (1982) and Chandrasekar (1983). In Chandrasekar and Kale (1984), it was shown that the Loewner ordering of $\text{eff}_\theta(g)$ was compatible with a partial ordering based either upon the trace or the determinant of $\text{eff}_\theta(g)$ in the sense that the following three conditions are equivalent:

1. $\text{eff}_\theta(g^*) - \text{eff}_\theta(g)$ is non-negative definite for all g and all θ;
2. $\text{trace}\{\text{eff}_\theta(g^*)\} \geq \text{trace}\{\text{eff}_\theta(g)\}$ for all g and all θ;
3. $\det\{\text{eff}_\theta(g^*)\} \geq \det\{\text{eff}_\theta(g)\}$ for all g and all θ.

2.5 A geometric interpretation of Godambe efficiency

Some of the motivation for the concept of Godambe efficiency can be obtained from geometric considerations. Readers who are unfamiliar with the methods of functional analysis and Hilbert space geometry can skip this section, as it is independent of subsequent material.

Let us restrict to the case of a one-parameter model, so that $k = 1$ and the estimating functions are real-valued. Note that the class of unbiased estimating functions forms a vector-space, because $ag(\theta) + bh(\theta)$ is an unbiased estimating function whenever g and h are unbiased estimating functions. The constants a and b can depend smoothly upon θ but cannot be random.

For any two estimating functions $g(\theta)$ and $h(\theta)$, let us define the inner product

$$\langle g, h\rangle_\theta = E_\theta\{g(\theta)h(\theta)\}.$$

A special case occurs when $g = h$. Following standard vector space convention, we shall write

$$||g||_\theta = \sqrt{\langle g, g\rangle_\theta}$$

Let us also define a linear functional ∇_θ on the set of estimating functions by

$$\nabla_\theta(g) = -E_\theta\{\dot{g}(\theta)\}.$$

The reason for the minus sign will become clearer later. The norm of the linear functional ∇_θ is then determined as

$$||\nabla||_\theta = \sup_g \frac{|\nabla_\theta(g)|}{||g||_\theta}. \tag{2.8}$$

Squaring both sides of this equation, we see that the right-hand side becomes $\sup_g \text{eff}_\theta(g)$. This provides us with an alternative interpretation of the problem of maximising the Godambe efficiency. The search for an estimating function which maximises the efficiency is also a calculation of the norm of a linear functional. Henceforth, let us call ∇_θ the *score functional*.

In a complete inner product space, the *Riesz Representation Theorem* tells us that a linear functional with finite norm can be represented as an inner product with respect to a fixed vector in the space. It is tempting to conclude immediately that this fixed vector should be the score function, so that

$$\nabla_\theta(g) = \langle g,\, u\rangle_\theta \tag{2.9}$$

where u is the score. We can check that this formula is correct by noting that

$$\begin{aligned}
\nabla_\theta(g) &= -\int_{\mathcal{Y}} \dot{g}(\theta, y)\, f(y;\theta)\, \mathrm{d}y \\
&= -\int_{\mathcal{Y}} \frac{\partial}{\partial\theta}\{g(\theta, y)\, f(y;\theta)\}\, \mathrm{d}y + \int_{\mathcal{Y}} g(\theta, y)\, \dot{f}(y;\theta)\, \mathrm{d}y \\
&= -\frac{\partial}{\partial\theta}\int_{\mathcal{Y}} g(\theta, y)\, f(y;\theta)\, \mathrm{d}y + \int_{\mathcal{Y}} g(\theta, y)\, \dot{f}(y;\theta)\, \mathrm{d}y \\
&= 0 + \int_{\mathcal{Y}} g(\theta, y)\, u(\theta, y)\, f(y;\theta)\, \mathrm{d}y \\
&= \langle g,\, u\rangle_\theta .
\end{aligned}$$

However, the Riesz Representation Theorem cannot be immediately invoked, because the space of estimating functions on which ∇_θ has been defined is not complete—the limit of a sequence of differentiable functions need not be differentiable. So in order to be able to apply the Riesz Representation Theorem, we have to remove the restriction of differentiability with respect to θ, and extend ∇_θ to the space of all unbiased estimating functions $g(\theta)$ such that $E_\theta\{g^2(\theta)\} < \infty$. We can extend $\nabla_\theta(g)$ to non-differentiable g by defining

$$\nabla_\theta(g) = \left\{\frac{\partial}{\partial\eta} E_\eta\, g(\theta)\right\}_{\eta=\theta} \tag{2.10}$$

Within the space of all unbiased, square integrable estimating functions, there are two subspaces of particular interest. The first of these subspaces is the set of all estimating functions $g(\theta)$ which are annihilated by ∇_θ. We say that $h(\theta)$ is a *first-order E-ancillary* function if $\nabla_\theta(h) = 0$ for all $\theta \in \Theta$. The complementary space is the set of all estimating functions which are orthogonal to all the first-order E-ancillary functions. That is, the set of all $g(\theta)$ such that $\langle g, h\rangle_\theta = 0$ for all first-order E-ancillary $h(\theta)$ and for all θ. The set of all $g(\theta)$ satisfying this property is called the set of *first-order E-sufficient functions*.

As we have seen, the Godambe efficient functions are multiples $a(\theta)\, u(\theta)$ of the score function $u(\theta)$, where $a(\theta)$ is not a function of y. These functions form a vector subspace, which is precisely the class of first-order E-sufficient estimating functions defined above. The complementary subspace of first order E-ancillary functions is the set of all estimating functions $h(\theta)$ such that $\langle h,\, u\rangle_\theta = 0$.

If maximising the Godambe efficiency is equivalent to calculating the norm of the score functional, then it is natural to consider how the concept of efficiency changes when we modify ∇_θ to some other functional. For example, consider the functional

$$D_{\theta,\epsilon}(g) = \epsilon^{-1} E_{\theta+\epsilon}\{g(\theta)\}.$$

It can be seen that ∇_θ is the limiting form of $D_{\theta,\epsilon}$ as $\epsilon \to 0$. If $||D_{\theta,\epsilon}|| < \infty$ then $D_{\theta,\epsilon}$ will have a Riesz representation.

Let us find this Riesz representation. Suppose

$$E_\theta \left\{ \frac{f(Y;\theta+\epsilon)}{f(Y;\theta)} \right\}^2 < \infty.$$

Define

$$u_\epsilon(\theta) = \frac{1}{\epsilon} \left\{ \frac{f(y;\theta+\epsilon)}{f(y;\theta)} - 1 \right\}.$$

Then

$$\begin{aligned}
\epsilon\, E_\theta \{u_\epsilon(\theta)\} &= \int_{\mathcal{Y}} \left\{ \frac{f(y;\theta+\epsilon)}{f(y;\theta)} - 1 \right\} f(y;\theta)\,\mathrm{d}y \\
&= \int_{\mathcal{Y}} f(y;\theta+\epsilon)\,\mathrm{d}y - \int_{\mathcal{Y}} f(y;\theta)\,\mathrm{d}y \\
&= 1 - 1 \\
&= 0.
\end{aligned}$$

So the function $u_\epsilon(\theta)$ is an unbiased square integrable estimating function. If we take the inner product of $u_\epsilon(\theta)$ with any other estimating function, we see that

$$\begin{aligned}
\langle g, u_\epsilon \rangle_\theta &= E_\theta\{g(\theta)\, u_\epsilon(\theta)\} \\
&= \epsilon^{-1} \int_{\mathcal{Y}} g(\theta) \left\{ \frac{f(y;\theta+\epsilon)}{f(y;\theta)} - 1 \right\} f(y;\theta)\,\mathrm{d}y \\
&= \epsilon^{-1} \int_{\mathcal{Y}} g(\theta)\, f(y;\theta+\epsilon)\,\mathrm{d}y - \epsilon^{-1} \int_{\mathcal{Y}} g(\theta)\, f(y;\theta)\mathrm{d}y \\
&= D_{\theta,\epsilon}(g).
\end{aligned}$$

This provides the Riesz representation for $D_{\theta,\epsilon}$.

Along with the linear functional $D_{\theta,\epsilon}$ and its Riesz representation, we have a corresponding efficiency criterion given by

$$\begin{aligned}
\mathrm{eff}_{\theta,\epsilon}(g) &= \frac{|D_{\theta,\epsilon}(g)|}{||g||_\theta} \\
&= \frac{|\epsilon^{-1}\, E_{\theta+\epsilon}\, g(\theta)|}{||g||_\theta}
\end{aligned}$$

which, in the limit as $\epsilon \to 0$, goes to the Godambe efficiency criterion.

An estimating function $h(\theta)$ which is annihilated by every functional $D_{\theta,\epsilon}$, in the sense that $D_{\theta,\epsilon}(h) = 0$ for all θ and ϵ is called an *E-ancillary* estimating function. The closure of the space of all such functions is called the *E-ancillary subspace*. Its orthogonal complement is the class of all estimating functions $g(\theta)$ such that

$$\langle g, h \rangle_\theta = 0$$

for every $\theta \in \Theta$ and every E-ancillary estimating function h. The subspace of all such functions is called the *complete E-sufficient subspace*. Together, the E-ancillary subspace and the complete E-sufficient subspace form an orthogonal decomposition of the space of all estimating functions: any estimating function can be uniquely represented as a sum of the form $g(\theta) + h(\theta)$, where $h(\theta)$ lies in the E-ancillary subspace, and $g(\theta)$ lies in the complete E-sufficient subspace. As can be seen, the decomposition is broadly analogous to *Basu's Theorem* which states that a complete sufficient statistic is independent of an ancillary statistic for a parameter. Indeed, in the special case where a complete sufficient statistic T exists, it can be shown that the complete E-sufficient subspace is the class of T-measurable estimating functions. However, the concept of complete sufficiency of a statistic and complete E-sufficiency of a subspace of estimating functions are not equivalent. The complete E-sufficient subspace exists in much greater generality than does a complete sufficient statistic for θ. Note that the first-order E-sufficient estimating functions are elements of the complete E-sufficient subspace. This follows from the fact that any E-ancillary estimating function is *a fortiori* a first-order E-ancillary estimating function.

In the presence of some regularity that we shall not develop here, it can be shown that the estimating functions generated by the likelihood ratios span the complete E-sufficient subspace. So any estimating function in the complete E-sufficient subspace can be represented as the limit of estimating functions of the form

$$g(\theta) = \sum_{j=1}^{m} a_j(\theta) \left\{ \frac{f(y; \theta + \epsilon_j)}{f(y; \theta)} - 1 \right\} \tag{2.11}$$

where $a_j(\theta)$ does not depend upon y, and the right-hand side is required to be square-integrable. The set of centred square-integrable likelihood ratios $f(y; \theta+\epsilon)/f(y; \theta) - 1$, for varying ϵ is called the *Bahadur basis* for the complete E-sufficient subspace.

Another basis for the complete E-sufficient subspace is obtained from a Taylor expansion of the numerator of a typical Bahadur basis estimating function. If $f(y; \theta)$ is analytic in θ, we can write each Bahadur basis function as

$$\begin{aligned} \frac{f(y; \theta + \epsilon) - f(y; \theta)}{f(y; \theta)} &= \frac{\epsilon\, \dot{f}(y; \theta) + (\epsilon^2/2)\, \ddot{f}(y; \theta) + \cdots}{f(y; \theta)} \\ &= \epsilon\, \frac{\dot{f}(y; \theta)}{f(y; \theta)} + \frac{\epsilon^2}{2} \frac{\ddot{f}(y; \theta)}{f(y; \theta)} + \cdots . \end{aligned}$$

The estimating functions

$$\frac{\dot{f}(y;\theta)}{f(y;\theta)}, \frac{\ddot{f}(y;\theta)}{f(y;\theta)}, \frac{\dddot{f}(y;\theta)}{f(y;\theta)}, \cdots \tag{2.12}$$

are called the *Bhattacharyya basis functions* for the complete E-sufficient subspace. It can be seen that the score function is the first of these basis functions. As with the Bahadur basis, the estimating functions of the Bhattacharyya basis can be used to expand or approximate elements in the complete E-sufficient subspace. For suitably regular models, the estimating functions in the complete E-sufficient subspace can be represented as limits of functions of the form

$$g(\theta) = \sum_{j=1}^{m} a_j(\theta) \frac{\partial^j/\partial\theta^j f(y;\theta)}{f(y;\theta)} \tag{2.13}$$

where $a_j(\theta)$ does not depend on y.

One of the advantages of the geometrical approach to estimating functions is that it is possible to extend the theory to multiple parameters without the need to use matrix algebra and partial orderings on matrices. For the geometrical approach we continue to regard an estimating function $g(\theta)$ as a real-valued function of a vector-valued parameter θ. (We can always assemble a vector-valued estimating function from a set of real-valued estimating functions by using the real-valued functions as the components of a vector.) An estimating function $h(\theta)$ will be said to be *first-order E-ancillary* if

$$E_\eta\, h(\theta) = o(||\eta - \theta||) \tag{2.14}$$

for all θ and all $\eta \to \theta$. An estimating function $g(\theta)$ is then said to be *first-order E-sufficient* if $\langle g, h\rangle_\theta = 0$ for all first-order E-ancillary functions $h(\theta)$ and for all $\theta \in \theta$. For regular models with a k-dimensional vector-valued parameter, the first-order E-sufficient subspace is spanned by the k linearly independent unbiased estimating functions which are the components of the score vector.

For more information about the geometry of estimating functions, we refer the reader to Small and McLeish (1994).

2.6 Types of estimating functions

2.6.1 QUASI-LIKELIHOOD AND SEMI-PARAMETRIC MODELS

Let us now consider the application of the theory of estimating functions to *semi-parametric models* in which the parametric assumptions are only partially specified. Suppose that Y is an $n \times 1$ vector of observations with mean $\mu(\theta) = E_\theta(Y)$ and covariance matrix $\Sigma(\theta)$. We shall suppose that $\mu(\theta)$ is a known function of the unknown parameter $\theta \in \Theta \subset \mathbb{R}^k$. Similarly, suppose that $\Sigma(\theta)$ is a known function of θ. With these assumptions, the observation vector Y is said to be governed by a semi-parametric model, in the sense that the first and second moments are specified by the parameter, while the distribution is otherwise unspecified.

With semi-parametric information about the distribution of Y, the class of estimating functions that can be guaranteed to be both unbiased and square-integrable is limited. A class of estimating functions which are both unbiased and square-integrable are those of the form $g(\theta) = a(\theta)\{Y - \mu(\theta)\}$, where $a(\theta)$ is a $k \times n$ matrix that does not depend on the observations.

Within this class we can look for that choice of $a(\theta)$ which maximises the value of $\mathrm{eff}_\theta(g)$. In this case, the score function will not be the general solution because it will not be a function of the required linear form in Y in general. The maximising value of $a(\theta)$ is found by setting $a^*(\theta) = \dot{\mu}^{\mathrm{t}}(\theta)\,\Sigma^{-1}(\theta)$. For a proof of this, we refer the reader to Heyde (1997, Theorem 2.3). The resulting estimating function

$$g(\theta) = \dot{\mu}^{\mathrm{t}}(\theta)\,\Sigma^{-1}(\theta)\,\{Y - \mu(\theta)\} \tag{2.15}$$

is called the *quasi-score function* and a solution to the equation $g(\hat{\theta}) = 0$ is called a *quasi-likelihood estimator*.

The quasi-score function shares many important properties with the score function but also has some important differences. The quasi-score and the score both have the properties of unbiasedness and information unbiasedness. However, an important difference between the score function $u(\theta)$ in any parametric model and the quasi-score function $g(\theta)$ in a semi-parametric model is found when we construct the matrix of derivatives for each function. Note that the matrix $\dot{u}(\theta)$ of partial derivatives of the components of u with respect to the components of θ is symmetric, because $\dot{u}(\theta)$ is actually the matrix of second partials of the log-likelihood. In contrast, the matrix $\dot{g}(\theta)$ is generally not symmetric. An important statistical consequence of this failure of symmetry comes when we examine an integrated quasi-score of the form

$$\int_{\hat{\theta}}^{\theta} g^{\mathrm{t}}(s)\,\mathrm{d}s = \sum_{j=1}^{k} \int_{\hat{\theta}_j}^{\theta_j} g_j(s)\,\mathrm{d}s_j$$

where $\mathrm{d}s = (\mathrm{d}s_1, \ldots, \mathrm{d}s_k)^{\mathrm{t}}$ is some vector differential along a smooth path from $\hat{\theta}$ to a given θ. The path integral under consideration will be path independent only if the matrix $\dot{g}(\theta)$ is symmetric. When $g(\theta)$ is the score function of a parametric model then the symmetry of $\dot{u}(\theta)$ makes this integral path independent. In fact, the integral becomes the difference in the log-likelihood:

$$\ell(\theta) - \ell(\hat{\theta}) = \int_{\hat{\theta}}^{\theta} u^{\mathrm{t}}(\theta)\,\mathrm{d}s. \tag{2.16}$$

However, the failure of $\dot{g}(\theta)$ to be symmetric means that there is generally no analog of equation (2.16) for the quasi-score. Since an analog of (2.16) does not exist, quasi-likelihood estimates cannot usually be interpreted as the stationary points (local maxima, minima, or saddle points) of some scalar objective function.

An obvious exception to this occurs when the parameter θ is real-valued. In this case, $\dot{g}(\theta)$ is a 1×1 matrix and trivially symmetric. The one-dimensional integral

$$\lambda(\theta) - \lambda(\hat{\theta}) = \int_{\hat{\theta}}^{\theta} g(s)\, ds \tag{2.17}$$

implicitly defines a function $\lambda(\theta)$, known as the *quasi-likelihood function*, which is determined by equation (2.17) except for an arbitrary additive constant.

To illustrate the use of quasi-likelihood functions, let us consider a data set analysed by Bissel (1972). This data set gives the number of defects found in 32 samples of cloth of various lengths, as shown in Figure 2.1.

At a first pass, it is tempting to treat the number of defects in various samples of cloth as a Poisson process. The number of defects appearing in a given piece of cloth would then have a Poisson distribution whose mean would be proportional to the length of the cloth. With this assumption, the numbers of defects Y_j for given cloth lengths L_j would be independent with Poisson distributions

$$Y_j \sim \text{Poiss}(\theta L_j), \quad j = 1, \ldots, n \quad (\text{where } n = 32),$$

and where θ is an unknown constant of proportionality. Fitting the MLE for θ gives

$$\hat{\theta} = \frac{\sum_{j=1}^{n} Y_j}{\sum_{j=1}^{n} L_j}$$

The mean number of defects plus and minus two standard deviations are plotted in Figure 2.1 for this Poisson model.

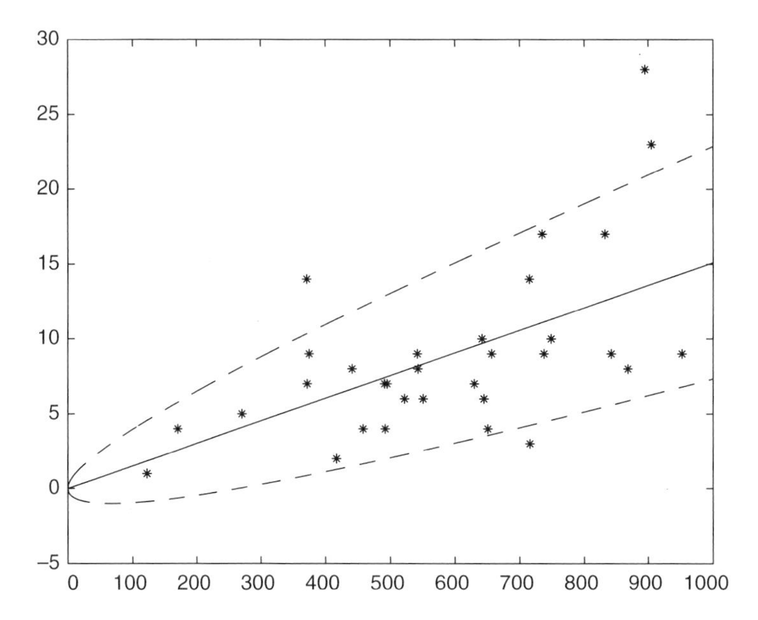

Fig. 2.1 The number of defects found in cloth of various lengths.

A glance at Figure 2.1 supports the idea that the mean number of defects is roughly proportional to the length. However, for Poisson variables, the variance equals the mean, and in this respect, the data look overdispersed. One possible cause of such overdispersion is some variation or heterogeneity in the cloth. From the standpoint of parametric inference, the next step is probably to find a model for the Y_j's which allows for greater dispersion than is possible in the Poisson family of distributions. For example, Bissel (1972) modelled the heterogeneity of the cloth by supposing that the parameter θ was distinct for each Y_j, and has a gamma distribution. However, we can avoid any additional parametric assumptions by using a semi-parametric model. Suppose we write

$$\mu_j = \theta\, L_j \quad \text{and} \quad \sigma_j^2 = \alpha\, \theta\, L_j = \alpha\, \mu_j$$

and assume that $Y_1, \ldots, Y_n$ are independent. Then the quasi-score equation (2.15) becomes

$$\begin{aligned} g(\hat{\theta}) &= \sum_{j=1}^{n} \frac{Y_j - \mu_j(\hat{\theta})}{\sigma_j^2(\hat{\theta})}\, \dot{\mu}_j(\hat{\theta}) \\ &= \sum_{j=1}^{n} \frac{Y_j - \hat{\theta}\, L_j}{\alpha\, \hat{\theta}\, L_j}\, L_j \\ &= 0. \end{aligned}$$

This leads to the same estimate for θ as that found for the Poisson model, and provides us with some confidence that a point estimator for θ is not unduly affected by overdispersion for this example. While the point estimate $\hat{\theta}$ is unchanged, the confidence interval for θ will be affected by the dispersion parameter α. In practice, α will be unknown, and will have to be estimated from the data. A naive moment estimator for α is given by

$$\hat{\alpha} = \sum_{j=1}^{n} \frac{(Y_j - \hat{\theta}\, L_j)^2}{(n-1)\, \hat{\theta}\, L_j}.$$

A rather different situation could arise if we were to consider modelling the variance as a non-linear function of θ and L_j. For example, if $\mu_j(\theta) = \theta L_j$ and $\sigma_j^2(\theta) = L_j^{\theta}$, then the efficient estimating equation for θ will not admit an explict solution. This problem, and the possible multiple solutions to estimating equations arising with certain types of non-linearity, shall be explored in future chapters.

Further consideration of this data set will take us beyond the scope of this book. The reader is referred to Bissel (1972) for additional details.

2.6.2 MARTINGALE ESTIMATING FUNCTIONS

Quasi-likelihood methodology has been found to be fruitful for the statistical inference on stochastic processes which are controlled by a stochastic differential equation involving one or more unknown parameters. In this section, we sketch the Hutton–Nelson theory of martingale estimating functions for inference of such stochastic processes. The reader who is unfamiliar with the theory of stochastic integration and semi-martingales is referred to Protter (1990). This application can be safely skipped by the reader who does not wish to invest the time on the theory of stochastic integration and semimartingale processes.

Let Y_t, $0 \leq t \leq T$, be a continuous time real-valued stochastic process defined on the time interval $[0, T]$, and adapted to a filtration $\mathcal{F}_t$. We shall assume that Y_t admits a decomposition into a sum of two stochastic processes

$$Y_t = A_t + M_t,$$

where A_t has *bounded variation* and M_t is a *continuous time martingale*. For stochastic processes of this kind, let $\langle\langle Y \rangle\rangle_t$ represent the predictable quadratic variation of the process Y_t.

There are many applications or interpretations of such a stochastic process, especially in finance where Y might describe the dynamic behaviour of a stock index, or in the biological sciences, where Y might denote the size of a population of organisms. We will assume that the stochastic behaviour of Y is controlled by an unknown parameter θ.

Martingale estimating functions are themselves martingales. This can be thought of as a dynamic version of the unbiasedness property of estimating functions.

Definition 2.7 *Let $Y_t, 0 \leq t \leq T$, be adapted to a right continuous filtration $\mathcal{F}_t$. A stochastic process $g_t(\theta)$ that is indexed by the parameter θ and adapted to $\mathcal{F}_t$ is said to be a* martingale estimating function *if $g_0(\theta) = 0$ and the martingale property*

$$E_\theta \left\{ g_{T_1}(\theta) \mid \mathcal{F}_{T_2} \right\} = g_{T_1 \wedge T_2}(\theta)$$

holds. (Here, $T_1 \wedge T_2$ is the minimum of T_1 and T_2.)

If $g_t(\theta)$ is a martingale estimating function, it follows that $E_\theta \, g_t(\theta) = 0$ for all t.

Suppose that we can decompose Y into the sum of two stochastic processes so that $Y_t = A_t + M_t$, where M_t is a continuous time martingale and A_t is a predictable process. As the distribution of Y is governed by the unknown value of θ, we should write more properly

$$Y_t = A_t(\theta) + M_t(\theta) \tag{2.18}$$

as the particular decomposition can only be performed if a value for θ is specified. The functions which have the form

$$g_T(\theta) = \int_0^T a_t(\theta) \, \mathrm{d}M_t(\theta) \tag{2.19}$$

where $a_t(\theta)$ is a given predictable process, are called *Hutton–Nelson martingale estimating functions*, or simply Hutton–Nelson estimating functions. It is easy to check that they satisfy Definition 2.7. In formula (2.19), the integral is an Itô stochastic integral. The integrand a_t may also depend upon Y, but does so as a predictable process. Some care must be taken in interpreting (2.19). The decomposition of Y into a predictable and a martingale component is performed under the stochastic model governed by θ. Thus different values of θ give different martingale components. In order for (2.19) to be meaningful, we must assume that $M_t(\theta)$ is a *semimartingale process* whatever the true value θ_0 of the parameter. In particular, this semimartingale becomes a martingale when $\theta = \theta_0$, a result that parallels the unbiasedness of estimating functions considered earlier.

Within the class of Hutton–Nelson estimating functions, it is possible to find one which maximises the Godambe efficiency. This is given by

$$g_T^*(\theta) = \int_0^T \left\{ \frac{-E_\theta\left(\mathrm{d}\dot{M}_t|\mathcal{F}_{t-}\right)}{\mathrm{d}\langle\langle M\rangle\rangle_t} \right\} \mathrm{d}M_t. \tag{2.20}$$

Example 2.8 Black–Scholes model

Consider an asset whose price Y_t at time t varies continuously according to the stochastic differential equation

$$\mathrm{d}Y_t = \theta\, Y_t\, \mathrm{d}t + \sigma\, Y_t\, \mathrm{d}W_t \tag{2.21}$$

where W_t is standard Brownian motion and $\sigma > 0$ is the volatility or the diffusion coefficient. According to the Black–Scholes model, the parameter $\theta = r - q$, where r is the risk-free interest rate, and q is a continous dividend yield. In the simplest form of the equation, the parameter θ is constant. If we wish to estimate θ to compare its estimated value with the constant provided by the Black–Scholes model, then we can use a martingale estimating function to do this. Setting $\mathrm{d}A_t(\theta) = \theta\, Y_t\, \mathrm{d}t$ and $\mathrm{d}M_t(\theta) = \mathrm{d}Y_t - \theta\, Y_t\, \mathrm{d}t$ gives us a class of estimating functions of the form

$$g_T(\theta) = \int_0^T a_t(\theta)\, \{\mathrm{d}Y_t - \theta\, Y_t\, \mathrm{d}t\}.$$

Note that W_t is not directly observed. So we cannot include it in the integrating function explicitly. Since $\mathrm{d}M_t = \mathrm{d}Y_t - \theta\, Y_t\, \mathrm{d}t$ we get $\mathrm{d}\dot{M}_t = -Y_t\, \mathrm{d}t$. Now $E(\mathrm{d}\dot{M}_t|\mathcal{F}_{t-}) = Y_t\, \mathrm{d}t$, because Y_t is a predictable process. Additionally, $\langle\langle M\rangle\rangle_t = Y_t^2\, \sigma^2\, \mathrm{d}t$. So the choice of $a_t(\theta)$ which maximises the Godambe efficiency is $a_t = 1/Y_t$, except for a constant factor that does not depend upon Y or t. Thus

$$g_T^*(\theta) = \int_0^T \left\{ \frac{\mathrm{d}Y_t}{Y_t} - \theta\, \mathrm{d}t \right\}.$$

Setting $g_T^*(\hat{\theta}) = 0$ gives

$$\hat{\theta} = \frac{1}{T} \int_0^T \frac{\mathrm{d}Y_t}{Y_t}.$$

This Itô integral cannot be evaluated as if it were a Riemann integral. Note that if $y_t = \log Y_t$, then Itô's Lemma implies that

$$\mathrm{d}y_t = \frac{\mathrm{d}Y_t}{Y_t} - \frac{\mathrm{d}\langle\langle Y_t \rangle\rangle}{2\, Y_t^2}$$

where $\mathrm{d}\langle\langle Y_t \rangle\rangle = \sigma^2\, Y_t^2\, \mathrm{d}t$. So

$$\begin{aligned} \int_0^T \frac{\mathrm{d}Y_t}{Y_t} &= \int_0^T \mathrm{d}y_t + \sigma^2 \int_0^T \frac{\mathrm{d}t}{2} \\ &= y_T + \frac{\sigma^2\, T}{2}. \end{aligned}$$

We can conclude that

$$\hat{\theta} = \frac{y_T}{T} + \frac{\sigma^2}{2}. \tag{2.22}$$

Some insight into the nature of this estimate can be obtained by noting that y_t obeys the stochastic differential equation

$$\mathrm{d}y_t = \left(\theta - \frac{\sigma^2}{2}\right) \mathrm{d}t + \sigma\, \mathrm{d}W_t$$

which is the equation for a Brownian motion with drift $\theta - \sigma^2/2$ and diffusion parameter σ. So Y_t is an example of *geometric Brownian motion*: an exponentiated Brownian motion. From this, it can be seen that $\hat{\theta}$ is a moment estimator for θ, which can also be obtained by solving $y_t = E_\theta(y_t)$.

To illustrate this formula, consider Figure 2.2, which shows the values of the Toronto Stock Exchange (TSE) Composite 300 Index from January 1984 to December 1987 inclusive. The data consist of 1009 observations of the index taken on a daily basis from January 3, 1984 to December 31, 1987.

Included in this picture is the famous drop in the market on October of 1987, which was dubbed 'Black-and-Blue Monday' by stock market analysts. If we treat the index as the realisation of a Black–Scholes process, then the parameter θ can be estimated by formula (2.22).

Figure 2.3 shows the process y_t. The increment in y_t over the observed period can be used to estimate θ provided σ is known. An important feature of such diffusion processes is that the value of σ need not be estimated, but can be completely determined from a path of the process. This follows from the fact that, in a certain sense, a single realisation of the process contains an infinite amount of information about σ.

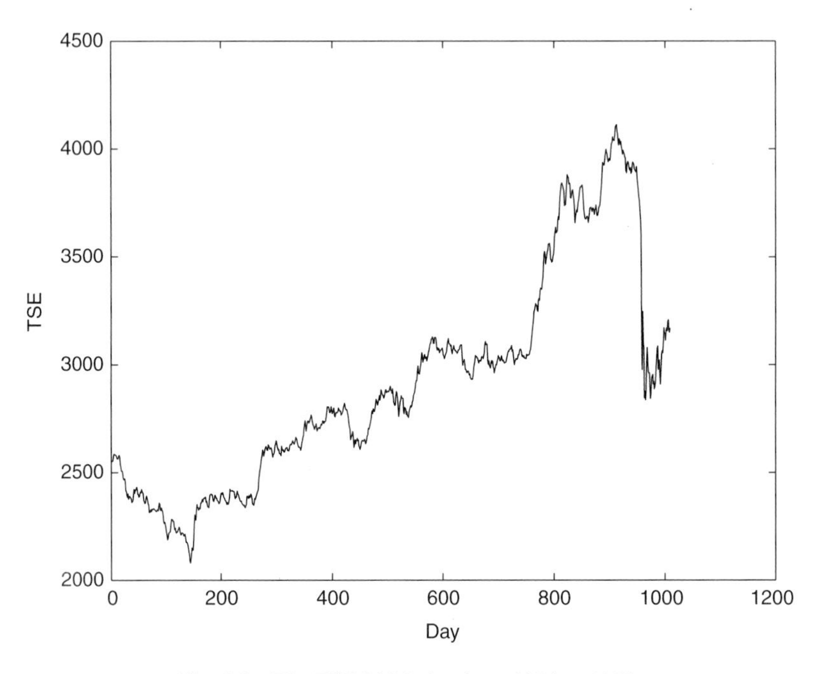

Fig. 2.2 The TSE 300 Index from 1984 to 1987.

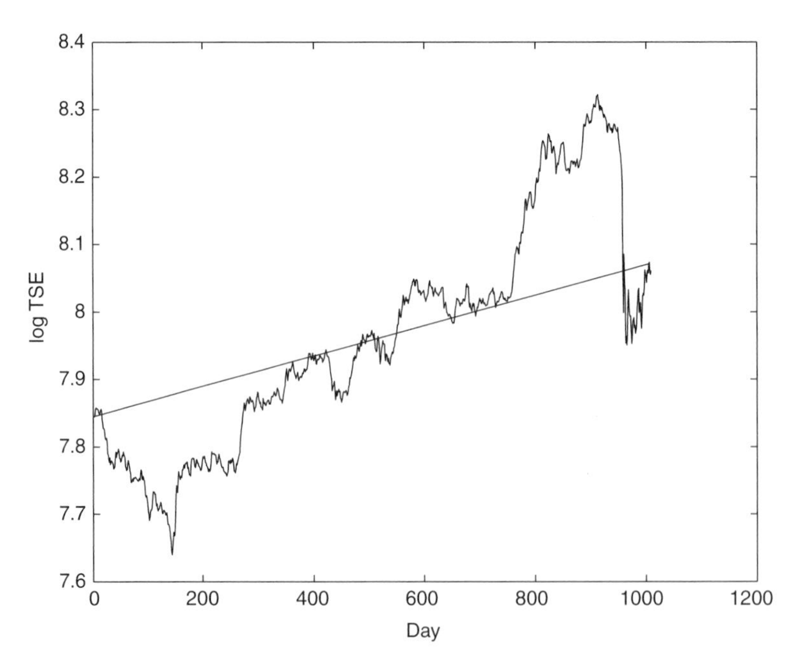

Fig. 2.3 The logarithm of the TSE 300 Index from 1984 to 1987.

From an observed realisation of the process, σ is determined by the formula

$$\sigma^2 = \lim \frac{1}{T} \sum_{j=1}^{n} (y_{t_j} - y_{t_{j-1}})^2$$

where the limit is taken as the mesh of the partition $0 = t_0 < t_1 < \cdots < t_n = T$ goes to zero. In practice, we can use a mesh no finer than one day, the gap between neighbouring observations of the index. This gives a value of σ of approximately 27.68, using one day as the unit of time. So $\hat{\theta} = 383.72$.

Examining the index, we can easily question the appropriateness of the Black–Scholes model. According to Black–Scholes theory, the increments of y_t should be normally distributed, while the increments of Y_t should have a log-normal distribution. In view of the large drop in 1987 and other increments at other times, we might be tempted to conclude that the increments of y_t have heavier distribution tails than can be explained by the normal assumption. We shall consider this possibility next as we look at estimating functions for symmetric stable distributions.

The reader who would like to find out more about martingale estimating functions is referred to Heyde (1997).

2.6.3 EMPIRICAL CHARACTERISTIC FUNCTIONS AND STABLE LAWS

Estimating function methodology is also useful for estimation in parametric models where the densities are difficult or intractable to calculate. One such example is the family of densities for *symmetric stable laws*. Stable laws arise as natural generalisations of the normal distribution, and include the Cauchy distribution as a special case. Suppose that Y_1 and Y_2 are independent random variables with common distribution F.

Definition 2.9 *The distribution F is said to be* stable *if for each pair of constants a and b, there exist constants c and d such that*

$$a\,Y_1 + b\,Y_2 \sim c\,Y_1 + d.$$

If F is a stable distribution, then it can be shown that there is a constant $0 < \alpha \leq 2$, called the *exponent* of F, such that $c^\alpha = a^\alpha + b^\alpha$. For example, a normal distribution is stable with exponent $\alpha = 2$, and a Cauchy distribution is also stable with exponent $\alpha = 1$. Stable laws are generalisations of the normal laws commonly used in statistical inference. Like the normal, the stable laws are closed as a family under linear combinations of independent random variables.

The symmetric stable laws form a special class of the stable laws, and are defined as follows.

Definition 2.10 *A random variable Y is said to have a* symmetric stable distribution *centred at θ if Y has a stable distribution and if, additionally, the random variables $Y - \theta$ and $\theta - Y$ have the same distribution.*

Symmetric stable laws are completely determined up to location and scale by the exponent α. For example, the normal distributions are characterised as the symmetric stable laws with exponent $\alpha = 2$. The exponent $\alpha = 1$ characterises the location-scale family of Cauchy distributions. The laws with exponent $\alpha < 2$ have heavier tails than the normal distribution. So they provide an interesting alternative for modelling error distributions when it is known that the tails of these distributions are heavy. Outlying observations are more likely to occur for stable laws with exponent $\alpha < 2$ than under the normal assumption. In fact, the normal distribution is the only member of the family with finite variance.

An example to illustrate this can be found in the fluctuations of the TSE 300 stock index that we considered earlier. Figures 2.4 and 2.5 show the values of the logarithmic differences $y_{t_{i+1}} - y_{t_i}$ of the TSE 300 index over time as measured in days. It is immediately evident that there is great volatility associated with the days around Black-and-Blue Monday. Clearly, a normal model for the increments $y_{t_{i+1}} - y_{t_i}$ will not be adequate, as Figure 2.5 shows. Note that in addition to the large drop of the stock market crash, there are also several large positive increments associated with the increased volatility of the market. How can we make inferences about the process in the presence of such high volatility? One approach is to model the events around Black-and-Blue Monday as anomalous. If this is truly the case, then the observations in the tails of the distribution in Figure 2.5 can be regarded as outliers.

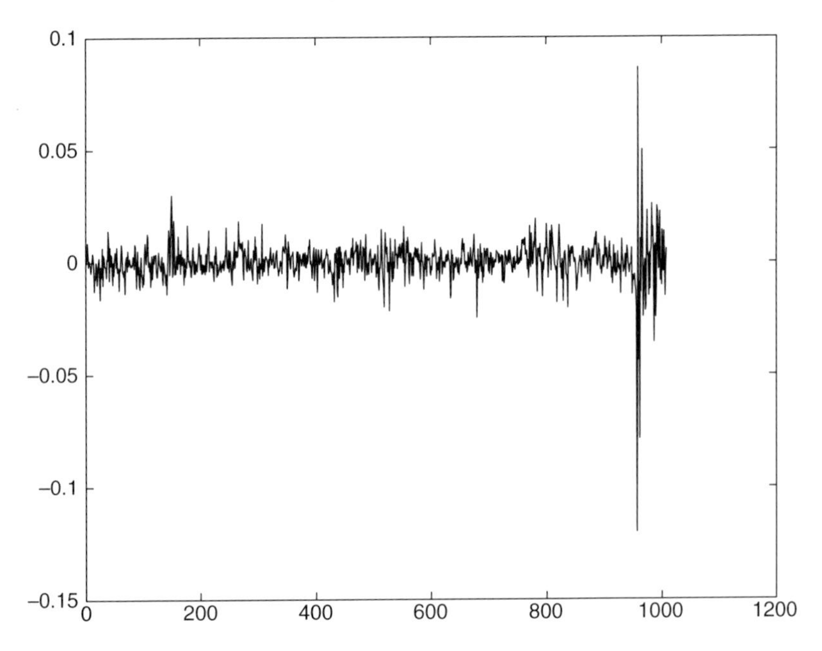

Fig. 2.4 Fluctuations of the logarithm of the TSE 300 index as a function of time in days.

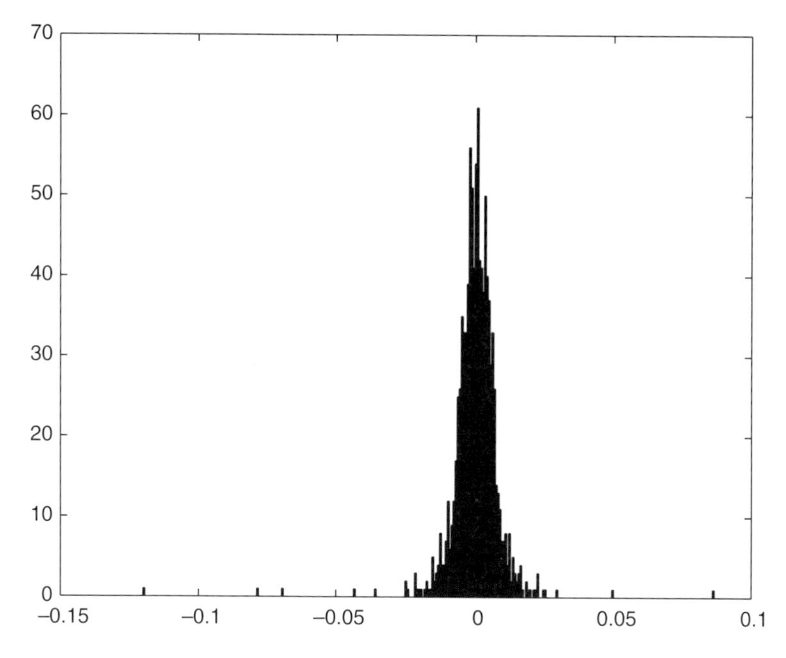

Fig. 2.5 Histogram of the fluctuations of y_t.

However, it is problematic to detect and trim outliers from data when the number of outliers is unknown. So we could construct a method of inference that is robust against the presence of outliers. A second approach to the presence of such volatility is to try to build a model which includes the possibility of high volatility. Modelling y_t as a *stable process*, i.e., a process of independent stable increments, has the advantage that it can explain the presence of extreme observations in Figure 2.5 and is in reasonable accord with our knowledge of market behaviour. For example, stable laws have a mathematical property in common with the normal distribution, namely that the linear combination of independent symmetric stable random variables with common exponent is also symmetric stable with the same exponent. What differs is the way in which we calculate the scale parameter for the linear combination. This property of symmetric stable laws is useful for continuous time stochastic processes that can be regarded as having certain self-similarity properties and independent increments. See Mandelbrot (1963). By matching quantiles, Small and McLeish (1994, section 6.7) obtained a good fit to a symmetric stable distribution with location parameter $\theta = 0.000016$, scale parameter $c = 0.0038$, and exponent $\alpha = 1.67$.

Although the density functions for the cases $\alpha = 1$, the Cauchy, and $\alpha = 2$, the normal, are fairly simple, the formulas for the densities of other symmetric stable laws cannot be written in simple form. So the likelihood equations for the location and scale parameters are just as difficult to write down. In contrast, the characteristic functions of symmetric stable laws are very simple. Suppose that Y has a symmetric stable law centred at θ with exponent α. Using Definition 2.9, it can be shown that

$$E\,\mathrm{e}^{\mathrm{i}tY} = \mathrm{e}^{\mathrm{i}\theta t - |ct|^{\alpha}} \tag{2.23}$$

where c is a scale parameter. Suppose that $w_1, w_2, \ldots, w_m$ are given complex numbers, and that $t_1, t_2, \ldots, t_m$ are given real numbers. We can construct an unbiased estimating function for the parameters of the symmetric stable law of Y, namely

$$g(\theta, c, \alpha, Y) = \sum_{j=1}^{m} w_j \left(\mathrm{e}^{\mathrm{i}t_j Y} - \mathrm{e}^{\mathrm{i}\theta t_j - |ct_j|^{\alpha}} \right).$$

If $Y_1, \ldots, Y_n$ is a random sample from a symmetric stable law, then we have an estimating function of the form

$$g(\theta, c, \alpha, Y) = \sum_{j=1}^{m} w_j \sum_{k=1}^{n} \left(\mathrm{e}^{\mathrm{i}t_j Y_k} - \mathrm{e}^{\mathrm{i}\theta t_j - |ct_j|^{\alpha}} \right) \tag{2.24}$$

$$= \sum_{j=1}^{n} w_j \left\{ \tilde{\phi}(t_j) - \phi(t_j;\, \theta,\, c,\, \alpha) \right\} \tag{2.25}$$

where $\phi(t; \theta, c, \alpha)$ is the characteristic function of the stable law, and $\tilde{\phi}(t)$ is the empirical characteristic function.

The fact that g is complex-valued introduces no additional complications here. We can return to real-valued estimating functions by taking the real and imaginary parts of g. Note also that the family of estimating functions can be extended by assuming that $w_1, \ldots, w_m$ are also functions of the unknown parameters θ, c and α. If this is the case, then the family of estimating functions so defined provides flexible opportunities for estimation, because the family of trigonometric polynomials span the space of all estimating functions with finite variance. Since the trigonometric polynomials are almost periodic functions, it is reasonable to expect the estimating functions defined by equation (2.25) to have multiple roots. We shall not attempt any solution to the multiple roots problems of (2.25), but shall defer this to later.

To illustrate how empirical characteristic functions can be used for estimation, let us consider the data set illustrated in Figure 2.5. We shall restrict attention to the estimation of θ and c, supposing for the moment that $\alpha = 1.67$ is given. Setting $m = 1$ in (2.25), we obtain the pair of estimating equations

$$\sum_k \sin\left\{ \frac{t(y_k - \hat{\theta})}{\hat{c}} \right\} = 0 \tag{2.26}$$

and

$$\sum_k \cos\left\{ \frac{t(y_k - \hat{\theta})}{\hat{c}} \right\} = n\,\exp(-t^{\alpha}). \tag{2.27}$$

In practice, it is convenient to allow different values of t in equations (2.26) and (2.27). Based upon the maximising of an expected information, Small and McLeish (1994, Section 6.7) recommended using $t = 0.49$ in equation (2.26) and $t = 0.73$ in (2.27). With these values, equations (2.26) and (2.27) yield the estimates $\hat{\theta} = 0.0004$ and $\hat{c} = 0.0104$. However, problems of multiple roots do arise. For example, Figure 2.6 shows the value of the estimating function in formula (2.26) when $c = \hat{c}$. The periodic nature of the sine function ensures the presence of multiple roots.

Fortunately, we can determine which root to choose from the quantile fit of the symmetric stable law mentioned above. As the value $\hat{\theta} = 0.0004$ is positive, we can see how the estimating function has discounted the effect of the large negative drop on Black-and-Blue Monday. However, for the particular interval of time under consideration, the value of θ is estimated to be very close to zero. This means that it would have been difficult for an investor whose portfolio resembled the TSE 300 stocks to make large amounts of money over the short term. Nevertheless, over a longer period of observation, a total of 1009 days, significant positive return was achievable.

In this section, we have emphasised the application of empirical characteristic estimating function methodology to symmetric stable laws whose density functions are difficult to obtain. It can be seen that these methods can be applied to

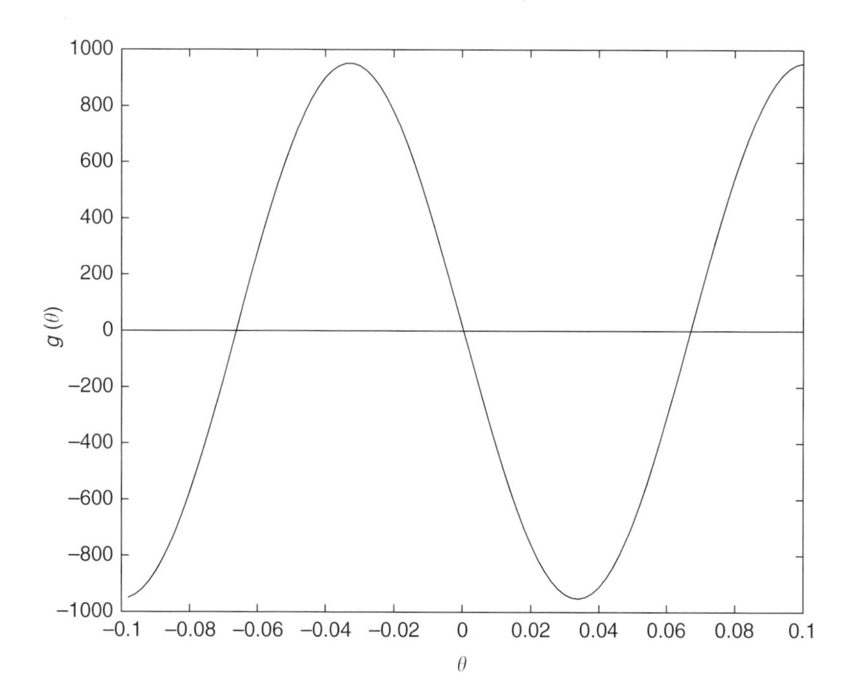

Fig. 2.6 A sine estimating equation for θ using the log-increments of the TSE 300 data.

other models whose characteristic functions are more tractable than their densities. In particular, distributions for data which form convolution families, but not of exponential families, are often of this type.

2.6.4 QUADRAT SAMPLING

Data in the forestry, geology and other biological and earth sciences can often be modelled as Poisson processes of particles scattered over a region. If the actual locations of the particles within a region can be determined with precision, then an exact likelihood function can be computed for inferential purposes. However, in a number of cases, the exact locations of particles of a Poisson process cannot be determined, and the researcher must rely on incomplete information about the process.

Data sets gathered in quadrat sampling are examples of this. If it is too time-consuming to record the exact locations of the particles of a point process, it may be more practical to divide the region of observation up into subregions and to count the number of particles in each subregion. Suppose, for example, that a region A is subdivided into n subregions $A_1, \ldots, A_n$. In each subregion A_j a total of Y_j particles is counted. Under the assumption that the individual particles are distributed as a Poisson process across A, the count statistics $Y_1, \ldots, Y_n$ will each have a Poisson distribution with parameters $\lambda|A_1|, \ldots, \lambda|A_n|$ respectively, where λ is the intensity parameter of the process. More generally, in a non-homogeneous Poisson process, $Y_j \sim \text{Poiss}(\mu_j(\theta))$, where

$$\mu_j(\theta) = \int_{x \in A_j} \lambda_\theta(x)\, dx$$

and θ is some parameter or vector of parameters determining the intensity function λ.

Provided that the regions A_j are disjoint, the counts $Y_1, \ldots, Y_n$ will be independent, and the MLE for θ will be found by solving

$$\sum_{j=1}^{n} \dot{\mu}_j(\hat{\theta}) \left\{ \frac{Y_j}{\mu_j(\hat{\theta})} - 1 \right\} = 0$$

where $\dot{\mu}(\theta)$ is the derivative of $\mu_j(\theta)$ with respect to θ. The sets A_j are called *quadrats*, the term deriving from the common procedure of dividing up a planar region into small squares or rectangles.

In general, it may not be practical to count particles across a region by dividing the region into disjoint sets. Figure 2.7 shows one such case in which the quadrats are overlapping strips. Sampling of this kind occurs when counts are taken by walking in a straight line through a region. If every individual 'particle' is counted if it is within a distance r of the observer, then the tallies for each person walking through the region will be a count of particles in a strip of width $2r$. When the quadrats overlap, the random variables $Y_1, \ldots, Y_n$ are no longer independent, and

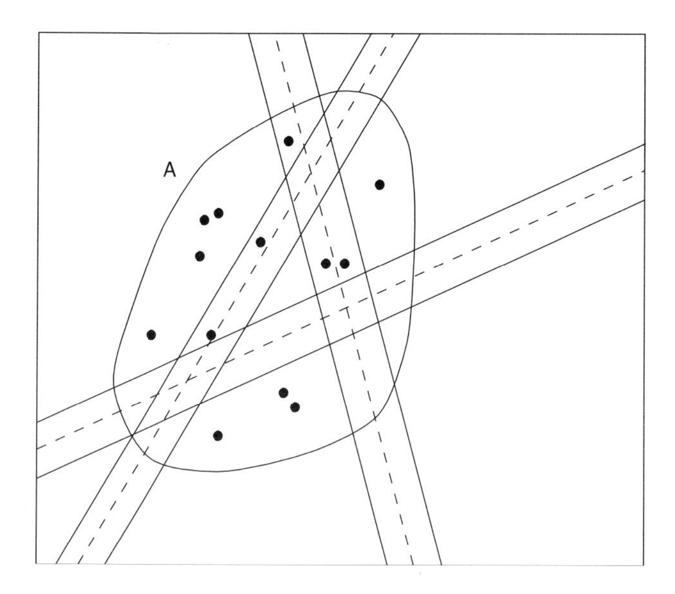

Fig. 2.7 Quadrat sampling of a region using strips as quadrats.

have a multivariate Poisson distribution. If the joint probability that $Y_j = y_j$ for $j = 1, \ldots, n$ can be calculated, then the likelihood function for θ can be found and maximised. However, as the number of quadrats increases, and as the count within each quadrat increases, the calculation of the likelihood function becomes increasingly difficult. So it is natural to consider an alternative approach to the estimation of θ which is more straightforward.

Since the first, second and product moments of the Y_j's can be calculated, a semi-parametric approach to estimation is sensible. As in the independent case, we have

$$\mu_j(\theta) = \int_{x \in A_j} \lambda_\theta(x)\,\mathrm{d}x. \tag{2.28}$$

We also have quite generally the covariance formula

$$\sigma_{jk} = \int_{x \in A_j \bigcap A_k} \lambda_\theta(x)\,\mathrm{d}x \tag{2.29}$$

for the covariance between Y_j and Y_k.

Let Y denote the $n \times 1$ column vector of observations, $\mu(\theta)$ the corresponding $n \times 1$ column vector of means, and $\Sigma(\theta)$ the $n \times n$ covariance matrix of the Y_j's. If the parameter θ is k-dimensional, then the derivative matrix $\dot{\mu}(\theta)$ will have dimension $n \times k$. As in equation (2.15), the quasi-likelihood estimate for θ will be the solution to the equation

$$\dot{\mu}^{\mathrm{t}}(\hat{\theta})\,\Sigma^{-1}(\hat{\theta})\{Y - \mu(\hat{\theta})\} = 0. \tag{2.30}$$

We shall develop some numerical methods for solving such vector-valued quasi-likelihood equations in the next chapter when we discuss Jacobi iteration.

2.7 Bibliographical notes

The theory of efficient estimating functions was proposed by Godambe (1960) and independently by Durbin (1960). The theory of quasi-likelihood was introduced by Wedderburn (1974, 1976), and was not initially derived using Godambe efficient estimating functions. Rather, it was through properties such as unbiasedness and information unbiasedness that it was recommended and motivated. However, subsequent research showed that quasi-likelihood methods share many of the properties of likelihood methods because they are both Godambe efficient within an appropriate class of estimating functions. Extensions of the quasi-likelihood methodology to semi-martingale processes using the Doob–Mayer decomposition were introduced by Hutton and Nelson (1986). The reader who is interested in a more complete description of quasi-likelihood methods is referred to Heyde (1997).

The geometric approach to estimating functions was proposed by Small and McLeish (1994). In particular, it was demonstrated that Godambe's efficiency criterion also had an interpretation as a measure of local sufficiency of the estimating function. For an alternative explanation of local E-sufficiency, the reader is referred to Heyde (1997).

3
Numerical algorithms

3.1 Introduction

For estimating equations that cannot be solved explicitly, it is useful to be able to find solutions using numerical methods. Such methods usually involve the construction of a sequence of approximations $\theta^{(1)}, \theta^{(2)}, \theta^{(3)}, \ldots$ to a root of the estimating function. Among available procedures, the Newton–Raphson algorithm is arguably the most popular and most successful method for finding roots. However, before choosing this, or any other method, the researcher is well served by performing a cost-benefit analysis of the available numerical methods.

A root-finding algorithm that is successful 99% of the time may seem quite satisfactory. However, if it is being used in a simulation study to determine the variance of the estimator defined by the root of an estimating function, it could be misleading. When algorithms fail to converge, they often produce values that are far from the root. So even a small set of cases where the algorithm fails can have a large effect on the sample variance determined by the simulation. Unless the researcher can detect those trials for which the algorithm fails, it is dangerous to tabulate a statistic which is so sensitive to outlying values.

Obviously, fast algorithms are better than slow ones. However, if time is not a serious factor in the analysis, then a slow algorithm that is guaranteed to converge may be more useful than a fast algorithm with occasional convergence problems. It is also useful to be able to provide a lower bound on the rate of convergence. Even very fast algorithms may have no guarantee that their rates of convergence are initially better than slow algorithms. To assess the rate of convergence of an algorithm we must consider the error as a function of the number of iterations the algorithm runs. In practice, it is impossible to give a sharp bound for the error of the approximation after a fixed number of iterations. The error of an algorithm can be assessed in two ways:

1. we can either determine the asymptotic size of the error, as the number of interations goes to infinity, or
2. we can try to determine an upper bound for the size of the error.

The following example illustrates how the asymptotic error can be obtained for the zero-truncated Poisson model.

Example 3.1 Truncated Poisson model

Consider the estimation of a parameter θ from the zero-truncated Poisson distribution having probability function

$$p(y;\theta) = \frac{\theta^y e^{-\theta}}{(1-e^{-\theta})y!}$$

where $\theta > 0$ and $y = 1, 2, 3, \ldots$. The zero-truncated Poisson distribution arises in the analysis of Poisson count data where a count is reported only when one or more events are observed. The likelihood equation for $\hat{\theta}$ based upon a random sample $Y_1, \ldots, Y_n$ reduces to

$$Y - \frac{n\hat{\theta}}{1-e^{-\hat{\theta}}} = 0 \tag{3.1}$$

where $Y = \sum Y_j$. To find a solution to equation (3.1), we could rewrite the equation as

$$\hat{\theta} = \bar{Y}\{1 - \exp(-\hat{\theta})\} \tag{3.2}$$

where $\bar{Y} = Y/n$. Thus $\hat{\theta}$ is seen to be a sample mean that is rescaled to adjust for zero-truncation. The reformulated identity suggests the iteration $\theta^{(1)} = \bar{Y}$ and

$$\theta^{(j+1)} = \bar{Y}\{1 - \exp(-\theta^{(j)})\} \tag{3.3}$$

for $j = 1, 2, 3, \ldots$. This iteration is illustrated geometrically in Figure 3.1.

First we need to check more formally whether this converges. Note that all iterates after the first lie in the bounded interval $(0, \bar{Y})$. Moreover, equation (3.2) tells us that $\hat{\theta}$ will also lie in this interval, where it is a fixed point of the iteration. Now

$$\begin{aligned}\theta^{(j+1)} - \hat{\theta} &= \bar{Y}\{\exp(-\hat{\theta}) - \exp(-\theta^{(j)})\} \\ &= \bar{Y}\exp(-\hat{\theta})\{(\theta^{(j)} - \hat{\theta}) + o(\theta^{(j)} - \hat{\theta})\}\end{aligned} \tag{3.4}$$

as $\theta^{(j)} \to \hat{\theta}$. So $\hat{\theta}$ will be an attractive fixed point of the iteration if $\bar{Y}\exp(-\hat{\theta}) < 1$. Next, let us define a real number α by writing $\hat{\theta} = \alpha\bar{Y}$. Since $\hat{\theta}$ lies strictly between 0 and $\bar{Y}$, the constant α must lie strictly between 0 and 1. It turns out that we can write both $\bar{Y}$ and $\hat{\theta}$ explicitly in terms of α. From equation (3.2) we find that

$$\bar{Y} = \frac{-\log(1-\alpha)}{\alpha} \quad \text{and} \quad \hat{\theta} = -\log(1-\alpha).$$

So,

$$\bar{Y}\exp(-\hat{\theta}) = \frac{(\alpha-1)\log(1-\alpha)}{\alpha}.$$

But it is easily checked that

$$0 < \frac{(\alpha-1)\log(1-\alpha)}{\alpha} < 1$$

for all $0 < \alpha < 1$. Thus $\hat{\theta}$ is an attractive fixed point for the iteration.

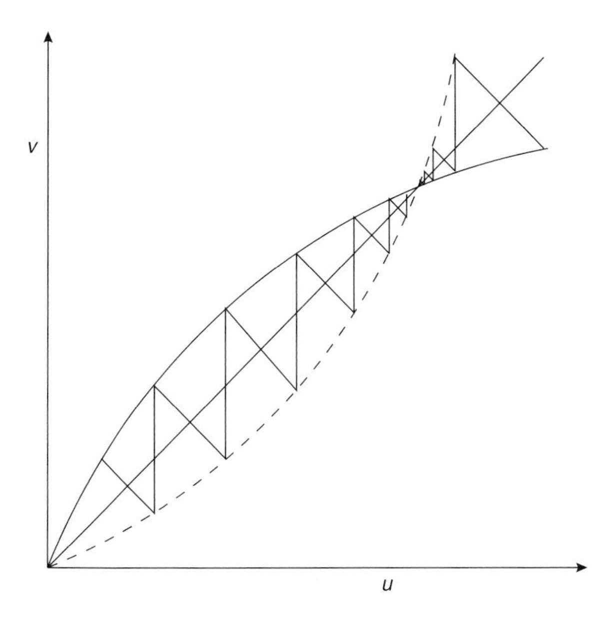

Fig. 3.1 Iterative substitution for the truncated Poisson model displayed geometrically. Here, the desired fixed point of the iteration is displayed as the intersection between two curves $v = u$ (diagonal line) and $v = \bar{Y}\{1 - \exp(-u)\}$ (plain curve). The dashed curve is the reflection of the plain curve about the line $v = u$. Two iterations are shown, one starting below the fixed point and one above. Both converge to the unique fixed point.

In a small neighbourhood of $\hat{\theta}$, equation (3.4) implies that the error $\theta^{(j)} - \hat{\theta}$ goes geometrically to zero with geometric ratio $\bar{Y}\exp(-\hat{\theta})$. It would be rather natural to say that this rate of convergence is geometric. However, this is not the standard terminology in the literature. Following standard terminology here, we shall say that the iterative substitution converges *linearly* to $\hat{\theta}$. The reason for this terminology is that the number of *correct decimal places* in the approximation is roughly a linear function of the *number of iterations*.

To illustrate this, let us consider the solution to equation (3.2) for $\bar{Y} = 3.0$, say. Using iterative substitution, we get Table 3.1 . The rightmost column of Table 3.1 shows the number of correct decimal places, which can be seen to be increasing at an approximately linear rate.

By determining the asymptotic nature of the size of an error in an algorithm, it is sometimes possible to speed up the algorithm. For example, when $\theta^{(j)} - \hat{\theta}$ converges geometrically to zero, approximately, then the values $\theta^{(j)}$ can be regarded as approximately equal to the partial sums of a geometric series. As infinite geometric series can be summed exactly it is natural to try to 'second guess' the outcome of the iteration by fitting a geometric series to the values $\theta^{(j)}$. We shall discuss this when we introduce Aitken acceleration of slow algorithms. Aitken acceleration can greatly improve the convergence of a slow algorithm.

Table 3.1 Correct decimal places when iterating using the substitution algorithm.

Step	θ	Decimal accuracy
1	3.0000000000	0
2	2.8506387949	1
3	2.8265778539	2
4	2.8223545493	2
5	2.8216027120	3
6	2.8214685358	4
7	2.8214445795	4
8	2.8214403020	4
9	2.8214395382	5
10	2.8214394018	6
11	2.8214393774	8
12	2.8214393731	8
13	2.8214393723	9

The second method of error analysis is to obtain an upper bound on the error which holds not just asymptotically but for all steps of the iteration. The advantage of being able to bound the error is that the bound can usually be transformed into a stopping rule for the algorithm. In the following example, we see how a stopping rule can be invoked for a polynomial estimating function from such a bound on the error.

Example 3.2 Error bound for polynomial estimating equation

Suppose that θ is real-valued and $g(\theta)$ can be written in the form

$$g(\theta) = a_0(y)\theta^m + a_1(y)\theta^{m-1} + \cdots + a_{m-1}(y)\theta + a_m(y)$$

with (complex-valued) roots $z_1(y), z_2(y), \ldots, z_m(y)$. Somewhere among these roots is the desired (real-valued) estimator for θ, namely $\hat{\theta}$. We can write

$$g(\theta) = a_0(y) \prod_{j=1}^{m} \{\theta - z_j(y)\} \tag{3.5}$$

when $a_0(y) \neq 0$. So

$$\frac{a_m(y)}{a_0(y)} = (-1)^m \prod_{j=1}^{m} z_j(y). \tag{3.6}$$

Suppose that we have an algorithm $\theta^{(j)}$, for $j = 1, 2, \ldots$ which provides successive approximations to the root $\hat{\theta}$. The particular form of this algorithm need not concern us here. Instead, the rule that the algorithm uses for termination will be of interest to us. Let us assume that the algorithm checks the value of $g(\theta^{(j)})$ at each step j. If the absolute value of $g(\theta^{(j)})$ is large, then it is reasonable to suppose that the algorithm is far from the root. On the other hand, when $g(\theta^{(j)})$ is close to zero the approximation $\theta^{(j)}$ can be presumed to be good enough. Therefore, the algorithm terminates when $|g(\theta^{(j)})|$ is less than or equal to some pre-specified value $\epsilon > 0$. We shall suppose that this occurs when $\theta^{(j)}$ equals some value θ^*, say. Is there any way that we can use the fact that $|g(\theta^*)| \leq \epsilon$ to get a bound on the error of the algorithm?

It turns out that the answer is yes. Using equations (3.5) and (3.6), we get

$$\frac{\epsilon}{|a_m(y)|} \geq \left|\frac{g(\theta^*)}{a_m(y)}\right| = \prod_{j=1}^{m}\left|1 - \frac{\theta^*}{z_j(y)}\right|.$$

We can take the m-th root of each side of this inequality. Since the minimum of m positive quantities is less than or equal to their geometric mean we obtain the inequality

$$\min_j \left|1 - \frac{\theta^*}{z_j(y)}\right| \leq \sqrt[m]{\frac{\epsilon}{|a_m(y)|}}. \tag{3.7}$$

Equation (3.7) can also be written as

$$\min_j \left|\frac{\theta^* - z_j(y)}{z_j(y)}\right| \leq \sqrt[m]{\frac{\epsilon}{|a_m(y)|}}. \tag{3.8}$$

Inequality (3.8) establishes a bound on the relative error of θ^* as an approximation to *some root* of the estimating equation. Note that we would prefer to have

$$\left|\frac{\theta^* - \hat{\theta}}{\hat{\theta}}\right| \leq \sqrt[m]{\frac{\epsilon}{|a_m(y)|}}. \tag{3.9}$$

However, this cannot be concluded directly unless we know that the algorithm is closer to the desired root $\hat{\theta}$ than to any other complex root of the polynomial equation. Only if the algorithm is known to converge to $\hat{\theta}$ can this be concluded for a sufficiently large number of iterations. Note also that there is no guarantee that the bound is close to the error. So it is problematic to use the bound as a method for terminating the algorithm; it could be that the true error is many orders of magnitude smaller than the calculated bound.

To illustrate the bound in (3.8), let us consider the roots of a simple cubic of the form

$$g(\theta) = \theta^3 - 7\theta + 5.$$

This has three real roots: one negative, one between 0 and 1, and one greater than 2. Let us suppose that the root in the interval $(0, 1)$ is of interest, and that an iterative substitution algorithm of the form

$$\theta^{(j+1)} = \frac{(\theta^{(j)})^3 + 5}{7}$$

Table 3.2 Error bounds for the substitution algorithm.

Step	θ	$\lvert g(\theta)\rvert$	Rel. error bound	Correct rel. error
1	0.00000000	5.000000	1.000000	1.000000
2	0.71428571	0.364431	0.417717	0.087543
3	0.76634736	0.085635	0.257767	0.021037
4	0.77858097	0.021900	0.163615	0.005410
5	0.78170953	0.005712	0.104540	0.001413
6	0.78252558	0.001498	0.066907	0.000371
7	0.78273952	0.000393	0.042840	0.000097
8	0.78279568	0.000103	0.027433	0.000026
9	0.78281043	0.000027	0.017568	0.000007
10	0.78281430	0.000007	0.011250	0.000002

is initialised at $\theta^{(1)} = 0$ to find this root. The error analysis for the algorithm is displayed in Table 3.2. The relative error bound is that given by the right-hand side of (3.8) using $|g(\theta)|$ for ϵ. The correct relative error is the value determined by the left-hand side of (3.9).

It can be seen from Table 3.2 that the bound is valid for all steps of the iteration shown. However, the quality of the bound as a guide to the correct relative error rapidly deteriorates. Nevertheless, the relative error bound proceeds geometrically to zero, although at a slower rate than the correct relative error.

The methods that we shall consider in this chapter are root-finding algorithms from the numerical analysis literature. While much of this methodology is directly applicable to estimating functions, there are some special features of estimating functions which require that the algorithms be modified in certain cases. For regular estimating functions, there is a particular root of interest, namely the one which is consistent and closest to the true value of the parameter for a sufficiently large sample size. A consequence of this fact is that statistical root searching is more specialised than the task of simply finding the roots of an equation. In most contexts, for example, the divergence of a root-search algorithm should be treated as a defect. However, the situation is not that clear in some statistical applications where the statistician is interested in finding the consistent root. An algorithm that fails to converge may provide better diagnostics for statistical purposes than one which converges to an inappropriate root.

It may be known that the consistent root will be close to some simple consistent estimator with high probability. If this is the case, the consistent estimator is a natural value for initialising a root search algorithm. For some applications involving location parameters, it will be known that all reasonable estimators will lie in the interval defined between the smallest and largest observation in the data set. Algorithms that require a search over successively smaller intervals can be initialised over the range of values defined by the data.

Knowledge about the ergodic properties of an estimating function can also be useful in the implementation of an algorithm. For example, an information unbiased estimating function should have sample covariance matrix and derivatives of opposite sign and approximately equal modulus at the consistent root. Such criteria can be built into a root searching method.

Another consideration that is particularly applicable to statistics is to avoid specious accuracy in algorithms. For example, if asymptotic theory tells us that the difference between the true value of a parameter and an estimate is likely to be of order 10^{-2}, there is little point in running an algorithm to generate accuracy to six decimal places for the root. Six digits may look very impressive in a report, but only the first three will have much relationship with reality.

3.2 The bisection method

Among all available methods for finding the root of a real-valued function of a single real variable, the bisection method is arguably the simplest and most reliable. It also yields a simple and useful bound on the error. On the negative side, the rate of convergence is not particularly fast, and cannot easily be accelerated. To use the bisection method, we need a few basic conditions (Figure 3.2). Suppose that g is continuous for all θ in the interval $[\theta^{(1)}, \theta^{(2)}]$, and that

$$g(\theta^{(1)})g(\theta^{(2)}) < 0.$$

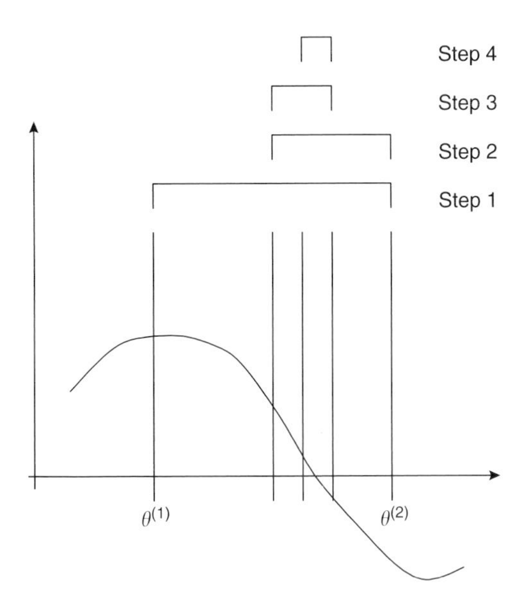

Fig. 3.2 The bisection method.

By the intermediate value theorem, a root must lie in the interval $(\theta^{(1)}, \theta^{(2)})$. The *bisection method* decreases the size of the interval by evaluating $g(\theta^{(3)})$, where

$$\theta^{(3)} = (\theta^{(1)} + \theta^{(2)})/2.$$

If this is non-zero, then the interval $(\theta^{(1)}, \theta^{(2)})$ is replaced by $(\theta^{(1)}, \theta^{(3)})$ or by $(\theta^{(3)}, \theta^{(2)})$, so that g undergoes a sign change over the new interval.

The algorithm proceeds in this fashion, replacing each interval with the new interval derived by bisecting the previous interval. At any stage, if $g(\theta^{(j)}) = 0$, the algorithm terminates. The given subinterval at each step of the algorithm provides a natural bound on the error of the algorithm. If the algorithm is terminated when the interval becomes sufficiently small, then the midpoint of the last interval can be used as the best approximation to the root.

While the bisection method is very reliable, it is possible for the practitioner to go astray if insufficient care is used. Most notably, when the function is not continuous on the interval, the bisection method can provide erroneous results. Note also that the method provides no simple guarantee of finding all the roots within a given starting interval $(\theta^{(1)}, \theta^{(2)})$. A sign change over an interval is sufficient to ensure the existence of a root for a continuous function, but is not necessary. To have a better chance to find all roots in a given interval, it is best to partition the original interval into more than two subintervals, and to check for sign changes in all intervals.

Example 3.3 Truncated Poisson model (continued)

We can apply the bisection method to the truncated Poisson in Example 3.1 as well. As noted above, the solution to the estimating equation

$$\bar{Y}\{1 - \exp(\hat{\theta})\} - \hat{\theta} = 0$$

lies in the interval $[0, \bar{Y}]$. The value $\theta = 0$ is an extraneous root that can be ignored. Taking this interval as a start for the bisection algorithm with $\bar{Y} = 3.0$ once again, and iterating we get Table 3.3.

The most obvious feature displayed in Table 3.3 is that convergence is very slow, even compared to the convergence of the substitution algorithm. In part this is due to the large size of the original interval which multiplies through as a scale factor in the sizes of subsequent intervals. However, the virtue of the bisection method is best seen when the rate of convergence is less important than the need to ensure a bound on the error that is guaranteed to go to zero. If speed is even moderately important for the algorithm, then other methods should be considered.

3.3 The method of false positions

The method of false positions is also known as the *regula falsi*. Consider a real-valued function, having a change of sign over an interval. If the absolute value of the function is large at one end of the interval and small at the other, then

Table 3.3 The bisection method for the truncated Poisson model.

Step	θ	Decimal accuracy
1	1.50000000	0
2	2.25000000	0
3	2.62000000	0
4	2.81250000	1
5	2.90625000	0
6	2.85937500	1
7	2.83593750	1
8	2.82421875	2
9	2.81835938	1

the root is usually closer to the endpoint where the function is small than to the endpoint where it is large. This intuition can be made rigorous if the function is approximately linear over the interval.

Suppose that g is a continuous real-valued estimating function defined for $\theta \in [a, b]$. Approximating g by a linear function passing though $(a, g(a))$ and $(b, g(b))$, we get

$$f(\theta) = g(a) + \frac{g(b) - g(a)}{b - a}(\theta - a)$$

which has a root at

$$\theta = \frac{ag(b) - bg(a)}{g(b) - g(a)}. \tag{3.10}$$

Equation (3.10) defines the iteration associated with the method of *false positions*. Suppose that $\theta^{(j)}$ and $\theta^{(j+1)}$ are two approximations to $\hat{\theta}$. We shall suppose that $\hat{\theta}$ lies in the interval between $\theta^{(j)}$ and $\theta^{(j+1)}$, and that g is positive on one of the two iterates, and negative on the other. We define

$$\theta^{(j+2)} = \frac{\theta^{(j)} g(\theta^{(j+1)}) - \theta^{(j+1)} g(\theta^{(j)})}{g(\theta^{(j+1)}) - g(\theta^{(j)})}. \tag{3.11}$$

To update the algorithm we drop the value at which g has the same sign as at $\theta^{(j+2)}$.

The method of false positions has been found to converge faster than the linear rate that is typical of iterative substitution methods such as given by equation (3.3). However, it can go quite wrong as illustrated in Figure 3.4. In this case, the slow convergence is due to the fact that the function g is fit rather poorly by a linear function on the given interval.

Example 3.4 Truncated Poisson model (continued)

To illustrate the method of false positions, let us consider the solution to

$$\bar{Y}\{1 - \exp(-\hat{\theta})\} - \hat{\theta} = 0$$

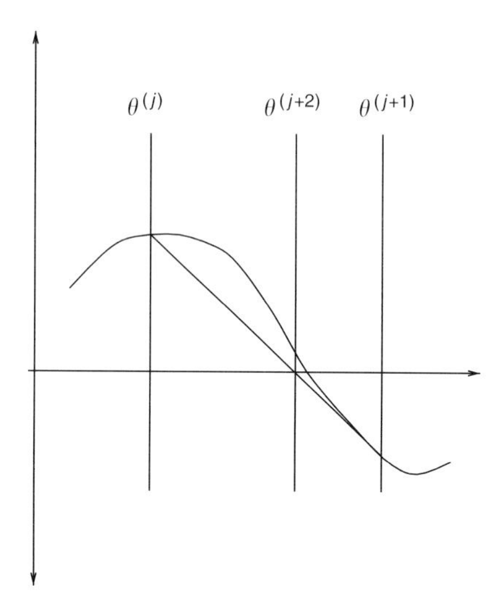

Fig. 3.3 The method of false positions.

Table 3.4 The method of false positions for the truncated Poisson model.

Step	θ	Decimal accuracy
1	1.500000000	0
2	3.000000000	0
3	2.771379081	0
4	2.820519493	2
5	2.821422832	4
6	2.821439075	6
7	2.821439367	7
8	2.821439372	9

from Example 3.1. To avoid convergence to the extraneous root $\theta = 0$, we shall start the iteration on the interval [1.5, 3.0] found after the first step of the bisection algorithm. The results are displayed in Table 3.4.

The convergence of the method of false positions has been found to be faster than the substitution method. Table 3.4 helps illustrate this fact. The method of false positions is guaranteed to provide a sequence of nested intervals containing a root if the function is continuous. Unfortunately, unlike the bisection method there is no guarantee that the size of the interval will decrease by a reasonable factor at each step. As Figure 3.4 illustrates, functions which are not approximately linear over the interval of consideration may require many steps for reasonable convergence. Nevertheless, the method of false positions is often very successful with a rapid convergence. This is particularly useful if it is difficult to calculate the derivative of a function that is needed for Newton–Raphson.

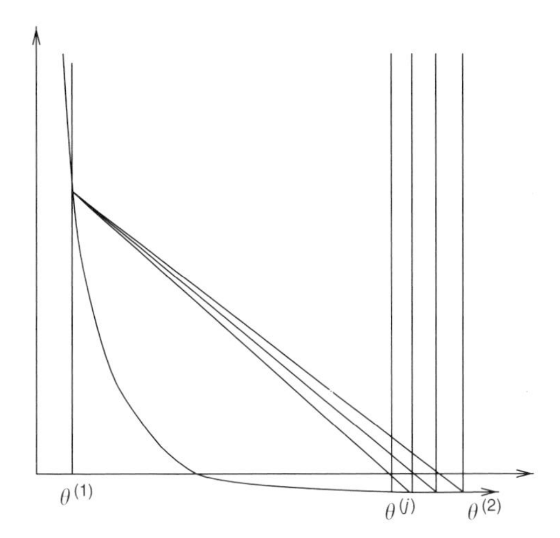

Fig. 3.4 An inappropriate application of the method of false positions.

A method that is closely related to the method of false positions is the *secant method*. It can be regarded as a relaxation of false positions to the case where the starting interval for approximation does not have an interval with a sign change of the function. In such cases, formula (3.11) can still be applied. The standard rule for such cases is to keep the most recent values of θ, i.e., $\theta^{(j+1)}$ and $\theta^{(j+2)}$ to compute the next value. Caution must be exercised when using the secant method on an estimating function which is not well approximated by a secant line.

In higher dimensions, the secant method is known as the *quasi-Newton method*. The reason for this terminology is that linear secant based upon the most recent iterates, $\theta^{(j)}$ and $\theta^{(j+1)}$, is an approximation to the tangent line through the most recent iterate $\theta^{(j+1)}$. So the secant method can also be interpreted as a Newton–Raphson algorithm with the derivative $\dot{g}(\theta^{(j+1)})$ replaced by a finite difference approximation. Quasi-Newton algorithms form a family of techniques that build up information about the derivative of the estimating function using the succession of iterates that have been computed at earlier stages.

We shall consider quasi-Newton algorithms in more detail in Section 3.6.

3.4 Muller's method

Muller's method is a natural extension of the method of false positions which can be used to resolve this difficulty. Since the linear approximation is a poor fit to the function in Figure 3.4, we can try to improve the approximation by fitting a quadratic to g using the three iterates $\theta^{(j)}$, $\theta^{(j+1)}$ and $\theta^{(j+2)}$ to generate a fourth. We shall omit some of the details involving fitting the required quadratic, and note that

$$\theta^{(j+3)} = \theta^{(j+2)} - \frac{2c_{j+2}(\theta^{(j+2)} - \theta^{(j+1)})}{b_{j+2} \pm \sqrt{b_{j+2}^2 - 4a_{j+2}c_{j+2}}} \tag{3.12}$$

where, letting

$$q_j = \frac{\theta^{(j)} - \theta^{(j-1)}}{\theta^{(j-1)} - \theta^{(j-2)}},$$

we have

$$a_j = q_j g(\theta^{(j)}) - q_j(1 + q_j) g(\theta^{(j-1)}) + q_j^2 g(\theta^{(j-2)}),$$
$$b_j = (2q_j + 1) g(\theta^{(j)}) - (1 + q_j)^2 g(\theta^{(j-1)}) + q_j^2 g(\theta^{(j-2)}),$$

and

$$c_j = (1 + q_j) g(\theta^{(j)}).$$

The troublesome aspect of equation (3.12) is that there are two solutions for $\theta^{(j+3)}$. The standard protocol for selecting a root is to choose that value of $\theta^{(j+3)}$ which is the closer of the two to $\theta^{(j+2)}$. However, for estimating functions, other selection criteria are worth considering. For example, if a $\sqrt{n}$-consistent estimator $\tilde{\theta}$ is available in closed form, then it may be reasonable to choose the solution to (3.12) which is closer to $\tilde{\theta}$ for the first few iterations, until the algorithm stabilises.

3.5 Iterative substitution and the contractive maps

Up to this point we have considered techniques for finding the root of a real-valued function of a real variable. However, in standard statistical applications, the parameter and the estimating function are vector-valued. Although many of the algorithms can be generalised into higher dimensions, such generalisations often lose the advantages of their lower-dimensional versions. For example, in dimension one, the bisection method and the method of false positions rely on the intermediate value theorem to guarantee convergence to a root. In higher dimensions there is no simple way that a root can be 'trapped' in a bounded region by evaluating a function at a finite set of points.

Suppose that g is a vector-valued estimating function of a vector parameter θ. Many algorithms to solve $g(\theta) = 0$ do so by constructing a function h and executing an iteration of the form $\theta^{(j+1)} = h(\theta^{(j)})$. Provided that the iteration converges and that the equations $\theta = h(\theta)$ and $g(\theta) = 0$ are equivalent, the algorithm will approximate roots with any required accuracy. For example, a simple substitution algorithm is obtained by setting $h(\theta) = \theta + g(\theta)$. However, there is no guarantee that the iteration

$$\theta^{(j+1)} = \theta^{(j)} + g(\theta^{(j)})$$

will converge. However, the following result is sufficient for the purpose.

Proposition 3.5 *Let $h : \Theta \to \Theta$ be a function defined on a closed domain Θ of $\mathbb{R}^k$. Suppose there exists a constant $c < 1$ such that*

$$\|h(\theta^{(1)}) - h(\theta^{(2)})\| \leq c\|\theta^{(1)} - \theta^{(2)}\| \tag{3.13}$$

for all $\theta^{(1)}$ and $\theta^{(2)}$ in Θ. Then

1. *the equation $\theta = h(\theta)$ has a unique solution $\hat{\theta} \in \Theta$;*
2. *for any $\theta^{(1)} \in \Theta$, the sequence $\theta^{(j)}$ defined by $\theta^{(j+1)} = h(\theta^{(j)})$ converges to $\hat{\theta}$; and*
3. *for all j, we have*

$$\|\theta^{(j)} - \hat{\theta}\| \leq \frac{c^j}{1-c}\|\theta^{(2)} - \theta^{(1)}\|.$$

To prove this result, we begin by noting that the sequence $\theta^{(j)}$ is well-defined from its starting point and lies inside Θ for all j. Proceeding by induction on equation (3.13), we get

$$\|\theta^{(j+1)} - \theta^{(j)}\| \leq c^j\|\theta^{(2)} - \theta^{(1)}\|.$$

For any j and k with $k > j$, we can write

$$\theta^{(k)} - \theta^{(j)} = \sum_{m=j}^{k-1}(\theta^{(m+1)} - \theta^{(m)}).$$

So by the triangle inequality, we have

$$\begin{aligned}\|\theta^{(k)} - \theta^{(j)}\| &\leq \sum_{m=j}^{k-1}\|\theta^{(m+1)} - \theta^{(m)}\| \leq \|\theta^{(2)} - \theta^{(1)}\|\sum_{m=j}^{k-1} c^m \\ &\leq \|\theta^{(2)} - \theta^{(1)}\|\frac{c^j}{1-c}.\end{aligned}$$

Taking the double limit as $j, k \to \infty$, we see that

$$\lim_{j,k}\|\theta^{(k)} - \theta^{(j)}\| \leq \|\theta^{(2)} - \theta^{(1)}\| \lim_j \frac{c^j}{1-c} = 0.$$

So the sequence $\theta^{(j)}$ satisfies the Cauchy criterion for convergence. Thus it has a limit in Θ, which we can call $\hat{\theta}$. Next, we need to show that $\hat{\theta}$ is a solution to the equation $\theta = h(\theta)$. To do this, we first note that h is a continuous function. This follows from equation (3.13), which is a *Lipschitz condition* for h. So

$$\lim_j h(\theta^{(j)}) = h(\lim_j \theta^{(j)}) = h(\hat{\theta}).$$

Therefore,

$$h(\hat{\theta}) = \lim_j h(\theta^{(j)}) = \lim_j \theta^{(j+1)} = \hat{\theta}$$

as required.

Finally, we need to show that $\hat{\theta}$ is the unique fixed point of h. Suppose that $\hat{\theta}$ and $\tilde{\theta}$ are two fixed points of h. By (3.13), we must have

$$\|h(\tilde{\theta}) - h(\hat{\theta})\| \leq c\|\tilde{\theta} - \hat{\theta}\|.$$

Since $\tilde{\theta}$ and $\hat{\theta}$ are both fixed points of h, this reduces to $\|\tilde{\theta} - \hat{\theta}\| \leq c\|\tilde{\theta} - \hat{\theta}\|$. But $c < 1$. Therefore, $\|\tilde{\theta} - \hat{\theta}\| = 0$. So $\tilde{\theta} = \hat{\theta}$, as was to be shown.

3.6 Newton–Raphson and its generalisations

3.6.1 NEWTON–RAPHSON AS A SUBSTITUTION ALGORITHM

The Newton–Raphson root-finding algorithm has been discussed earlier, and needs little introduction here. At this stage, we shall consider its relationship to iterative substitution methods. We shall also see how to find flexible families of methods which bridge the gap, so to speak, between slow and reliable methods and fast methods like Newton–Raphson. The goal is to build iterative methods which are both reliable—in the sense of being guaranteed to converge—and fast.

If we write an iterative substitution algorithm for an equation of the form $g(\theta) = 0$, we obtain

$$\theta^{(j+1)} = \theta^{(j)} + cg(\theta^{(j)}) \tag{3.14}$$

which could possibly diverge or converge at a linear rate. If convergence does occur, it is usually because the iteration is locally a contractive mapping within some neighbourhood of the root in the sense of (3.13). The constant c can be chosen in this iteration to ensure that the iteration is contractive. Note that (3.14) can include the multiparameter case. In such situations the constant c will be a square matrix of appropriate dimensions.

An extension of the class of iterations defined by (3.14) is

$$\theta^{(j+1)} = \theta^{(j)} + c(\theta^{(j)})g(\theta^{(j)}). \tag{3.15}$$

With this added flexibility, we are in a position to ask how the multiplying function $c(\theta)$ should be chosen so as to ensure that the convergence in (3.15) is most rapid. We will not discuss the details of the argument at this stage. However, it is possible to show that the rate of convergence is maximised locally about $\hat{\theta}$ if we choose

$$c(\theta) = -\{\dot{g}(\theta)\}^{-1}$$

which corresponds to Newton–Raphson iteration. With this choice, the convergence can be shown to be quadratic provided $\ddot{g}(\theta)$ exists and is continuous.

The following modifications of Newton–Raphson are often useful:

- *Whittaker's method*, a modified Newton–Raphson algorithm, is based upon the fact that in the later stages of iteration, the quantity $\dot{g}(\theta^{(j)})$ does not change substantially when j is sufficiently large. If this is the case, it can be useful not to waste time updating $\dot{g}$ for iterations beyond some integer m. While the rate of convergence of the algorithm

$$\theta^{(m+j+1)} = \theta^{(m+j)} - \{\dot{g}(\theta^{(m)})\}^{-1} g(\theta^{(m+j)})$$

is usually linear, it may be faster in practice than Newton–Raphson. This is because Newton–Raphson, whose rate of convergence is quadratic, may not be as fast as Whittaker's method in the early stages of iteration. Therefore, Whittaker's method can sometimes reach the required accuracy earlier than the former method. The most obvious difficulty with Whittaker's method is that it is likely to be difficult to determine the best value of m to use.

- Sometimes

$$c(\theta) = -\{E_\theta \dot{g}(\theta)\}^{-1}$$

is easier to compute or more stable than $-\{\dot{g}(\theta)\}^{-1}$. When g is the score function, this method is known as *Fisher scoring*. The rationale behind Fisher scoring is similar to Whittaker's method. Like Whittaker's method, it converges linearly to the root, but can be faster in practice than Newton–Raphson for similar reasons. It is particularly advantageous to use Fisher scoring early in the iteration if the algorithm is started in a domain over which $g(\theta)$ is not approximately linear. If this is the case, then early iterations of Newton–Raphson may be far less efficient in converging to a point close to the root. The advantages of Fisher scoring are most evident when the average value of $\dot{g}(\theta)$ over the domain of the iteration is better represented by $E_\theta\{\dot{g}(\theta)\}$ than by $\dot{g}(\theta^{(1)})$. Fisher scoring also offers advantages if each step can be computed more easily than each step of Newton–Raphson. This will often be the case with likelihood equations where the information function can be computed in simple form. However, there are numerous examples of models whose information functions cannot be expressed analytically. For these models, the particular advantages of Fisher scoring may be offset by particular computational difficulties.

- Figure 3.5 shows what happens when Newton–Raphson is used to find a root $\hat{\theta}$ where $\dot{g}(\hat{\theta})$ is infinite. To compensate for this instability, it is natural to try to shrink the value of $\theta^{(j+1)}$ back towards $\theta^{(j)}$. The choice

$$c(\theta) = \alpha\{\dot{g}(\theta)\}^{-1}$$

with $0 < \alpha < 1$, attempts to provide the right compromise. If α is too close to one, the iteration will be unstable as shown. However, a value of α close to zero, while stable, will converge very slowly.

- The opposite problem can occur if $\dot{g}(\hat{\theta}) = 0$, as shown in Figure 3.6. In this case, we must use $\alpha > 1$ as the desired scaling factor. Theory shows that when $\dot{g}(\hat{\theta}) = 0$, and $\ddot{g}(\hat{\theta}) \neq 0$, the choice $\alpha = 2$ is best, and achieves quadratic convergence.

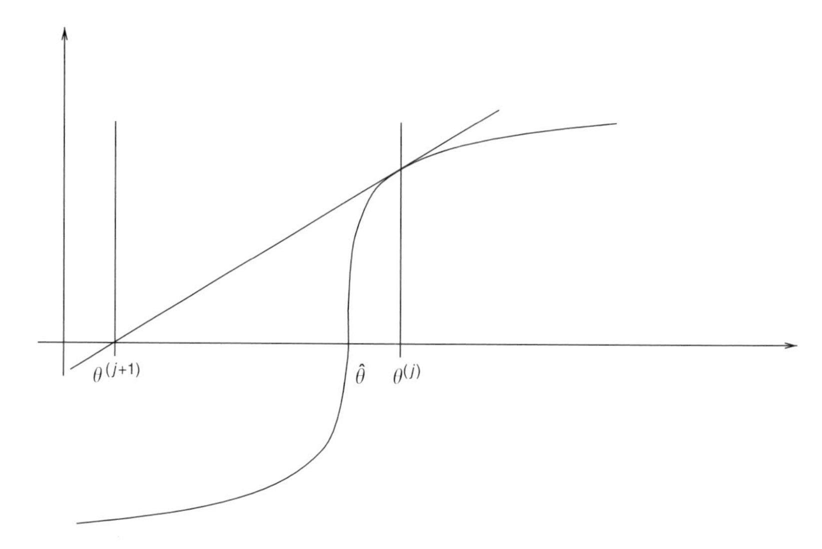

Fig. 3.5 One step of Newton–Raphson iteration when $\dot{g}(\hat{\theta})$ is infinite.

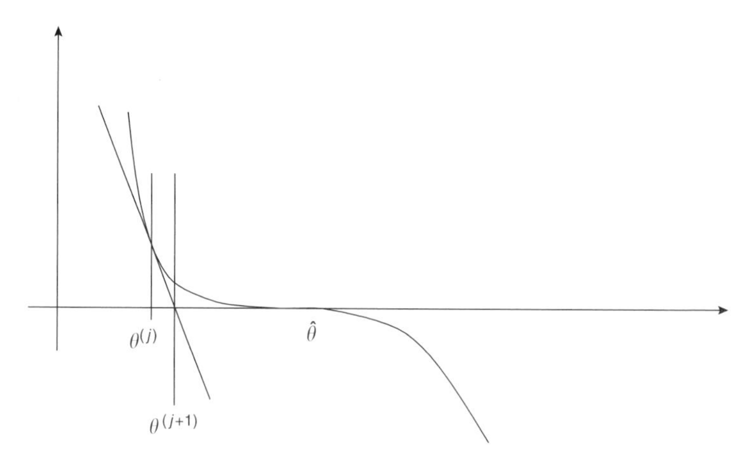

Fig. 3.6 One step of Newton–Raphson iteration when $\dot{g}(\hat{\theta})$ is zero.

3.6.2 QUASI-NEWTON ALGORITHMS

The next modification of the basic Newton–Raphson algorithm that we shall consider is the quasi-Newton family of algorithms. We have already seen an example of this. The secant method for finding a root can be interpreted as an approximation to the Newton–Raphson algorithm with the derivative $\dot{g}(\theta^{(j)})$ replaced by a finite difference. Quasi-Newton algorithms are extensions of this idea into higher dimensions. Suppose that $\theta \in \mathbb{R}^k$ is a vector, and that $g(\theta)$ is a vector-valued function taking values in $\mathbb{R}^k$ as well. As usual, we represent θ and

$g(\theta)$ as column vectors. A linear approximation to the function $g(\theta)$ at the point $\theta^{(j)}$ will take the form

$$g(\theta^{(j)}) + a(\theta - \theta^{(j)}) \tag{3.16}$$

where a is some matrix of dimension $k \times k$. When $a = \dot{g}(\theta^{(j)})$, then the solution $\theta^{(j+1)}$ of the linear approximation yields Newton–Raphson iteration. If $\dot{g}(\theta^{(j)})$ is not available, then the natural analog of the secant method is to choose a matrix a which solves

$$g(\theta^{(j-1)}) = g(\theta^{(j)}) + a(\theta^{(j-1)} - \theta^{(j)}). \tag{3.17}$$

This equation is known as the *secant equation*. Unfortunately, there can be more than one solution to the secant equation when θ and $g(\theta)$ are vector-valued. It might be tempting to try to solve this problem by incorporating earlier iterates. However, if the most recent iterates $\theta^{(j)}$, $\theta^{(j-1)}$, $\theta^{(j-2)}$, and so on, lie in a flat set of low dimension, the matrix a will continue to be under-specified.

A solution to this problem was proposed by Broyden (1965). Suppose that we have some value $a^{(j-1)}$ which was used to compute $\theta^{(j)}$, and we wish to update $a^{(j-1)}$ to $a^{(j)}$ so that we can compute $\theta^{(j+1)}$ using the linear approximation in equation (3.16). Broyden's recommendation was to use that solution $a^{(j)}$ to the secant equation (3.17) whose difference $a^{(j)} - a^{(j-1)}$ is minimised in the sense that its Frobenius norm is minimised. The particular choice that minimises the Frobenius norm is

$$a^{(j)} = a^{(j-1)} + \frac{[\{g(\theta^{(j)}) - g(\theta^{(j-1)})\} - a^{(j-1)}(\theta^{(j)} - \theta^{(j-1)})](\theta^{(j)} - \theta^{(j-1)})^{\mathrm{t}}}{(\theta^{(j)} - \theta^{(j-1)})^{\mathrm{t}}(\theta^{(j)} - \theta^{(j-1)})} \tag{3.18}$$

which is called the *Broyden update*.

This still leaves the problem of finding the initial matrix $a^{(1)}$. The recommended strategy is to approximate $\dot{g}(\theta^{(1)})$ using finite differences and to use this approximation as $a^{(1)}$.

We shall consider additional modifications of Newton–Raphson in Chapter 5. These examples indicate that while the justification for Newton–Raphson is asymptotic in the number of iterations, the algorithm used in practice will terminate after finitely many steps. During early stages of the iteration, Newton–Raphson can be unstable, in the sense of diverging from roots, or at least highly inefficient in step size. The methods above offer excellent practical alternatives to Newton–Raphson, but leave many unanswered questions about how to fine tune the methods to particular estimating functions.

3.7 The E–M algorithm

3.7.1 THE E–M ALGORITHM FOR LIKELIHOOD EQUATIONS

In some estimation problems, the ideal estimating equation is hard to solve because its computation involves integration or a difficult summation over cases. For example, suppose that $Y = (V, W)$ is a full data set for a statistical problem, but we only observe V. The marginal likelihood function for $V = v$ can be written as

$$L(\theta; v) = \int L(\theta; y) dw$$

where $L(\theta; y)$ is the full likelihood based upon $Y = y$.

To compute the marginal likelihood requires an integration that may not be simple to solve or expressible in analytically closed form. The E–M algorithm is a tool for maximising likelihoods and solving more general equations for problems where the solution would be easier if we knew the value of W.

Suppose we let

$$L(\theta; y|v) = L(\theta; y)/L(\theta; v)$$

denote the conditional likelihood for θ given v. Taking logarithms and differentiating with respect to θ, we find that the marginal score function for θ is

$$u(\theta; v) = u(\theta; y) - u(\theta; y|v) \tag{3.19}$$

where $u(\theta; y)$ and $u(\theta; y|v)$ are scores based upon y and y given v, respectively. Under standard regularity, the conditional score will be conditionally unbiased. Therefore,

$$E_\theta\{u(\theta; Y|v)|V = v\} = 0.$$

Therefore, the marginal score based upon v can be calculated conditionally as

$$u(\theta; v) = E_\theta\{u(\theta; Y)|V = v\}. \tag{3.20}$$

The E–M algorithm for solving the equation $u(\hat{\theta}; v) = 0$, is based upon this identity. We start with an initial approximation $\theta^{(1)}$ to the solution. The algorithm computes each value $\theta^{(j+1)}$ from the previous $\theta^{(j)}$ using a two-step procedure. For the *E-step* of the algorithm, we calculate the function

$$\mathcal{E}_j(\theta) = E_{\theta^{(j)}}\{u(\theta; Y)|V = v\} \tag{3.21}$$

for each value of θ. For the *M-step* of the algorithm, we solve the equation

$$\mathcal{E}_j(\theta^{(j+1)}) = 0. \tag{3.22}$$

The immediate rationale behind the algorithm can be seen by noting that if $\hat{\theta}$ is a fixed point of the algorithm, then $\hat{\theta}$ solves the equation $u(\hat{\theta}; v) = 0$. While this algorithm has been used in specific applications for many years, the terminology *E–M algorithm* derives from Dempster *et al.* (1977) in which a general framework was developed for models in the exponential family.

Example 3.6 Truncated Poisson model (continued)

To illustrate the E–M algorithm, let us return to the zero-truncated Poisson distribution considered in Example 3.1. We can regard the zero-truncated sample y

as a marginal data set, with the full data set being (y, m), where m is the number of zeros in the data. To make the augmented model precise, we shall suppose that sample of Poisson variables is collected using inverse sampling until precisely n non-zero values are obtained, where n is a fixed number. The number M of zeros in the data then has a negative binomial distribution.

The likelihood function for the augmented data is therefore

$$L(\theta; y, m) = \exp\{-(n+m)\theta\} \prod_{j=1}^{n} \frac{\theta^{y_j}}{y_j!}.$$

So the score function for the augmented data is

$$u(\theta; y, m) = -m + \sum_{j=1}^{n} \left(\frac{y_j}{\theta} - 1\right).$$

Taking the conditional expectation with respect to y, we see that

$$u(\theta; y) = E_\theta\{u(\theta; y, m)|y\} = -E_\theta(M|\, y) + \sum_{j=1}^{n} \left(\frac{y_j}{\theta} - 1\right)$$

$$= -E_\theta(M) + \sum_{j=1}^{n} \left(\frac{y_j}{\theta} - 1\right).$$

The last step follows from the fact that Y and M are independent. Since M has a negative binomial distribution, we can calculate its expectation to be

$$E_\theta(M) = \frac{n \exp(-\theta)}{1 - \exp(-\theta)}.$$

The E–M algorithm proceeds from $\theta^{(j)}$ to $\theta^{(j+1)}$ using the formula

$$-\frac{n \exp(-\theta^{(j)})}{1 - \exp(-\theta^{(j)})} + \sum_{j=1}^{n} \left(\frac{y_j}{\theta^{(j+1)}} - 1\right) = 0. \tag{3.23}$$

When we solve for $\theta^{(j+1)}$, we find that this is precisely the algorithm proposed in (3.3).

In the next example, we shall consider the application of the E–M algorithm to calculating the maximum likelihood estimator for a mixture model.

Example 3.7 Normal mixture

Let ϕ denote the probability density function for the standard normal distribution. We shall consider the mixture density

$$f(y; \theta) = \tfrac{1}{2}\phi(y - \theta + c) + \tfrac{1}{2}\phi(y - \theta - c)$$

obtained by averaging two normal distributions centred at $\theta - c$ and $\theta + c$, respectively. Let $Y_1, \ldots, Y_n$ be a random sample from $f(y; \theta)$.

After some algebraic manipulation, we can show that the likelihood equation for θ based upon $y_1, \ldots, y_n$ reduces to

$$\sum_{i=1}^{n}[w_i(\hat{\theta})\{y_i - \hat{\theta} + c\} + \{1 - w_i(\hat{\theta})\}\{y_i - \hat{\theta} - c\}] = 0 \qquad (3.24)$$

where

$$w_i(\theta) = \frac{\phi(y_i - \theta + c)}{\phi(y_i - \theta + c) + \phi(y_i - \theta - c)}.$$

The likelihood equation can be interpreted as a weighted average of the likelihood equations for each of the normal components. However, the weights themselves depend upon θ. Thus there is no simple explicit representation of $\hat{\theta}$ as a weighted mean. Nevertheless, we can write

$$\hat{\theta} = \bar{y} + \frac{c}{n}\sum_{i=1}^{n}\{2w_i(\hat{\theta}) - 1\}. \qquad (3.25)$$

This equation suggests the following algorithm for finding a root. We use the values $w_i(\theta^{(j)})$ as weights for a new weighted average of the data. So

$$\theta^{(j+1)} = \bar{y} + \frac{c}{n}\sum_{i=1}^{n}\{2w_i(\theta^{(j)}) - 1\}. \qquad (3.26)$$

It turns out that this algorithm can be obtained as an E–M algorithm. To see this, we augment the data set to include a set of independent binary random variables $Z_1, \ldots, Z_n$. Suppose

$$Z_i = \begin{cases} -1 & \text{with probability } 1/2 \\ +1 & \text{with probability } 1/2 \end{cases}$$

Generate Y_i so that

$$(Y_i | Z_i) \sim N(\theta + cZ_i, 1)$$

with all variables independent except where explicitly indicated above. The augmented data set consists of a set of n pairs $(Y_1, Z_1), \ldots, (Y_n, Z_n)$. The score function for the augmented data is

$$u(\theta; y, z) = \sum_{i=1}^{n}\left\{\frac{1 - z_i}{2}(y_i - \theta + c) + \frac{1 + z_i}{2}(y_i - \theta - c)\right\}. \qquad (3.27)$$

So the E–M algorithm is determined by the equation

$$E_{\theta^{(j)}}\{u(\theta^{(j+1)}; y, Z) | y\} = 0. \qquad (3.28)$$

Solving for $\theta^{(n+1)}$ we obtain formula (3.26).

Conditions for ensuring that $L(\theta^{(j+1)}) \geq L(\theta^{(j)})$ were given by Dempster *et al.* (1977). Conditions for convergence to the maximum likelihood estimate were studied by Boyles (1983).

While the E–M algorithm is quite reliable, its rate of convergence is usually only linear. This is not surprising, because the E–M algorithm is based upon the same kind of substitution method that was used to define simple iterative substitution earlier in the chapter. In the next section, we shall consider a method for accelerating the convergence of linear iterations including the E–M algorithm.

3.7.2 THE E–M ALGORITHM FOR OTHER ESTIMATING EQUATIONS

Wang and Pepe (2000) argued that the E–M algorithm has wider applicability than solving the likelihood equations. In this section we consider the case when $u(\theta; V)$ is replaced by any unbiased estimating function $g(\theta, V)$, where v is the observed or incomplete data. Let $g(\theta, Y)$ denote the unbiased estimating function based on the hypothetical complete data Y. Suppose that the estimating function based upon V and the estimating function based upon the complete data Y are related by the equation

$$g(\theta, V) = E_\theta\{g(\theta, Y)|V\}$$

which is similar to the likelihood case described above. This equation may hold even when the complete data estimating function $g(\theta, Y)$ is not a score function. Formally extending the ideas of last section, we solve the estimating equation $g(\theta, v) = 0$ by utilising the *expected estimating equation* (Wang and Pepe, 2000)

$$\mathcal{E}_{\theta^{(j)}}(\theta, v) = E_{\theta^{(j)}}\{g(\theta, Y)|V = v\} = 0. \tag{3.29}$$

The generalised E–M algorithm now consists of the repeated use of the following two steps until convergence:

- *Generalised E-step:* Compute the expected estimating function $\mathcal{E}_{\theta^{(j)}}(\theta, v)$ for a given value $\theta^{(j)}$;
- *Generalised M-step:* Solve the expected estimating equation

$$\mathcal{E}_{\theta^{(j)}}(\theta^{(j+1)}, v) = 0$$

for $\theta^{(j+1)}$ using the current value $\theta^{(j)}$.

Example 3.8 Longitudinal data analysis

Wang and Pepe (2000) applied the above ideas to longitudinal data analysis when measurement errors are present. Suppose that a sample consists of n individuals who are studied over time. A response measurement is recorded for individual i at time t_j. Suppose Y_{ij} denotes the response measurement of individual i at time t_j, where $i = 1, \ldots, n$ and $j = 1, \ldots, n_j$. Let X_{ij} be a covariate for individual i

at time t_j. It may happen that X_{ij} cannot be directly measured with precision, and only estimated with some quantity

$$\widehat{X}_{ij} = \sum_k W_{ijk}/m_{ij},$$

where W_{ijk}, for $k = 1, \ldots, m_{ij}$ are a sample of approximate measurements of X_{ij}.

With the true covariate data, suppose estimation of θ is based upon

$$g(\theta) = \sum_{i=1}^{n} \sum_{j=1}^{n_i} g(\theta, X_{ij}, Y_{ij}) = 0.$$

If only $(\widehat{X}_{ij}, Y_{ij})$ are observed, then this estimating equation cannot be used directly. We can consider the X_{ij} to be nuisance parameters for the model with observations $(\widehat{X}_{ij}, Y_{ij})$. Unfortunately, the number of such nuisance parameters goes proportionally with the size of the sample. However, provided that each variate $\widehat{X}_{ij}$ is a sufficient statistic for the quantity X_{ij} that it estimates, it will follow that the expected estimating equation:

$$\sum_{i=1}^{n} \sum_{j=1}^{n_i} E_\theta\{g(\theta, X_{ij}, Y_{ij}) | \widehat{X}_{ij}, Y_{ij}\} = 0$$

will be functionally independent of the unknown covariates X_{ij}. More generally, we can suppose that the measurements

$$W_{ij} = (W_{ij1}, W_{ij2}, \ldots, W_{ijm_{ij}})$$

are a random sample whose distribution is governed by the parameter X_{ij}. If this is the case, then the expected estimating equation

$$\sum_{i=1}^{n} \sum_{j=1}^{n_i} E_\theta\{g(\theta, X_{ij}, Y_{ij}) | W_{ij}, Y_{ij}\} = 0$$

will be functionally free of the unknown covariates.

Wang and Pepe (2000) have described an E–M algorithm for estimating θ in a number of applications of this kind based upon the solution to this expected estimating equation. We refer the reader to this paper for a number of examples.

3.8 Aitken acceleration of slow algorithms

We have seen that iterative substitution methods and the E–M algorithm usually converge at a linear rate, while Muller's method and Newton–Raphson are faster. In this section we shall consider a method for 'accelerating' the linear convergence of some algorithms to a rate that is closer to that of Newton–Raphson.

To motivate Aitken's formula, let us consider an iterative method for finding the root of an estimating function for a one-parameter model. Suppose that the method produces values according to the formula

$$\theta^{(j)} = h(\theta^{(j-1)}).$$

We shall also suppose that h has continuously nonvanishing derivative on some interval I, and that $h(\theta) \in I$ for all $\theta \in I$. For some unique value $\hat{\theta}$ in the interval I, we shall assume that $\hat{\theta} = h(\hat{\theta})$, and that this is the root to be sought. The following result will be useful.

Proposition 3.9 *If the iteration starts at some value in I distinct from $\hat{\theta}$, then for all subsequent steps j, the difference $\theta^{(j)} - \hat{\theta}$ will be non-zero.*

That this is true follows easily from the mean value theorem, because

$$\theta^{(j+1)} - \theta^{(j)} = h(\theta^{(j)}) - h(\theta^{(j-1)}) = \dot{h}(\theta^*)(\theta^{(j)} - \theta^{(j-1)})$$

for some value θ^* strictly between $\theta^{(j-1)}$ and $\theta^{(j)}$. If $\theta^{(j)} \neq \theta^{(j-1)}$, it follows that $\theta^{(j+1)} \neq \theta^{(j)}$, because $\dot{h}(\theta^*) \neq 0$. If $\theta^{(j-1)} - \hat{\theta}$ is non-zero, then $\theta^{(j-1)}$ is not a fixed point of h on I. It follows that $\theta^{(j)}$ is not a fixed point of h. Therefore, we can conclude that $\theta^{(j)} - \hat{\theta}$ is also non-zero.

Next, let us consider the limiting form of the error, under the assumption that $\theta^{(j)} \to \hat{\theta}$. Using the mean value theorem again, we see that

Proposition 3.10 *Under the assumptions of Proposition 3.9, and assuming additionally that $\theta^{(j)}$ converges to $\hat{\theta}$, we obtain*

$$\lim_{j\to\infty} \frac{\theta^{(j+1)} - \hat{\theta}}{\theta^{(j)} - \hat{\theta}} = \dot{h}(\hat{\theta}). \tag{3.30}$$

This follows from the fact that θ^* must converge to $\hat{\theta}$. As $\dot{h}$ is continuous on I, it follows that $\dot{h}(\theta^*)$ converges to $\dot{h}(\hat{\theta})$.

Another way to write (3.30) is

$$\theta^{(j+1)} - \hat{\theta} = \{\dot{h}(\hat{\theta}) + \epsilon_j\}(\theta^{(j)} - \hat{\theta}) \tag{3.31}$$

where ϵ_j goes to zero as $j \to \infty$. Applying one more iteration, we get the equation

$$\theta^{(j+2)} - \hat{\theta} = \{\dot{h}(\hat{\theta}) + \epsilon_{j+1}\}(\theta^{(j+1)} - \hat{\theta}). \tag{3.32}$$

Now let us make a simple approximation by setting $\epsilon_j, \epsilon_{j+1} \approx 0$. Subtracting (3.31) from (3.32) we get

$$\dot{h}(\hat{\theta}) \approx \frac{\theta^{(j+2)} - \theta^{(j+1)}}{\theta^{(j+1)} - \theta^{(j)}}. \tag{3.33}$$

If this is plugged into (3.31), we can solve for $\hat{\theta}$ to obtain

$$\hat{\theta} \approx \theta^{(j)} - \frac{(\theta^{(j+1)} - \theta^{(j)})^2}{\theta^{(j+2)} - 2\theta^{(j+1)} + \theta^{(j)}}. \tag{3.34}$$

The right-hand side of equation (3.34) is the *Aitken acceleration* of the sequence $\theta^{(j)}$, $\theta^{(j+1)}$, and $\theta^{(j+2)}$. A convenient representation of this formula uses the shift operator

$$\Delta\theta^{(j)} = \theta^{(j+1)} - \theta^{(j)}.$$

Applying this operator twice gives us

$$\Delta^2\theta^{(j)} = \theta^{(j+2)} - 2\theta^{(j+1)} + \theta^{(j)}.$$

So the Aitken acceleration of the original sequence can also be written as

$$\theta^{(j)\prime} = \theta^{(j)} - \frac{\{\Delta\theta^{(j)}\}^2}{\Delta^2\theta^{(j)}}. \tag{3.35}$$

Example 3.11 Truncated Poisson model (continued)

To illustrate Aitken acceleration, let us return to Example 3.1 and apply the acceleration method to the sequence there. The original sequence and the accelerated sequence are shown in Table 3.5.

An examination of Table 3.5 shows that the rate of covergence is quite fast. The six steps of the accelerated iteration require eight steps of the original algorithm. Thus the convergence to nine decimal places has been obtained at roughly the same

Table 3.5 Aitken acceleration of the substitution algorithm.

Step	θ	θ'
1	3.0000000000	2.8219575215
2	2.8506387949	2.8214554351
3	2.8265778539	2.8214398819
4	2.8223545493	2.8214393384
5	2.8216027120	2.8214393726
6	2.8214685358	2.8214393723
7	2.8214445795	
8	2.8214403020	
9	2.8214395382	
10	2.8214394018	
11	2.8214393774	
12	2.8214393731	
13	2.8214393723	

place as the method of false positions discussed above. As no additional information about the estimating function has been incorporated into the algorithm, it can be argued that the increased rate of convergence has been obtained virtually for free.

A pleasant feature of Aitken acceleration can be observed if we rewrite equation (3.35) as

$$\theta^{(j)\prime} = \frac{\theta^{(j)}\theta^{(j+2)} - (\theta^{(j+1)})^2}{\theta^{(j+2)} - 2\theta^{(j+1)} + \theta^{(j)}}. \tag{3.36}$$

Note that this is formula remains unchanged if the order of $\theta^{(j)}$, $\theta^{(j+1)}$, and $\theta^{(j+2)}$ is reversed. This means that if Aitken acceleration is applied to a sequence that is diverging away from a desired point $\hat{\theta}$, the accelerated value will be identical to that produced by the sequence converging towards $\hat{\theta}$ in the reverse order.

Equation (3.33) can be used to provide an approximation to the derivative of the function h. This is particularly useful when finding the maximum likelihood estimate. Suppose that $u(\theta)$ is the usual score function for θ and $h(\theta) = \theta + u(\theta)$. Then $\hat{\theta}$ is a fixed point of the iteration $\theta^{(j+1)} = h(\theta^{(j)})$ although not necessarily an attractive fixed point. Applying Aitken acceleration to the iteration, we find that we can write

$$\theta^{(j)\prime} = \theta^{(j)} - \left(\theta^{(j+1)} - \theta^{(j)}\right)\left(\frac{\theta^{(j+2)} - \theta^{(j+1)}}{\theta^{(j+1)} - \theta^{(j)}} - 1\right)^{-1}.$$

However, $\theta^{(j+1)} - \theta^{(j)} = u(\theta^{(j)})$, and

$$\frac{\theta^{(j+2)} - \theta^{(j+1)}}{\theta^{(j+1)} - \theta^{(j)}} - 1 \approx \dot{h}(\theta^{(j)}) - 1 = \dot{u}(\theta^{(j)}).$$

Therefore,

$$\theta^{(j)\prime} \approx \theta^{(j)} - \frac{u(\theta^{(j)})}{\dot{u}(\theta^{(j)})} \tag{3.37}$$

which is a step in Newton–Raphson iteration. As the denominator in this iteration is approximately the negative of the observed information about θ, we are also simultaneously calculating the approximate observed information about the parameter.

While Newton–Raphson requires knowledge of the derivative of a function, this accelerated method seems to have come with no cost other than the calculation of a few steps in a slow algorithm. The acceleration in speed is real, but does have its price: we are assuming that the score function is approximately linear.

We shall close this section with a modification of Aitken acceleration known as *Steffensen's method*. Suppose an iteration using the function h converges linearly. Applying Aitken's method to three iterates, we obtain θ'. It seems more sensible to use this as the new starting point for iteration by h, rather than any of the three original values. After three iterations starting from θ', we can apply Aitken's formula again to obtain θ'', as shown below:

$$\begin{array}{ccccccc} \theta & & \theta' & & \theta'' & & \\ h(\theta) & \nearrow & h(\theta') & \nearrow & h(\theta'') & \nearrow & \cdots \\ h(h(\theta)) & & h(h(\theta')) & & h(h(\theta'')) & & \end{array}$$

The sequence

$$\theta \to \theta' \to \theta'' \to \cdots$$

is called *Steffensen iteration*. Under fairly general conditions, it will converge to the same set of fixed points as those of h.

3.9 Bernoulli's method and the quotient-difference algorithm

In this section and the next, we shall consider root-finding methods that are particularly applicable to polynomial estimating functions.

Suppose that we have a polynomial estimating function $g(\theta)$ of degree m for a real-valued parameter θ. By dividing $g(\theta)$ by its leading coefficient, we obtain an estimating function of the form

$$p(\theta) = \theta^m + a_1(y)\theta^{m-1} + a_2(y)\theta^{m-2} + \cdots + a_m(y) \tag{3.38}$$

which is equivalent to $g(\theta)$ in the sense that it has the same roots.

Consider a sequence of values, $\ldots Z_{-2}, Z_{-1}, Z_0, Z_1, Z_2, \ldots$, where $Z_k = 0$ for $k < 0$, $Z_0 = 1$, and

$$Z_k = -a_1(y)Z_{k-1} - a_2(y)Z_{k-2} - \cdots - a_m Z_{k-m}. \tag{3.39}$$

Suppose that the sequence of quotients $q_k = Z_{k+1}/Z_k$ converges to some value r, say. Then it is easy to see that $p(r) = 0$. That is, r must be a root of the polynomial estimating equation.

Next, suppose that the roots of (3.38) are the m distinct complex numbers $z_1, z_2, \ldots, z_m$. Furthermore, let us suppose that one of these roots, say z_1, has a modulus that is *strictly larger* than the moduli of all other roots. That is,

$$|z_1| > \max(|z_2|, \ldots, |z_m|).$$

It follows from this that z_1 is a real, because each non-real root is paired with its complex conjugate, which is also a root. To see which root r is, we note that equation (3.39) is a linear difference equation of order m for the sequence Z_k, and (3.38) is the characteristic polynomial for this equation. Therefore, the sequence Z_k is representable in the form

$$Z_k = c_1 z_1^k + \cdots + c_m z_m^k. \tag{3.40}$$

Here, the coefficients are generally complex numbers. We shall not go into the proof of this result, but shall refer the reader to the theory of linear difference equations. So

$$\frac{Z_{k+1}}{Z_k} = \frac{c_1 z_1^{k+1} + \cdots + c_m z_m^{k+1}}{c_1 z_1^k + \cdots + c_m z_m^k}.$$

As z_1 has a strictly larger modulus than $z_2, \ldots z_m$, it follows that this ratio converges to $r = z_1$ as $k \to \infty$, provided $c_1 \neq 0$.

The algorithm that we have just described is called *Bernoulli's method.* It has the advantage over many algorithms that it is guaranteed to converge to a root provided the assumptions are satisfied. Note also that we do not have to specify an interval of parameter values in which the root search is to be conducted. However, there are some difficulties with the algorithm as a root search method for polynomial estimating functions. First, and most importantly, there is no reason to think that the consistent root of an estimating function is the one with the largest modulus. For example, if the true value of a parameter is set to zero in a simulation study, the closest root to zero will *a fortiori* be of smallest modulus. A second problem that needs consideration is that the argument above gives us no immediate insight into the rate of convergence of the algorithm.

To attack the first problem, we can generalise Bernoulli's method to the *quotient-difference algorithm*, which simultaneously finds all the roots of the polynomial. Rather than having a sequence of quotients q_k, as in Bernoulli's algorithm, the quotient-difference algorithm produces a set of $2m - 1$ sequences labelled

$$q_k^{(1)}, e_k^{(1)}, q_k^{(2)}, e_k^{(2)}, \ldots, q_k^{(m-1)}, e_k^{(m-1)}, q_k^{(m)}$$

for $k = 1, 2, 3, \ldots$. The sequence $q_k^{(1)}$ is simply the sequence of quotients q_k produced by Bernoulli's method. The other sequences are computed using the formulas $e_k^{(0)} = 0$ and

$$e_k^{(j)} = (q_{k+1}^{(j)} - q_k^{(j)}) + e_{k+1}^{(j-1)}$$

and

$$q_k^{(j+1)} = \frac{e_{k+1}^{(j)}}{e_k^{(j)}} q_{k+1}^{(j)}.$$

Under certain regularity conditions more general than those of Bernoulli's method, the sequences $q_k^{(j)}$, $j = 1, \ldots, m$ will converge to the roots of $p(\theta)$. The proof of this and the various types of regularity required are beyond the scope of this book. See Henrici (1958).

If the quotient-difference algorithm is initialised with real values, it is clear that it cannot converge to complex roots off the real axis. However, it is reasonable to hope that the real roots can be found as limits of the algorithm. Under certain conditions, this turns out to be true. The following proposition is useful. Suppose that the complex roots of $p(\theta)$ are ordered so that

$$|z_1| \geq |z_2| \geq |z_3| \geq \cdots \geq |z_m|.$$

Then we have the following proposition.

Proposition 3.12 *For every j such that $|z_{j+1}| > |z_j| > |z_{j-1}|$,*

$$\lim_{n \to \infty} q_k^{(j)} = z_j.$$

Furthermore, for every j such that $|z_j| > |z_{j+1}|$,

$$\lim_{n\to\infty} e_k^{(j)} = 0.$$

Note that if z_j has a distinct modulus, it follows that it is real. The proposition tells us that z_j can be obtained as the limit of a real sequence such as $q_k^{(j)}$. However, the converse is not true: a real root need not have a distinct modulus. The convergence properties of the algorithm are difficult to determine, and will not be discussed here.

The other problem that we shall consider here is the rate of convergence of the quotient-difference algorithm. It has been found that the algorithm converges rather slowly to the roots compared to Newton–Raphson, say, when the latter is initialised close to a root. For this reason, it is best to use the quotient-difference method to find rough approximations to the roots by running it sufficiently long so that the algorithm stabilises. After that, Newton–Raphson can be used to 'fine-tune' the digits of the numerical approximation. Generally, the quotient-difference method and Newton–Raphson work in complementary ways. Newton–Raphson has good local properties close to a root, but behaves badly with some functions when initialised far from a root. The reverse is the case for the quotient-difference algorithm.

3.10 Sturm's method

Suppose we wished to run a simulation study on an estimating function for a single parameter. Let us suppose that the purpose of the study is to construct a probability histogram for the distribution of the root(s) of the function. We might begin by dividing the real line into intervals or bins $(a_j, b_j]$, $j = 0, \pm 1, \pm 2, \ldots$, on which the histogram is to be constructed. To determine the number of roots in any interval and any trial of the simulation, it is unnecessary to compute the precise value of a root. Knowing the approximate value of a root will only determine the interval to which it belongs if the error in the approximation is smaller than the distance to the endpoints of the interval. So many of the algorithms discussed so far will be inefficient for this purpose.

If it is known that the estimating function is continuous, then a change of sign over the interval is sufficient to ensure that there is a root in the interval. However, if there is an even number of roots in the interval, then the function will not change sign over the interval. Thus there is no guarantee that roots can be detected by this method. In some cases, there is additional information available at the endpoints of the intervals, because we can also differentiate the function. This fact forms the basis for *Sturm's method* for polynomial estimating functions.

We begin with the original estimating function $g(\theta)$, where $\theta \in \mathbb{R}$, and shall assume that g is a polynomial of degree m. Suppose that we wish to find all roots in some interval $[a, b]$. If $g(b) = 0$, then we can replace g by $(\theta - b)^{-1} g(\theta)$, which will also be a polynomial. Continuing in this way, we eventually have $g(b) \neq 0$.

A similar argument can be used for $g(a)$. Set $g_0(\theta) = g(\theta)$, and define

$$g_1(\theta) = \dot{g}(\theta).$$

From there we proceed to define $g_j(\theta)$ recursively. Divide $g_{j-1}(\theta)$ by $g_{j-2}(\theta)$, leaving remainder $-g_j(\theta)$, and so on. That is, we can write

$$g_{j-2}(\theta) = q_{j-1}(\theta)g_{j-1}(\theta) - g_j(\theta) \tag{3.41}$$

where $q_{j-1}(\theta)$ is a polynomial in θ and $\deg g_j < \deg g_{j-1}$. As each function g_j is calculated, it can also be divided by the absolute value of its leading coefficient as this has no effect on sign changes. This often prevents overflow or underflow for polynomials of high degree.

We now have the following proposition, which we state without proof.

Proposition 3.13 *The number of distinct roots—ignoring multiplicities—of $g(\theta)$ in (a, b) is given by the number of sign changes in the sequence*

$$g_0(a), g_1(a), \ldots, g_m(a)$$

minus the number of sign changes in the sequence

$$g_0(b), g_1(b), \ldots, g_m(b).$$

For a proof of this result, the reader is referred to Barbeau (1989, p. 175).

The next example illustrates how the method can be used in the statistical analysis of epidemics.

Example 3.14 Measles epidemiology

In a family with n children, the following stochastic model is assumed for the spread of measles among the children if one of the children comes home with measles. Each of the other $n - 1$ children has probability θ of catching measles directly from the infected child. If no other child gets measles from that child the epidemic stops. If j children catch measles from the original infection, then the probability that an uninfected child will subsequently catch measles from one of these j is $1 - (1 - \theta)^j$. In turn, a certain number of children, say k, will be infected from the $n - j - 1$ children still at risk, and so on. Let Y be the total number of children in the family who eventually get measles. Set $p_{in} = P(Y = i)$.

For $n = 2$, we have

$$p_{12} = 1 - \theta, \quad p_{22} = \theta.$$

For $n = 3$, the formulas can be shown to be

$$p_{13} = (1 - \theta)^2, \quad p_{23} = 2\theta(1 - \theta)^2, \quad p_{33} = \theta^2(3 - 2\theta).$$

As n increases, the calculation of probabilities becomes more complex. For $n = 4$, we get

$$p_{14} = (1-\theta)^3, \quad p_{24} = 3\theta(1-\theta)^4, \quad p_{34} = 3\theta^2(1-\theta)^3 + 6\theta^2(1-\theta)^4$$

and

$$p_{44} = 1 - (p_{14} + p_{24} + p_{34}) = 1 - (1+3\theta^2)(1-\theta)^3 - 3\theta(1+2\theta)(1-\theta)^4.$$

Suppose that in a study of 100 families, each with four children, it was found that exactly one child caught measles in 32 cases, exactly two in 28 cases, three in 25 cases and four in 15 cases. Such a sample can be regarded as the outcome from 100 multinomial trials, with four possible outcomes on each trial. So the likelihood function is

$$\{p_{14}(\theta)\}^{32}\{p_{24}(\theta)\}^{28}\{p_{34}(\theta)\}^{25}\{p_{44}(\theta)\}^{15}.$$

Upon simplification, the score function is found to be

$$u(\theta) = \frac{5712\,\theta^5 - 32562\,\theta^4 + 72396\,\theta^3 - 77245\,\theta^2 + 37886\,\theta - 5904}{\theta\,(-16 + 33\,\theta - 24\,\theta^2 + 6\,\theta^3)\,(-3 + 2\,\theta)\,(-1 + \theta)}$$

and the likelihood equation simplifies to

$$5712\,\hat\theta^5 - 32562\,\hat\theta^4 + 72396\,\hat\theta^3 - 77245\,\hat\theta^2 + 37886\,\hat\theta - 5904 = 0 \qquad (3.42)$$

where $0 < \theta < 1$.

A general quintic equation can have as many as five roots in a given interval. So it is worth checking to see if there is a unique solution to this equation for θ between 0 and 1. The calculation of the Sturm sequence is routine and easily implemented in a symbolic computing package such as MAPLE or MATHEMATICA. For example, MAPLE provides the following sequence from the command *sturmseq*:

$$g_0(\theta) = \theta^5 - \frac{5427}{952}\theta^4 + \frac{6033}{476}\theta^3 - \frac{11035}{816}\theta^2 + \frac{18943}{2856}\theta - \frac{123}{119}$$

$$g_1(\theta) = \theta^4 - \frac{5427}{1190}\theta^3 + \frac{18099}{2380}\theta^2 - \frac{2207}{408}\theta + \frac{18943}{14280}$$

$$g_2(\theta) = \theta^3 - \frac{6301723}{1470498}\theta^2 + \frac{58533895}{8822988}\theta - \frac{3616229}{980332}$$

$$g_3(\theta) = \theta^2 - \frac{270761746753}{540814611780}\theta - \frac{89916695331}{60090512420}$$

$$g_4(\theta) = -\theta + \frac{11160795036204153801}{7441682496972499633}$$

and

$$g_5(\theta) = -1.$$

The MAPLE command *sturm* can then be used to calculate the number of roots in a given interval. It can be checked that there is exactly one root in the interval $(0, 1)$, and that this root is approximately $\hat\theta = 0.2748$.

3.11 Roots and eigenvalues

The roots of a polynomial function, and the search for them, can also be related to the eigenvalues of an appropriate matrix. Suppose that

$$g(\theta) = \theta^m + a_1(y)\theta^{m-1} + a_2(y)\theta^{m-2} + \cdots + a_m(y) = 0 \tag{3.43}$$

is an estimating equation whose roots are to be determined. If we construct the matrix

$$A = \left(\begin{array}{cccc|c} -a_1(y) & -a_2(y) & \cdots & -a_{m-1}(y) & -a_m(y) \\ \hline 1 & 0 & \cdots & 0 & 0 \\ 0 & 1 & \cdots & 0 & 0 \\ \vdots & \vdots & \ddots & \vdots & \vdots \\ 0 & 0 & \cdots & 1 & 0 \end{array}\right) \tag{3.44}$$

then the eigenvalues of A are determined by its *characteristic equation* $\det(A - \theta I) = 0$, where I is the $m \times m$ identity matrix. Note that this characteristic equation is equivalent to (3.43).

Although the solutions to the equations are formally equivalent, there are many methods for finding eigenvalues which do not use the characteristic equation. For example, it is possible to transform A to another matrix $B = S^{-1}AS$ which is similar to A and has a convenient canonical form. Such a transformation preserves the eigenvalues. A popular representation is upper Hessenberg form. A matrix $B = (b_{ij})$ is said to be in *upper Hessenberg form* if $b_{ij} = 0$ for all $j \leq i - 2$. The advantage of the upper Hessenberg form is that it permits a QR decomposition of the matrix to be performed relatively quickly.

The *QR algorithm* for computing the eigenvalues of A is an iterative method. The algorithm starts with the matrix A itself, which we presume has been converted into upper Hessenberg form. Suppose that at the kth stage of the algorithm, a matrix $A^{(k)}$ has been generated. The QR decomposition allows us to write $A^{(k)} = Q^{(k)}R^{(k)}$, where $Q^{(k)}$ is orthogonal, and $R^{(k)}$ is upper triangular. The QR algorithm then sets $A^{(k+1)} = R^{(k)}Q^{(k)}$. As eigenvalues are preserved under orthogonal transformations, and

$$A^{(k+1)} = R^{(k)}Q^{(k)} = [Q^{(k)}]^{-1}A^{(k)}Q^{(k)}$$

the step $A^{(k)} \rightarrow A^{(k+1)}$ preserves eigenvalues. Each step has the effect of transferring mass from the subdiagonal part of the matrix to the superdiagonal part. The limit matrix, if it exists, is upper triangular, with the eigenvalues appearing as the elements along the main diagonal. This algorithm for finding the roots of polynomials is now standard using the function *roots* available within MATLAB.

Example 3.15 Solving quadratic equation

As an easy example, we consider the equation $\theta^2 - 4\theta + 3 = 0$ with roots $\hat{\theta} = 1$ and $\hat{\theta} = 3$. The matrix

$$A = \begin{pmatrix} 4 & -3 \\ 1 & 0 \end{pmatrix}$$

is already in upper Hessenberg form. Performing a QR decomposition of $A = A^{(0)}$, we get from the MATLAB function *qr*

$$Q^{(1)} = \begin{pmatrix} -0.9701 & -0.2425 \\ -0.2425 & 0.9701 \end{pmatrix}, \quad R^{(1)} = \begin{pmatrix} -4.1231 & 2.9104 \\ 0 & 0.7276 \end{pmatrix}.$$

Therefore,

$$A^{(1)} = R^{(1)}Q^{(1)} = \begin{pmatrix} 3.2941 & 3.8235 \\ -0.1765 & 0.7059 \end{pmatrix}.$$

Continuing in this way, we have

$$A^{(2)} = \begin{pmatrix} 3.0919 & -3.9514 \\ 0.0486 & 0.9081 \end{pmatrix}, \quad A^{(3)} = \begin{pmatrix} 3.0300 & 3.9847 \\ -0.0153 & 0.9700 \end{pmatrix}$$

and so on. Of course, all these steps can be combined together. For example, the MATLAB function *roots* operates on the coefficient vector $[1, -4, +3]$ to give

```
>> roots([1,-4,+3])
ans=
     3
     1
```

with the solution in the form of a column vector. Note that in addition to the steps shown, the function *roots*, which uses the MATLAB function *eig*, attempts to balance the matrix when computing its eigenvalues. Balancing a matrix serves to minimise the effect of roundoff error which can make large changes to eigenvalues.

3.12 The Nelder–Mead algorithm

The *Nelder–Mead simplex algorithm*, also known as the *amoeba algorithm*, falls within a class of simplex search methods that search for an optimal value by moving the vertices of a simplex around in a region according to prescribed rules. It should not be confused with the simplex method of linear programming with which it has little in common.

The Nelder–Mead algorithm has become enormously popular in certain applications and is incorporated as a standard algorithm in packages such as MATLAB. Part of the reason for its popularity is that it is quite reliable, even for searches in high dimensional spaces. However, it is not a fast algorithm. So it is to be preferred for problems in which there is little known about the nature of the objective function to be minimised or maximised. For example, it may be difficult or impossible to compute the derivative matrix of a score function—that is, the matrix of second partials of the log-likelihood—so that Newton–Raphson cannot be implemented. Indeed the log-likelihood might have a singularity set where the derivative matrix is not defined. If this is the case, the Newton–Raphson algorithm cannot be routinely applied.

In the standard formulation of the algorithm, a scalar-valued objective function is minimised. As we shall be more properly concerned with the maximisation of an objective function such as a likelihood, log-likelihood or quasi-likelihood, etc., we shall formulate the algorithm in the reverse direction from the standard formulation. The modification is trivial, of course.

Suppose that Θ is a k-dimensional parameter space, and $\ell(\theta)$, $\theta \in \Theta$, is a real-valued objective function (which need not be a likelihood or log-likelihood) which is to be maximised. Each iteration of the algorithm begins with a simplex in Θ determined by $k+1$ vertices $\theta^{(1)}, \ldots, \theta^{(k+1)}$. Based on the evaluation of ℓ at these vertices, one or more vertices are computed, and a new simplex is constructed.

In order to implement the Nelder–Mead algorithm, four tuning parameters must be chosen (Figure 3.7). These are coefficients of

- reflection $\rho > 0$ (usually set to 1);
- expansion $\chi > 1$ (usually 2);

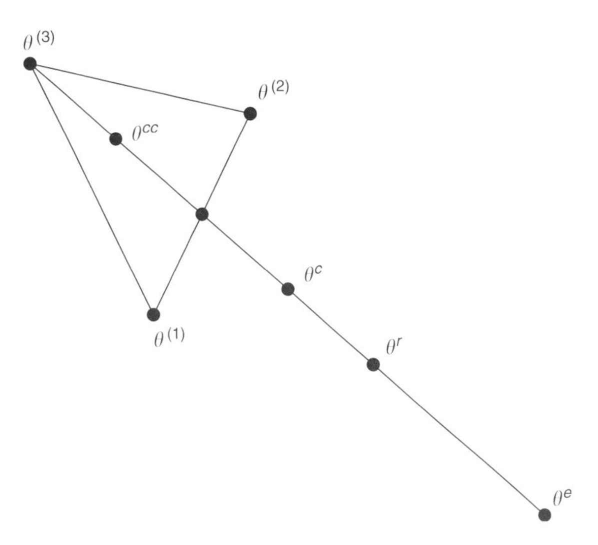

Fig. 3.7 The reflection, expansion and contraction steps in the Nelder–Mead algorithm in two dimensions.

- contraction $0 < \gamma < 1$ (usually $\frac{1}{2}$); and
- shrinkage $0 < \sigma < 1$ (usually $\frac{1}{2}$).

At any step of the iteration, a non-degenerate simplex with vertices $\theta^{(1)}, \ldots, \theta^{(k+1)}$ is determined. We assume that these vertices are ordered in decreasing values of the objective function ℓ, so that

$$\ell^{(1)} \geq \ell^{(2)} \geq \cdots \geq \ell^{(k+1)}$$

where $\ell^{(j)} = \ell(\theta^{(j)})$. Since the task is to maximise ℓ, we call $\theta^{(1)}$ the *best point* and $\theta^{(k+1)}$ the *worst point*. The algorithm seeks to replace the worst vertex by a more appropriate one. To this end, we begin by *reflecting* $\theta^{(k+1)}$ about the point which is the *centroid* of the other k vertices. That is, we define θ^r to be

$$\theta^r = \bar{\theta} + \rho(\bar{\theta} - \theta^{(k+1)})$$

where $\bar{\theta} = (\theta^{(1)} + \cdots + \theta^{(k)})/k$. If $\ell^{(k)} < \ell^r \leq \ell^{(1)}$, then we have improved on the second worst vertex. The point θ^r replaces $\theta^{(k+1)}$ and the iteration is complete.

On the other hand, if $\ell^{(1)} < \ell^r$, then it is appropriate to try to do even better than this reflected point. We calculate the *expansion step*

$$\theta^e = \bar{\theta} + \chi(\theta^r - \bar{\theta}).$$

If θ^e is better than θ^r we accept θ^e and the iterative step is complete with θ^e replacing $\theta^{(k+1)}$. If θ^r is better than θ^e, then θ^r replaces $\theta^{(k+1)}$, and once again the iterative step is complete.

The case where the reflection actually makes things worse or does not improve matters enough occurs when $\ell^r \leq \ell^{(k)}$. In this case, we attempt the point which is the *contraction* of $\bar{\theta}$ towards the better of $\theta^{(n+1)}$ and θ^r. If $\ell^{(k+1)} < \ell^r \leq \ell^{(k)}$, then an *outside contraction* is performed, namely

$$\theta^c = \bar{\theta} + \gamma(\theta^r - \bar{\theta}).$$

If $\ell^r \leq \ell^{(k+1)}$ then an *inside contraction*

$$\theta^{cc} = \bar{\theta} - \gamma(\bar{\theta} - \theta^{(k+1)})$$

is performed. Either way, we end up with a contracted point, namely θ^c or θ^{cc}. If an outside contraction is performed and $\ell^c \geq \ell^r$, or if an inside contraction is performed and $\ell^{(k+1)} < \ell^{cc}$, then the iterative step is terminated and the contracted point replaces $\ell^{(k+1)}$.

However, the step of the iteration does not necessarily end with this contracted point. If the contracted point does not improve enough to satisfy the relevant inequality, we proceed to the final step which is *shrinkage*. We evaluate ℓ at the k points determined by shrinking $\theta^{(2)}, \ldots, \theta^{(k+1)}$ towards $\theta^{(1)}$ using

$$\eta^{(j)} = \theta^{(1)} + \sigma(\theta^{(j)} - \theta^{(1)}).$$

The vertices of the next simplex are $\theta^{(1)}, \eta^{(2)}, \ldots, \eta^{(k+1)}$ in no particular order (Figure 3.8).

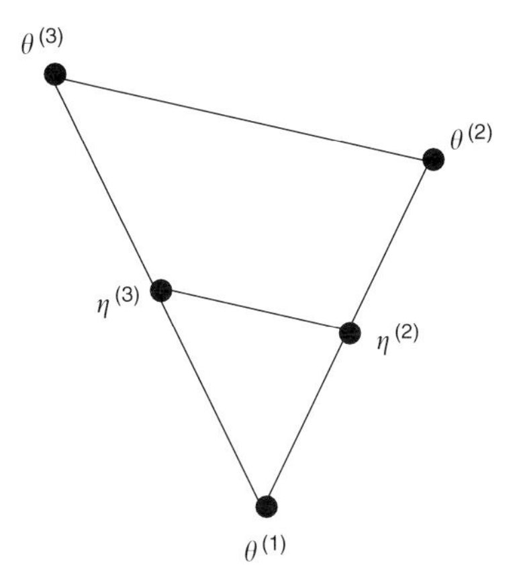

Fig. 3.8 The final shrinkage step in the Nelder–Mead algorithm in two dimensions.

The MATLAB command *fmins* conducts a minimisation routine using the Nelder–Mead algorithm. As such, it is appropriate for optimising a wide variety of functions. The initial simplex for the algorithm is constructed internally by *fmins* so that the user need only specify the objective function and an initial starting point in the domain of the function. The default accuracy of *fmins* is 10^{-4}, which can be reset when calling the function. If convergence to this accuracy has not occurred after 400 iterations, the algorithm terminates. Once again, this default value can be changed. The function can be called with a statement of the form

```
>>xmin=fmins('function',x0,[display, acc],[],
             p1,p2,p3,p4,...)
```

where *function* is the objective function, `x0` is the initial value for the algorithm, *display* is a real number which causes intermediate steps to be shown when *display* is non-zero, `acc` is a real number defining the required accuaracy for the algorithm, and `p1`, `p2`, and so on, are real-valued parameters that can be passed to *function*. The value `xmin` that is the output from `fmins` is the smallest domain variable found by the search algorithm.

While the Nelder–Mead algorithm does not require the calculation of gradients, it can fail to perform well if the objective function has multiple extrema or is piecewise constant on certain regions. In the former case, it can converge to a locally optimal point that is not the global solution. Similarly in the latter case, the piecewise constancy of an objective function may 'trick' the algorithm into terminating prematurely.

3.13 Jacobi iteration for quasi-likelihood

In this section, we shall consider a method for inverting matrices that is useful for solving quasi-likelihood equations.

Let Y denote an $n \times 1$ column vector of observations, $\mu(\theta)$ the corresponding $n \times 1$ column vector of means, and $\Sigma(\theta)$ the $n \times n$ covariance matrix of the Y_j's. If the parameter θ is k-dimensional, then the derivative matrix $\dot{\mu}(\theta)$ will have dimension $n \times k$. As in equation (2.15), the quasi-likelihood estimate for θ will be obtained by solving the equation

$$\dot{\mu}^t(\hat{\theta})\Sigma^{-1}(\hat{\theta})\{Y - \mu(\hat{\theta})\} = 0. \tag{3.45}$$

Two computational problems arise in solving equation (3.45). The first is that the matrix $\Sigma(\theta)$ apparently needs to be inverted. The second problem is that, even with this inversion in hand, the equation may not have an explicit solution for $\hat{\theta}$. We shall concentrate on the problem of matrix inversion below.

The coefficient matrix $a(\theta) = \dot{\mu}^t(\hat{\theta})\Sigma^{-1}(\hat{\theta})$ can be regarded as the solution to the set of simultaneous linear equations provided by

$$a(\theta)\Sigma(\theta) = \dot{\mu}^t(\theta). \tag{3.46}$$

Gaussian elimination can be used to find $a(\theta)$ directly, and this is more computationally efficient than inverting Σ in equation (3.45) above. A simple alternative to Gaussian elimination is *Jacobi iteration*. Let us write $\Sigma(\theta) = \Delta(\theta) + \Gamma(\theta)$ where $\Delta = \mathrm{diag}(\sigma_{11}, \ldots, \sigma_{nn})$. So, applying this identity, we can rewrite equation (3.46) as

$$a(\theta) = \{\dot{\mu}^t(\theta) - a(\theta)\Gamma(\theta)\}\Delta^{-1}(\theta).$$

The form of this identity suggests the iteration

$$a^{(j+1)}(\theta) = \{\dot{\mu}^t(\theta) - a^{(j)}(\theta)\Gamma(\theta)\}\Delta^{-1}(\theta). \tag{3.47}$$

Jacobi iteration consists of the application of this recursion to some starting matrix $a^{(0)}$. An application of the contractive mapping theorem shows that the iteration will converge provided that the *spectral radius* of $\Gamma\Delta^{-1}$ is less than one. When Y is bivariate ($n = 2$), then the covariance inequality can be used to prove that this is always the case. However, when $n > 2$ and the observations Y_j are highly correlated, Jacobi iteration can diverge for some starting values.

To estimate θ and to invert Σ simultaneously, one can iterate jointly over a and θ. For example, the iteration $(a^{(j)}, \theta^{(j)}) \to (a^{(j+1)}, \theta^{(j+1)})$ that is defined by

$$a^{(j)}\mu(\theta^{(j+1)}) = a^{(j)}Y \tag{3.48}$$

$$a^{(j+1)} = \left\{\dot{\mu}^t(\theta^{(j+1)}) - a^{(j)}\Gamma(\theta^{(j+1)})\right\}\Delta^{-1}(\theta^{(j+1)}) \tag{3.49}$$

can be used. The solution to (3.48) may be immediate or quite difficult, depending upon the particular model.

Example 3.16 Data from quadrat sampling

In the following example, we shall consider the solution to the particular quasi-likelihood equation that was proposed in Section 2.6.4. We shall further restrict to the case where the Poisson process is homogeneous with intensity parameter θ. Then $\mu_j(\theta) = \theta|A_j|$, and $\sigma_{jk}(\theta) = \theta|A_j \cap A_k|$, where $|B|$ represents the area of the set B. So (3.48) and (3.49) become

$$\theta^{(j+1)} = (a^{(j)}Y)/(a^{(j)}\nu) \tag{3.50}$$

and

$$a^{(j+1)} = 1_{1\times n} - a^{(j)}T \tag{3.51}$$

where $\nu = (|A_1|, \ldots, |A_n|)^t$, $1_{1\times n}$ is an n-dimensional row vector of ones, and

$$T = \begin{pmatrix} 0 & \frac{|A_1 \cap A_2|}{|A_2|} & \cdots & \frac{|A_1 \cap A_n|}{|A_n|} \\ \frac{|A_2 \cap A_1|}{|A_1|} & 0 & \cdots & \frac{|A_2 \cap A_n|}{|A_n|} \\ \vdots & \vdots & \ddots & \vdots \\ \frac{|A_n \cap A_1|}{|A_1|} & \frac{|A_n \cap A_2|}{|A_2|} & \cdots & 0 \end{pmatrix}.$$

Some insight into the Jacobi iteration here can be obtained if we set $a^{(0)} = 1_{1\times n}$. Then

$$a^{(n)} = 1_{1\times n}\left\{I + \sum_{j=1}^{n}(-1)^j T^j\right\} \tag{3.52}$$

which is the binomial expansion of $1_{1\times n}(I + T)^{-1}$.

3.14 Bibliographical notes

Many of the properties of Newton–Raphson iteration can be found in Dennis (1996). Particular attention is paid in that book to quasi-Newton algorithms both for solving non-linear equations and unconstrained minimisation problems. A survey of traditional numerical analysis can be found in Henrici (1964). While the literature on numerical analysis is extensive, there are fewer books which concentrate on the use of numerical methods specifically within statistics. The reader is referred to Lange (1999) for the applications of numerical analysis to a variety of statistical problems.

The Nelder–Mead algorithm was introduced by Nelder and Mead (1965). While the Nelder–Mead algorithm is not fast compared to the Newton–Raphson algorithm, it has proved to be very reliable. Despite the limited number of theoretical results showing that it converges successfully, the practical results have been mostly positive.

4
Working with roots

4.1 Introduction

In this chapter, we will consider a number of examples where estimating functions lead to multiple roots. As we shall see, an estimating function can have multiple roots for a variety of reasons. So it is not surprising that—to use a medical analogy—the appropriate prescription for a multiple root problem requires a careful analysis of its aetiology. In each of the examples below, we must consider several questions:

1. Is the existence of multiple roots in an estimating equation a symptom of a problem in the choice of *model* for the data?
2. Is the existence of multiple roots symptomatic of a problem in the choice of the *estimating function* for analysing the data?
3. Are multiple roots symptomatic of problems in the *data themselves*, i.e., that the data have little information about the parameter?

We will not be able to answer all these questions at this stage. However, to continue in a medical vein, good practice will require an extensive case book of patients on which a variety of remedies have been tried.

4.2 Non-identifiable parameters in mixture models

A parameter or parametrisation is said to be identifiable if, for all pairs of distinct parameter values, say θ and θ', the induced distributions on the data $Y \in \mathbb{R}^n$ are also distinct. More precisely, we have the following.

Definition 4.1 *A parametrisation is said to be* identifiable *if, for any distinct parameter values θ and θ', there is some (measurable) set A such that the probabilities of the event $Y \in A$ assuming θ and θ' are different.*

For any parametric model, we can partition the parameter space into equivalence classes, with the elements of each class being equivalent in the sense that they induce the same distribution on the data Y. Identifiable parametrisations are those whose equivalence classes are singleton sets. The likelihood function will always be constant on an equivalence class of parameter values. So if a maximum likelihood estimate lies in an equivalence class with more than one element, it will not be unique.

If a model has a non-identifiable parametrisation, it is possible to find an identifiable submodel. However, as the following example shows, even when the problems of non-identifiability are eliminated, the likelihood may remain multimodal.

Example 4.2 A mixture model for fish populations

Individual fish of a particular species caught in a river live for a maximum of two years. Up to the time of death, the growth rate for the fish is approximately linear, so that the size of the fish is roughly proportional to its age. Thus the average size of fish spawned in the year of capture is $c\theta$, where c is the time, measured in months since the spawning season and θ is the rate of growth. The average size of fish spawned in the previous season is $(c + 12)\theta$. A possible model for the distribution of size for a randomly selected fish from the population is therefore

$$p\mathcal{N}(c\theta, \sigma^2) + (1 - p)\mathcal{N}(c\theta + 12\theta, \sigma^2)$$

where the parameter σ represents individual variation within age group and the weight p represents the proportion of the population of fish spawned in the season of capture. The growth rate parameter θ should be constrained to be a positive real number, as should the fish sizes themselves. Therefore, we must assume that the positive probabilities assigned to negative sizes by the model are negligible for practical purposes.

If p is assumed to lie in the closed interval $[0, 1]$, then a problem of parameter non-identifiability occurs. This is easily seen by noting that the parameter assignment $p = 0$ and $\theta = \theta_0$ leads to the same distribution as $p = 1$ and $\theta = (1 + 12c^{-1})\theta_0$. So for any data set, the likelihoods will be the same. For example, in Figure 4.1, the log-likelihood function is graphed for a special case of three observations artificially set to be $y_j = 7$, for $j = 1, 2, 3$. To illustrate the behaviour of the likelihood, the parameter space has been reduced to the set where $\sigma = 1$, $c = 11$, and $\theta = 0.3(1 + 12p/c)$. In this submodel, parametrized by the single parameter p, the endpoints $p = 0$ and $p = 1$ impose identical distributions on the data, and therefore are not identifiable. Correspondingly, the log-likelihood has two equal maxima at the endpoints of the interval. In Figure 4.2, the log-likelihood is plotted as a function of both p and θ, with c and y_j as before.

It is tempting to try to restore identifiability to the parameter space by 'pasting' the parameter points with $p = 0$ to those with $p = 1$, so to speak, thereby declaring the parameter space to be the set of its equivalence classes. While this makes good sense for the distribution of the data, it is not justified in the context of the application: the cases $p = 0$ and $p = 1$ represent quite different situations in the population biology of the fish species.

The next example that we shall consider is drawn from a study of haemophilia patients as described by Basford and McLachlan (1985).

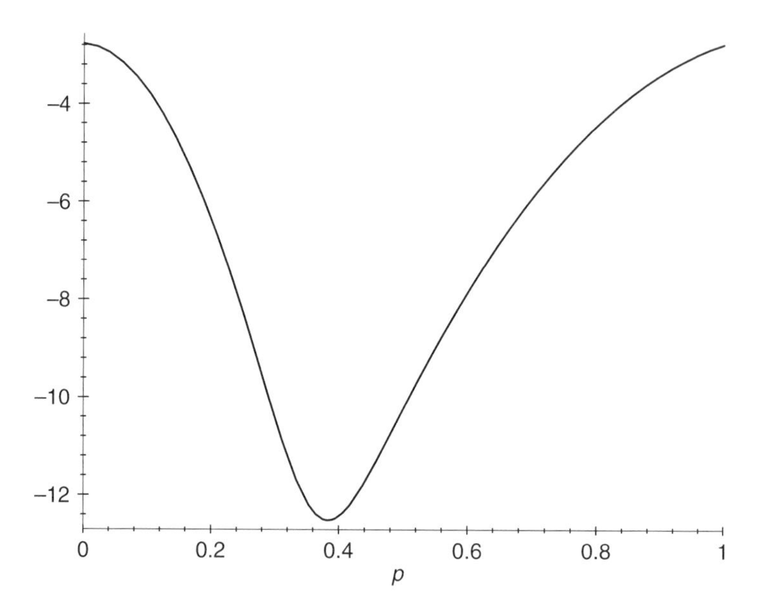

Fig. 4.1 The log-likelihood function for the weight parameter p in a submodel with $\theta = 0.3(1 + 12p/c)$ and three observations.

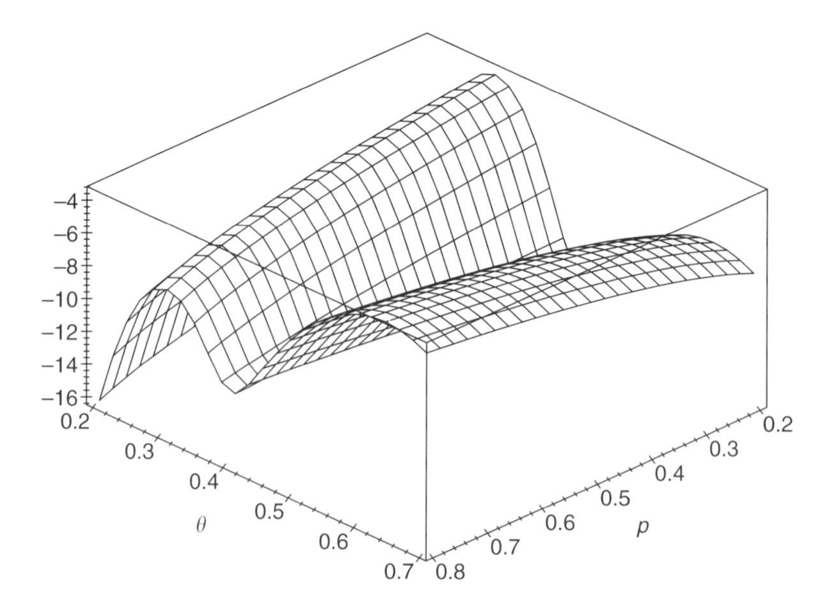

Fig. 4.2 The log-likelihood function Y as a function of both p and θ, with c and y_i.

Example 4.3 Haemophilia study

Habbema *et al.* (1974) considered the problem of discriminating between normal women and *haemophilia A* carriers based upon *AHF activity* and *AHF-like antigen*. On each of 75 subjects the variables

$$y_1 = \log_{10}(\text{AHF activity}) \quad y_2 = \log_{10}(\text{AHF-like antigen})$$

were determined. Let y_{j1} and y_{j2} be these two variables respectively for subject j, where $j = 1, \ldots, 75$. A scatterplot of the 30 known non-carriers and 45 known obligatory carriers is shown in Figure 4.3.

Pooling samples of non-carriers and carriers together, Basford and McLachlan analysed the data with an IID homoscedastic mixture of normals of the form

$$pN(\mu_1, \Sigma) + (1 - p)N(\mu_2, \Sigma)$$

where μ_i is a 1×2 row vector representing the expectation of $y = (y_1, y_2)$ assuming that it comes from population $i = 1, 2$, respectively, and Σ is the *common* 2×2 covariance matrix for the two populations. The parameter vector for this model is

$$\theta = (p, (\mu_1)_1, (\mu_1)_2, (\mu_2)_1, (\mu_2)_2, \Sigma_{11}, \Sigma_{22}, \Sigma_{12}).$$

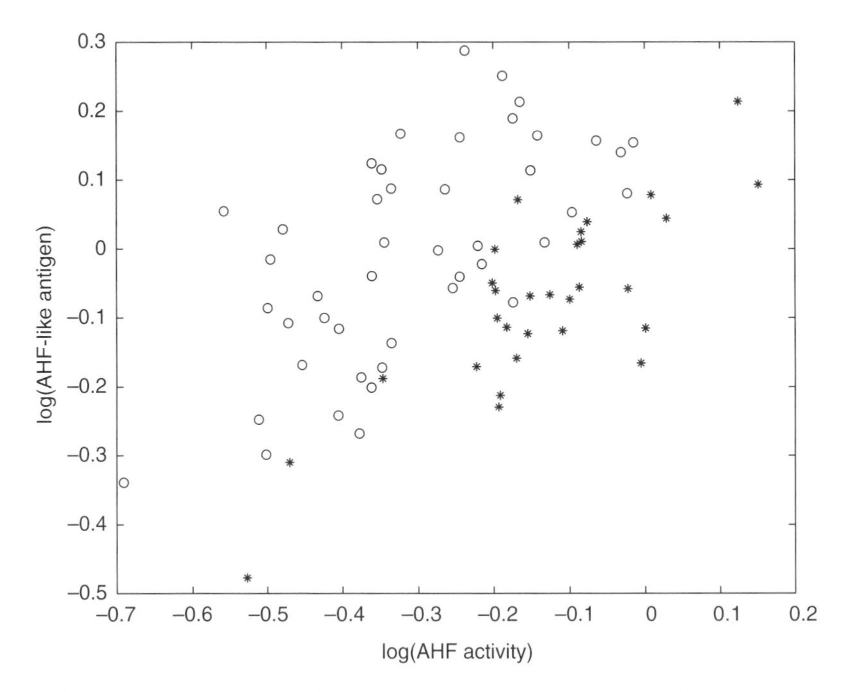

Fig. 4.3 Scatterplot of y_1 and y_2 for the 30 non-carriers ($*$) and 45 carriers (o) of haemophilia A.

So the parameter space has eight dimensions. In the more general heteroscedastic model, the dimensionality of the parameter space increases to 11.

As we noted earlier for such mixture models, the parameters are not fully identifiable. Under the symmetry $\theta \leftrightarrow \theta'$, where

$$\theta = (p, (\mu_1)_1, (\mu_1)_2, (\mu_2)_1, (\mu_2)_2, \Sigma_{11}, \Sigma_{22}, \Sigma_{12})$$
$$\theta' = (1 - p, (\mu_2)_1, (\mu_2)_2, (\mu_1)_1, (\mu_1)_2, \Sigma_{11}, \Sigma_{22}, \Sigma_{12}),$$

the density is invariant. This implies that $L(\theta) = L(\theta')$ at such pairs of points in the parameter space. An artificial constraint that restores parameter identifiability is the restriction $(\mu_1)_1 > (\mu_2)_1$, which shall be assumed henceforth.

Even with the problems of identifiability removed, the problems of likelihood multimodality remain. Basford and McLachlan discovered three distinct local maxima of the likelihood. As the mixture model is obtained by pooling together the two samples of the data set, the three local MLE's can be compared with the unique vector of estimates for the two-sample data under correct allocation as shown in Figure 4.3. Table 4.1 shows the roots found by Basford and McLachlan.

The first row of Table 4.1 shows estimates for means and covariances for the two-sample (ts) model using the correct allocation of observations. The remaining three rows show the three local maxima of the likelihood for the mixture model as discovered by Basford and McLachlan. The first of these three is the global maximum. Note that the MLE for the two-sample model is based upon the complete data, whereas the MLE's for the mixture model are based upon marginal data with the indicator variable for the allocation discarded. Therefore, the two-sample mean and variance estimates are to be preferred over the three mixture estimates. A perusal of this table offers little reassurance that the global MLE is to be preferred over the other two local maxima.

An important question is left unanswered in Basford and McLachlan (1985). Since the parameter space for the homoscedastic model is eight dimensional, how can we be sure that all the local maxima of the likelihood have been discovered? The dimensionality of the parameter space makes the search for local maxima like wandering in a maze. One could miss a maximum completely if its domain of attraction in a search algorithm is very small. With eight dimensions, we might

Table 4.1 Local maxima of the mixture likelihood found by Basford and McLachlan (1985).

$\hat{\theta}$	$\hat{p}$	$(\hat{\mu}_1)_1$	$(\hat{\mu}_1)_2$	$(\hat{\mu}_2)_1$	$(\hat{\mu}_2)_2$	$\hat{\Sigma}_{11}$	$\hat{\Sigma}_{22}$	$\hat{\Sigma}_{12}$	$\ell(\hat{\theta})$
ts	0.400	−0.135	−0.078	−0.308	−0.006	0.0226	0.0216	0.0154	
1	0.716	−0.206	−0.080	−0.321	0.079	0.0265	0.0171	0.0158	75.00
2	0.528	−0.121	−0.019	−0.370	−0.052	0.0137	0.0220	0.0100	73.49
3	0.681	−0.153	0.012	−0.420	−0.135	0.0138	0.0175	0.0035	73.29

conceivably need millions of iterations with different starting points to find all maxima. One solution to this problem is based upon the following principle:

- Each local maximum of the likelihood is usually supported by a *subset* of observations for which the marginal MLE is unique and close to that local MLE for the full data set.

We shall examine this principle, which cannot be applied too generally, in Section 4.10. On the basis of this principle we can examine subsets of six observations, say, at a time. By dividing the six observations into two groups of three,

$$y_{i_j} = (y_{i_j 1}, y_{i_j 2}), \quad j = 1, 2, 3, 4, 5, 6$$

we obtain starting estimates for the parameters of the heteroscedastic model by setting

$$\mu_1^{(1)} = \frac{y_{i_1} + y_{i_2} + y_{i_3}}{3}, \quad \mu_2^{(1)} = \frac{y_{i_4} + y_{i_5} + y_{i_6}}{3},$$

$$\Sigma_1^{(1)} = \frac{y_{i_1}^t y_{i_1} + y_{i_2}^t y_{i_2} + y_{i_3}^t y_{i_3} - (\mu_1^{(1)})^t \mu_1^{(1)}}{2},$$

and

$$\Sigma_2^{(1)} = \frac{y_{i_4}^t y_{i_4} + y_{i_5}^t y_{i_5} + y_{i_6}^t y_{i_6} - (\mu_2^{(1)})^t \mu_2^{(1)}}{2}.$$

If $\mu_{11}^{(1)} < \mu_{21}^{(1)}$, we switch the groups. The simplest way to initialise p is to let $p^{(1)}$ be chosen randomly from the interval $(0, 1)$. Then with this random choice, the pooled covariance for the homoscedastic model can be initialised at

$$\Sigma^{(1)} = p^{(1)} \Sigma_1^{(1)} + (1 - p^{(1)}) \Sigma_2^{(1)}.$$

From the initial parameter estimates, a variety of iterative methods are available to climb the likelihood function. Basford and McLachlan used the E–M algorithm, which is a slow but reliable method. Newton–Raphson is faster but less reliable. Also to be considered is the Nelder–Mead algorithm, which also reliably increases the likelihood.

Iterating from the initial estimates using the Nelder–Mead algorithm, we find that there is an additional mode of the likelihood that is distinct from the ones found by Basford and McLachlan (1985). Our expanded table of estimates becomes Table 4.2. As the fourth root is close to the global maximum, its presence may have been 'masked' by the global maximum. Are there any other local maxima? While this investigation did not find any others in the interior of the parameter space, the existence of yet more maxima cannot be ruled out. We shall consider the question of undiscovered roots and local maxima in Section 4.10.

Table 4.2 Expanded list of local maxima for Basford and McLachlan (1985).

$\hat{\theta}$	$\hat{p}$	$(\hat{\mu}_1)_1$	$(\hat{\mu}_1)_2$	$(\hat{\mu}_2)_1$	$(\hat{\mu}_2)_2$	$\hat{\Sigma}_{11}$	$\hat{\Sigma}_{22}$	$\hat{\Sigma}_{12}$	$\ell(\hat{\theta})$
ts	0.400	−0.135	−0.078	−0.308	−0.006	0.0226	0.0216	0.0154	
1	0.716	−0.206	−0.080	−0.321	0.079	0.0265	0.0171	0.0158	75.00
2	0.528	−0.121	−0.019	−0.370	−0.052	0.0137	0.0220	0.0100	73.49
3	0.681	−0.153	0.012	−0.420	−0.135	0.0138	0.0175	0.0035	73.29
4	0.890	−0.212	−0.009	−0.454	−0.247	0.0235	0.0167	0.0064	73.01

What are we to conclude from the proliferation of roots in this model? The global maximum of the likelihood is quite distant from the two-sample MLE which may be considered definitive for the data set—at least for the homoscedastic model. The presence of multiple maxima suggests the need to analyse the data with a heteroscedastic model. In this we concur with Basford and McLachlan (1985). However, even the heteroscedastic model has multiple maxima, with quite distinct parameter values. It seems likely that, fundamentally, the presence of multiple maxima points to the lack of clear information in the pooled data set about the model parameters.

4.3 Estimation of the correlation coefficient

Consider a set of independent bivariate observations (x_i, y_i), $i = 1, \ldots, n$, from a bivariate normal distribution which is standardised to have means $\mu_x = \mu_y = 0$ and variances $\sigma_x^2 = \sigma_y^2 = 1$. We assume that there is an unknown correlation coefficient ρ between any x_i and y_i. The likelihood equation $\dot{\ell}(\rho) = 0$, where $\ell(\rho) = \log L(\rho)$, reduces to

$$P(\rho) = \rho(1-\rho^2) + (1+\rho^2)\frac{\sum xy}{n} - \rho\left\{\frac{\sum(x^2+y^2)}{n}\right\} = 0 \tag{4.1}$$

which can have as many as three real roots in the interval $(-1, 1)$. If three roots are present, then these will correspond to two relative maxima and one relative minimum of the likelihood.

A sufficient (but not necessary) condition for the cubic equation $P(\rho)$ to have a unique root is that it be monotone. The cubic $P(\rho)$ will be monotone, and therefore have a unique real root, when the equation $\dot{P}(\rho) = 0$ has at most one real solution. In turn, this will be true if the discriminant of the quadratic

$$D = 4\left(\frac{\sum xy}{n}\right)^2 + 12\left\{1 - \frac{\sum(x^2+y^2)}{n}\right\}$$

is zero or strictly negative. From the law of large numbers, we see that D converges to $4\rho^2 - 12$ as $n \to \infty$. So with probability converging to one, the likelihood equation will have a unique root for large sample sizes.

To analyse this cubic equation further, let us define

$$S_1 = \frac{\sum_{j=1}^{n} x_j y_j}{n} \qquad S_2 = \frac{\sum_{j=1}^{n} (x_j^2 + y_j^2)}{n}.$$

The pair (S_1, S_2) forms a minimal sufficient statistic for the estimation of ρ. Next, we perform a location shift $z = \rho - S_1/3$. Equation (4.1) reduces to the depressed equation

$$z^3 + az + b = 0 \tag{4.2}$$

where the coefficients

$$a = S_2 - \frac{S_1^2}{3} - 1$$

and

$$b = \frac{S_1 S_2}{3} - \frac{4S_1}{3} - \frac{2S_1^3}{27}$$

are functions of the data through the sample moments S_1 and S_2. We can study the multiple solutions to this equation by plotting in $\mathbb{R}^3$ all points (a, b, z), where z is a root of equation (4.2) with given coefficients a and b. Figure 4.4 shows the surface so obtained. The resulting surface is an example of the well-known *cusp catastrophe*. With this interpretation, the coefficients a and b represent *control parameters* for the cusp catastrophe. In the control space, the projection of the folds of the surface defines the *separatrix*, whose equation in (a, b) is

$$4a^3 + 27b^2 = 0.$$

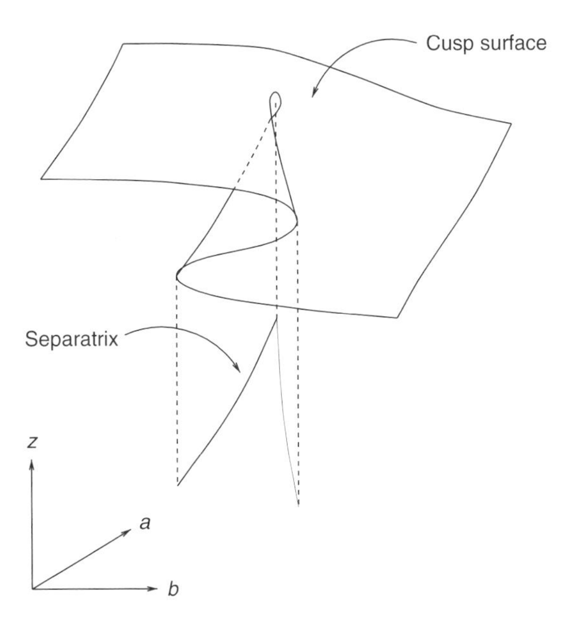

Fig. 4.4 The cusp surface with the roots as functions of the control parameters.

The separatrix divides the control space into two regions. In the first region,

$$4a^3 + 27b^2 > 0$$

and there is a single root. In the other region where

$$4a^3 + 27b^2 < 0$$

the surface folds back on itself, so that there are three roots. The point $a = b = 0$ defines the control parameters of the cusp catastrophe where the two fold lines of the separatrix meet (see Figure 4.5). The reader is referred to Gilmore (1981, p. 61) for more on the theory of the cusp catastrophe.

The detailed analysis of the likelihood function and its extrema can be found in Stuart and Ord (1991).

If we examine the necessary and sufficient conditions in for the uniqueness of a root, we see that multiple roots will occur if and only if $4a^3 + 27b^2 < 0$. Equivalently, in S_1 and S_2 we have

$$4S_2^3 - S_2^2 S_1^2 - 12S_2^2 - 16S_2 S_1^2 + 12S_2 + 4S_1^4 + 44S_1^2 - 4 < 0, \quad -\frac{S_2}{2} \leq S_1 \leq \frac{S_2}{2}. \tag{4.3}$$

The constraint that $2|S_1| \leq S_2$ holds for all data sets. Setting $U = S_1^2$, we can rewrite (4.3) as

$$4U^2 + (44 - 16S_2 - S_2^2)U + (4S_2^3 - 12S_2^2 + 12S_2 - 4) < 0$$

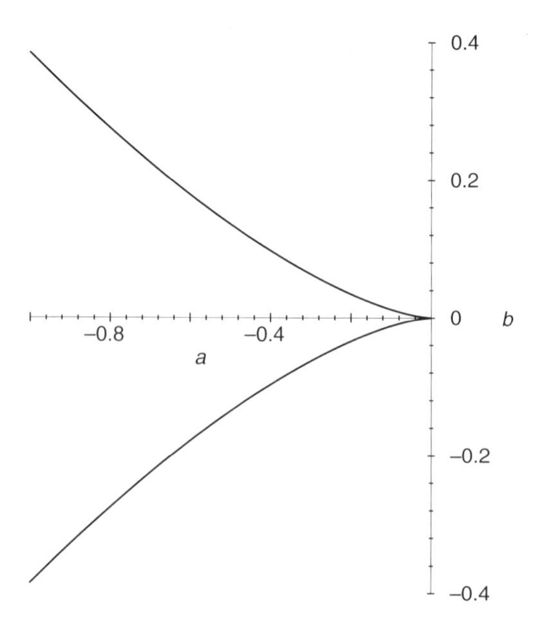

Fig. 4.5 The separatrix dividing the control space into two regions.

which is a quadratic in U. So it will have negative values for certain choices of S_1 when the discriminant of this quadratic is positive. As a function of S_2, the discriminant in the variable S_1^2 is

$$\begin{aligned} D &= 2000 - 1600S_2 + 360S_2^2 - 32S_2^3 + S_2^4 \\ &= (S_2 - 2)(S_2 - 10)^3. \end{aligned}$$

From this factorisation, we see that D will be positive when S_2 lies outside the interval $[2, 10]$. So multiple roots can occur when $S_2 < 2$ or $S_2 > 10$. The case where $S_2 > 10$, need not concern us, because the spurious roots of the likelihood equation are outside the interval $(-1, +1)$. So we can consider three cases:

1. $S_2 > 2$. The likelihood equation has a unique solution in the interval $(-1, +1)$.
2. $S_2 = 2$. The likelihood equation has a unique and explicit solution $\hat{\rho} = S_1$. Note that $E(S_2) = 2$ for all values of ρ. So the assignment $S_2 = 2$ most clearly represents the case where the data are in agreement with model assumptions.
3. $S_2 < 2$. The likelihood equation can have multiple solutions in the interval $(-1, +1)$.

The value $S_2 = 2$ is clearly the critical cutoff, both for multiple roots and for obtaining $\hat{\rho} = S_1$, as the explict and standard estimate for the correlation. This suggests that a perturbation analysis of the likelihood equation around $\rho = S_1$ and $S_2 = 2$ would be helpful. Equation (4.1) can be rewritten as

$$\rho = S_1 + \frac{\rho}{1 + \rho^2}(2 - S_2). \tag{4.4}$$

The accompanying iterative substitution is

$$\rho^{(j+1)} = S_1 + \frac{\rho^{(j)}}{1 + (\rho^{(j)})^2}(2 - S_2)$$

with $\rho^{(1)} = S_1$. This can be Aitken accelerated.

4.4 The Cauchy distribution and stable laws

The previous example may suggest that multiple root problems are small sample issues which will disappear for sufficiently large sample sizes. This commonly held belief is unfortunately too optimistic, as the following example illustrates.

Definition 4.4 *Suppose a sample of n random variables is drawn from a Cauchy distribution with density function*

$$f(y; \xi, c) = \frac{1}{\pi} \frac{c}{c^2 + (y - \xi)^2}. \tag{4.5}$$

Then the n variables are said to come from a Cauchy location-scale model with location parameter $\xi \in (-\infty, +\infty)$ and scale parameter $c > 0$.

This family of densities is characterised as the family of non-degenerate symmetric stable laws with exponent one. In the special case where $c = 1$, the densities

$$f(y;\xi) = \frac{1}{\pi}\frac{1}{1+(y-\xi)^2} \tag{4.6}$$

are said to form the Cauchy location model.

The Cauchy distribution is often considered mainly a source for counter-examples and pathologies in statistical inference. However, it can arise in financial applications as the following example involving limit pricing illustrates.

Example 4.5 Limit pricing

Suppose R_t and S_t are two stocks whose fluctuations are modelled as diffusions which are functions of time t. In particular, we suppose that $Z_t = \log R_t$ and $W_t = \log S_t$ are Brownian motions with zero drift and diffusion parameters σ_1 and σ_2 respectively. An order is placed to sell stock S at any time τ that S_τ hits a given price. On the logarithmic scale, we might suppose that the order to sell S occurs at the earliest time τ such that $W_\tau = a$, say. The recurrence property of Brownian motion in one dimension with zero drift tells us that τ will be finite—that is, the stock will eventually be sold—with probability one (Figure 4.6). What distribution does $Y = Z_\tau$ have when this stock is sold?

The answer depends upon the joint distribution of the two stocks. In many cases, the pair (Z_t, W_t) can be modelled as a correlated Brownian motion in the plane. This means there exists a linear transformation

$$Z_t = Z'_t + \beta W_t$$

where Z'_t and W_t are independent Brownian motions and Z'_t also has zero drift.

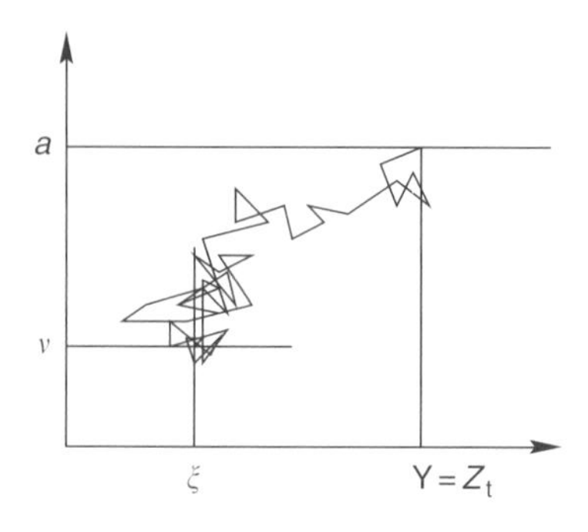

Fig. 4.6 A Cauchy variable Y generated from a hitting time of Brownian motion Z_t.

Suppose that $Z_0 = \xi$, and $W_0 = \nu$, say. For any b, let $\tau(b)$ be the time until W_t hits the value $\nu + b$. The strong Markov property of Brownian motion implies that when $b, c > 0$ or when $b, c < 0$, the distribution of $\tau(b + c)$ is the convolution of the distributions of $\tau(b)$ and $\tau(c)$. Additionally, since the increments of Brownian motion scale as the square root of the time variable t, it follows that

$$\tau(b) \sim b^2 \tau(1).$$

In particular, $\tau \sim (a - \nu)^2 \tau(1)$. These considerations imply that the waiting time τ has a stable distribution with exponent $1/2$ that is concentrated on the positive axis. Using the reflection principle for the hitting times of random walks, we can show that

$$P(\tau < t) = 2P(W_t > a) = 2\left\{1 - \Phi\left(\frac{a - \nu}{\sigma_2 \sqrt{t}}\right)\right\}$$

where Φ is the CDF for the standard normal distribution. Differentiating with respect to the variable t, we obtain the density function for τ, namely

$$f_\tau(t) = \frac{a - \nu}{t^{3/2}\sigma_2\sqrt{2\pi}} \exp\left\{-\frac{(a - \nu)^2}{2\sigma_2^2 t}\right\}, \qquad t > 0.$$

In a similar way, it can be shown that the distribution of $Z_{\tau(b+c)}$ (resp. $Z'_{\tau(b+c)}$) is the convolution of the distributions of $Z_{\tau(b)}$ (resp. $Z_\tau(b)'$) and $Z_{\tau(c)}$ (resp. $Z'_{\tau(c)}$), with scaling law

$$Z_{\tau(b)} \sim b Z_{\tau(1)} \quad (\text{resp. } Z'_{\tau(b)} \sim b Z'_{\tau(1)}).$$

So Z_τ and Z'_τ have stable distributions. The scaling law in this case implies that Z_τ and Z'_τ are stable with exponent one. As Z'_τ has a distribution that is symmetrical about $\xi - \beta\nu$ and

$$Z_\tau = Z'_\tau + \beta(\nu + b)$$

it follows that the distribution of Z_τ is symmetric about $\xi + \beta b$. However, the only symmetric stable law with exponent one is the Cauchy distribution.

Let us turn now to the estimation of parameters for the Cauchy model. Inferentially, the Cauchy location model is quite distinct from the location-scale model. So we shall consider these two models separately. We begin with the location model, and shall set $c = 1$. If c is known and $c \neq 1$, the general comments below hold true with appropriate modification of the formulas. A sample of variables from a Cauchy location model presents a number of problems for likelihood inference. The likelihood function for location parameter ξ is

$$L(\xi) = \prod_{j=1}^{n} \frac{1}{1 + (y_j - \xi)^2}$$

and the likelihood equation reduces to

$$\sum_{j=1}^{n} \frac{y_j - \hat{\xi}}{1 + (y_j - \hat{\xi})^2} = 0. \tag{4.7}$$

Equation (4.7) can be rearranged so that any solution $\hat{\xi}$ can be represented as a weighted average of the data:

$$\hat{\xi} = \sum_{j=1}^{n} w_j(\hat{\xi}) y_j \tag{4.8}$$

where the weights

$$w_j(\xi) = \frac{\{1 + (y_j - \xi)^2\}^{-1}}{\sum_{k=1}^{n} \{1 + (y_k - \xi)^2\}^{-1}}$$

are positive and sum to one. The representation of $\hat{\xi}$ as a weighted average suggests the iterative substitution algorithm

$$\xi^{(k+1)} = \sum_{j=1}^{n} w_j(\xi^{(k)}) y_j$$

for finding roots. Convergence will be linear, and can be improved with Aitken acceleration.

How many roots does the likelihood equation have? Upon taking a common denominator in equation (4.7), we see that the solution set for this equation is equivalent to that of

$$\sum_{i=1}^{n} \left\{ (y_i - \xi) \prod_{j \neq i} [1 + (y_j - \xi)^2] \right\} = 0$$

which is a polynomial equation of degree $2n - 1$. Since the degree of the equation is odd, it will have at least one real root, and if multiplicities of the roots are counted, the total number of real roots will always be odd number between 1 and $2n - 1$. It is not hard to check that these extremes can be attained for certain data sets. For example, if $y_j = y$, for $j = 1, \ldots, n$, then $\hat{\xi} = y$ will be the only root. For extreme configurations of highly separated variates, the polynomial equation admits a full $2n - 1$ distinct solutions, corresponding to n relative maxima and $n - 1$ relative minima of the likelihood. Fortunately, as Reeds (1985) has shown, this extreme situation is rare. As $n \to \infty$, the asymptotic distribution of the number of local maxima of the Cauchy likelihood converges to that of $1 + M$, where M has a Poisson distribution with mean $1/\pi$. A consequence of this is that the number

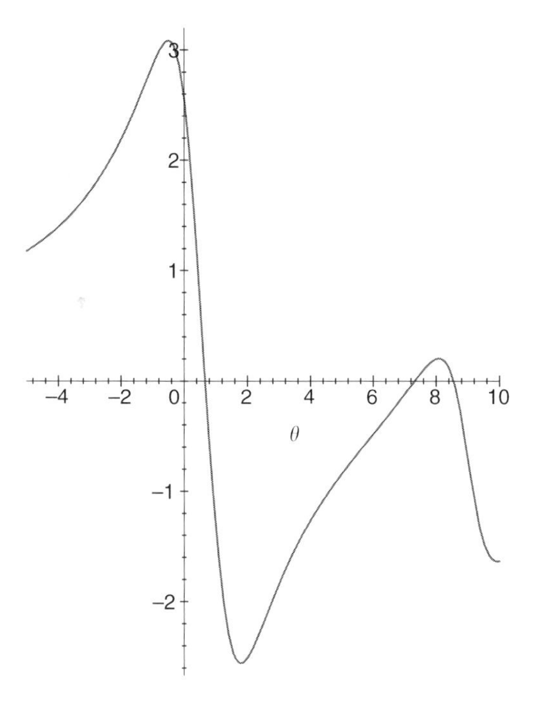

Fig. 4.7 The score function for the Cauchy location-model with an outlier producing extraneous roots.

of extraneous local maxima of the likelihood will be positive with an asymptotic probability given by

$$1 - e^{-1/\pi} \approx 0.2726$$

which is less than one time in three. However, the probability that extraneous roots occur does not go to zero as the sample size gets large (Figure 4.7).

We shall now consider the Cauchy location-scale model with unknown parameters ξ and c. Unlike the location model, the problem of multiple roots does not arise in this case. Copas (1975) showed that with probability one there is a unique solution to the simultaneous likelihood equations

$$\frac{\partial L(\xi, c)}{\partial \xi} = 0 \quad \frac{\partial L(\xi, c)}{\partial c} = 0.$$

The solution to these equations fails to be unique only if exactly 50% of the data values are coincident at some y_1 and the other 50% are coincident at some other y_2. The other case worthy of special consideration occurs when more than 50% of the values are coincident at a point y. In this case there is no solution to the likelihood equations, and the likelihood is maximised on the boundary of the parameter space with $\hat{\xi} = y$ and $\hat{c} = 0$. Both of these special cases have probability zero.

After some rearrangement, the likelihood equations reduce to

$$\hat{\xi} = \sum_{j=1}^{n} w_j(\hat{\xi}, \hat{c}) y_j, \tag{4.9}$$

$$\hat{c}^2 = \sum_{j=1}^{n} w_j(\hat{\xi}, \hat{c})(y_j - \hat{\xi})^2 \tag{4.10}$$

where

$$w_j(\xi, c) = \frac{\{c^2 + (y_j - \xi)^2\}^{-1}}{\sum_{k=1}^{n}\{c^2 + (y_k - \xi)^2\}^{-1}}.$$

These equations suggest the iterative substitution algorithm

$$\xi^{(k+1)} = \sum_{j=1}^{n} w_j(\xi^{(k)}, c^{(k)}) y_j, \tag{4.11}$$

$$c^{(k+1)^2} = \sum_{j=1}^{n} w_j(\xi^{(k)}, c^{(k)})(y_j - \xi^{(k)})^2 \tag{4.12}$$

which can be Aitken accelerated.

The Cauchy distribution is a special case in the family of symmetric stable distributions, to which we now turn. A general representation for symmetric stable distributions is provided by the following construction.

Proposition 4.6 *Suppose that $X_1, \ldots, X_n$ are independent random variables that are uniformly distributed in the interval $[-n, +n]$. Let $p > 1/2$. Define*

$$Y_n = \sum_{j=1}^{n} \operatorname{sgn}(X_j)|X_j|^{-p} \tag{4.13}$$

where $\operatorname{sgn}(X)$ is the sign of X, which is $+1$, 0, or -1 as $X > 0$, $X = 0$ or $X < 0$, respectively. Then in the limit as $n \to \infty$, Y_n converges in distribution to a symmetric stable law with exponent $\alpha = 1/p$, centered at zero with scale parameter

$$c = \left\{\alpha \int_0^\infty u^{-\alpha-1}(1 - \cos u)\, du\right\}^{1/\alpha}.$$

To prove this result, we shall examine the characteristic function of Y_n. Note first that

$$E(e^{i\lambda \operatorname{sgn}(X)|X|^{-p}}) = n^{-1} \int_0^n \cos(\lambda x^{-p})\, dx. \tag{4.14}$$

So we can write

$$\begin{aligned}
E(\mathrm{e}^{\mathrm{i}\lambda Y_n}) &= \left\{n^{-1}\int_0^n \cos(\lambda x^{-p})\,\mathrm{d}x\right\}^n \\
&= \left[1 - n^{-1}\int_0^n \{1-\cos(\lambda x^{-p})\}\,\mathrm{d}x\right]^n \\
&\to \exp\left[-\int_0^\infty \{1-\cos(\lambda x^{-p})\}\,\mathrm{d}x\right] \\
&= \exp\left[-\frac{|\lambda|^{1/p}}{p}\int_0^\infty u^{-(1+p)/p}(1-\cos u)\,\mathrm{d}u\right] \\
&= \exp[-|c\lambda|^\alpha].
\end{aligned}$$

It can be checked that the integral converges for $p > 1/2$. This last expression is the characteristic function for a symmetric stable law centred at zero with scale c and exponent α.

The density functions for symmetric stable laws other than the normal and the Cauchy cannot be expressed in closed form with elementary functions. This creates major obstacles for implementing a likelihood analysis for data from a symmetric stable distribution. The density function for these laws can be expressed through an infinite series. Bergström (1952) gives the formulas for the CDF

$$F(y;\alpha) = \frac{1}{2} + \frac{1}{\pi\alpha}\sum_{k=1}^{\infty}(-1)^{k-1}\frac{\Gamma((2k-1)/\alpha)}{(2k-1)!}y^{2k-1} \tag{4.15}$$

and

$$F(y;\alpha) = 1 + \left\{\frac{1}{\pi}\sum_{k=1}^{\infty}(-1)^k\frac{\Gamma(\alpha k)}{k!\,y^{\alpha k}}\sin\left(\frac{k\pi\alpha}{2}\right)\right\} \tag{4.16}$$

for the special case $\xi = 0$ and $c = 1$. Formula (4.15) is formally convergent for all y when $\alpha > 1$, while formula (4.16) converges for all $y \neq 0$ when $\alpha < 1$. These series are easily modified to account for general scale and location parameters.

Both equations can also be formally differentiated with respect to y to obtain expansions for the density, namely

$$f(y;\xi,c,\alpha) = \frac{1}{\pi\alpha c}\sum_{k=0}^{\infty}(-1)^k\frac{\Gamma((2k+1)/\alpha)}{(2k)!}\left(\frac{y-\xi}{c}\right)^{2k} \tag{4.17}$$

and

$$f(y;\xi,c,\alpha) = \frac{1}{\pi c}\sum_{k=1}^{\infty}(-1)^{k+1}\frac{\alpha\Gamma(\alpha k)}{(k-1)!}\left|\frac{y-\xi}{c}\right|^{-\alpha k-1}\sin\left(\frac{k\pi\alpha}{2}\right). \tag{4.18}$$

Once again, (4.17) converges when $\alpha > 1$ and (4.18) converges when $\alpha < 1$. Rather surprisingly, the formal convergence properties of these series are not closely related to the practical utility of the series for approximating the density functions. For example, when $|y - \xi|$ is large and $\alpha > 1$, the terms in (4.17) will only decrease to zero when n is very large. Before that, they can increase monotonically by many orders of magnitude. On the other hand, if $|y - \xi|$ is small and $\alpha < 1$, the series, although formally divergent, appears to converge for many terms. Similar remarks hold for formula (4.18). When $|y - \xi|$ is small and $\alpha < 1$, the series appears to diverge although it is formally convergent. When $|y - \xi|$ is large and $\alpha > 1$, the series appears to converge but is formally divergent.

Some explanation for this phenomenon can be found in the fact that in those cases when the series diverges, the desired density behaves as an *anti-limit* of the series rather than as a limit. An anti-limit can be regarded as a point of repulsion of the sequence of partial sums just as a limit is a point of attraction for the partial sums. For practical purposes, (4.18) works well when $|y - \xi|$ is large for all α, while (4.17) works well when $|y - \xi|$ is small. In cases where the series are formally divergent, it is important to only use terms which are decreasing early in the sequence.

There remains the difficult problem of determining which series to use for a given value of the density. In fact there may be values of y for which neither series works in a practical sense. To ensure the best convergence results, it is helpful to accelerate the convergence properties of such series, whenever possible. In the following example, we shall consider an acceleration method known as the *Euler transformation*.

Example 4.7 Let

$$s = a_1 - a_2 + a_3 - a_4 + \cdots \tag{4.19}$$

be a convergent alternating series with positive a_j decreasing monotonically to zero. We write

$$\begin{aligned} 2s &= a_1 + (a_1 - a_2) - (a_2 - a_3) + (a_3 - a_4) - \cdots \\ &= a_1 + \Delta a_1 - \Delta a_2 + \Delta a_3 - \cdots \end{aligned}$$

where $\Delta a_j = a_j - a_{j-1}$ is the usual first difference operator. Therefore,

$$s = \frac{a_1}{2} + \frac{1}{2}(\Delta a_1 - \Delta a_2 + \Delta a_3 - \cdots).$$

But the expression in parentheses is another alternating series. Therefore the same operation can be applied again. Continuing in this fashion we obtain the expression

$$s = \frac{a_1}{2} + \frac{\Delta a_1}{4} + \frac{\Delta^2 a_1}{8} + \frac{\Delta^3 a_1}{16} + \cdots . \tag{4.20}$$

Here,

$$\begin{aligned} \Delta^2 a_1 &= \Delta(a_1 - a_2) \\ &= (a_1 - a_2) - (a_2 - a_3) \\ &= a_1 - 2a_2 + a_3 \end{aligned}$$

and, in general

$$\Delta^j a_1 = \sum_{k=0}^{j} (-1)^k \binom{j}{k} a_{1+k}.$$

Equation (4.20) is the *Euler transformation*. As the coefficients of the difference operators sum to zero, the numerators $\Delta^j a_1$ do not grow large, and the transformed series behaves approximately like a geometric series.

The Euler transformation is so successful at accelerating convergence, that considerable attention has been paid to converting series to an alternating form. One such method is *Van Wijngaarden's transformation* which converts positive series to alternating series. Suppose that

$$a_1 + a_2 + a_3 + \cdots$$

is a convergent series with positive terms. Define

$$a_k^* = a_k + 2a_{2k} + 4a_{4k} + \cdots$$

for $k = 1, 2, 3, \ldots$. Then

$$a_1 + a_2 + a_3 + a_4 + \cdots = a_1^* - a_2^* + a_3^* - a_4^* + \cdots . \tag{4.21}$$

The Van Wijngaarden transformation coupled with the Euler transformation often produces rapid convergence. For example, the series $\sum n^{-2}$ becomes

$$\begin{aligned}
\sum_{n=1}^{\infty} \frac{1}{n^2} &= \sum_{n=1}^{\infty} \left\{ (-1)^{n-1} \frac{2}{n^2} \right\} \qquad \text{(Van W.)} \\
&= \sum_{n=1}^{\infty} \left\{ \frac{\sum_{j=0}^{n-1} (-1)^j \binom{n-1}{j} (1+j)^{-2}}{2^{n-1}} \right\} \qquad \text{(Euler)} \\
&= 1 + \frac{3}{8} + \frac{11}{72} + \frac{25}{384} + \frac{137}{4800} + \cdots .
\end{aligned}$$

An additional advantage of the Euler transformation is that it can increase the radius of convergence of some power series. For example, the series $1 - x^2 + x^4 - \cdots$ converges for x in the interval $(-1, +1)$. Under the Euler transformation, the series becomes

$$\frac{1}{2} + \frac{1 - x^2}{4} + \frac{(1 - x^2)^2}{8} + \cdots ,$$

which converges for x in $(-\sqrt{3}, +\sqrt{3})$. So the transformation is often useful for computing the anti-limits of series.

One last issue that needs consideration is the effect of an Euler transformation on series which are not alternating. In such cases, the transformation usually leads to an equivalent series. If the original series has a large number of sign changes among neighbouring terms, then the Euler transformation usually improves the rate of convergence. However, for series with few sign changes, the rate of convergence may not be as good.

Returning to the problem of calculating symmetric stable densities, we can apply the Euler transformation to formula (4.17). This increases the region of satisfactory convergence sufficiently to allow the tail formula in (4.18) to finish the job. In Figure 4.8, we see the effect of applying the Euler transformation to the series expansion with eight terms for $\alpha = 1.5$. The expansion in (4.17) with eight terms works well when $|y| < 2$, but the approximation degenerates rapidly outside that region. On the other hand, the tail formula (4.18), although formally divergent, works well when $|y| > 3$ using eight terms. Neither formula captures the 'shoulders' of the distribution between 2 and 3. However, by applying the Euler transformation, the region of satisfactory convergence for (4.17) is extended to $|y| < 4$, which is more than sufficient to cover all choices of y. It is encouraging to note that the density can be accurately approximated piecewise with only eight terms of the appropriate series for all real y.

Figure 4.9 shows the effect of the Euler transformation on the tail formula in (4.18) for the sample distribution. Here the improvement is more modest, but is still quite clear. The ability to accelerate the convergence from a few terms of the

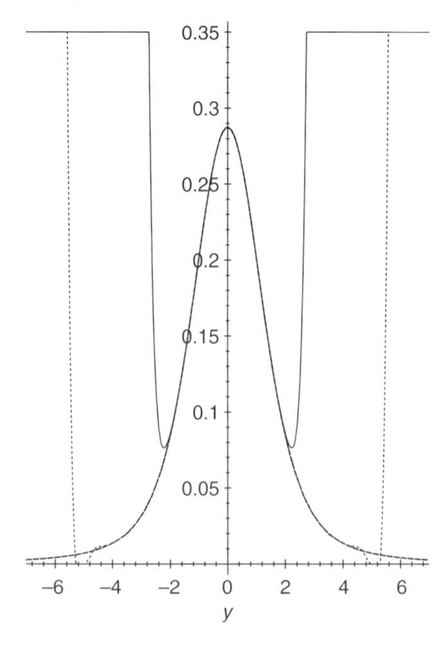

Fig. 4.8 The improvement of the Euler transformation on series (4.17). The plot shows series (4.17) and the Euler transformed improvement of (4.17) with eight terms for each. Parameters $\xi = 0$, $c = 1$ and $\alpha = 1.5$ have been used. The large deviations of each series have been truncated in the ordinate to ensure appropriate scaling. The correct density is also plotted for comparison purposes.

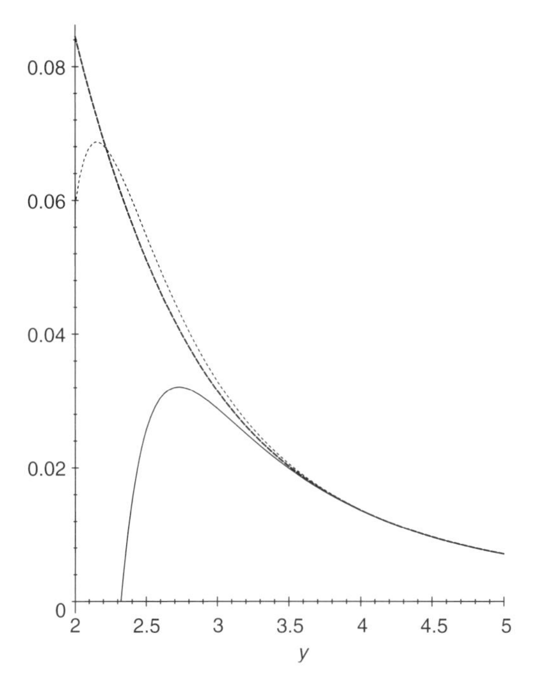

Fig. 4.9 The improvement of the Euler transformation on series (4.18). The plot shows the series (4.18) and its Euler transformed improvement with eight terms for each. Parameters $\xi = 0$, $c = 1$ and $\alpha = 1.5$ have been used. The large deviations of each series have been truncated in the ordinate to ensure appropriate scaling. The correct density is also plotted for comparison purposes.

tail formula is especially useful in (4.18) because the density is the anti-limit of the formula, and cannot be approximated with a large number of terms.

By patching the formulas together piecewise, it is possible to plot the densities completely. For example, Figure 4.10 shows the densities for values of α between 0.7 and 1.5 with $\xi = 0$ and $c = 1$.

Once densities are calculated, maximum likelihood estimation becomes possible. In much the same way as equations (4.9) and (4.10), the likelihood equations for ξ and c can be represented as weighted least squares equations. Let $\psi(y) = \log f(y; 0, 1, \alpha)$. Then the likelihood equations for ξ and c transform to

$$\sum_{j=1}^{n} w_j(\hat{\xi}, \hat{c}, \alpha)(y_j - \hat{\xi}) = 0 \tag{4.22}$$

$$\sum_{j=1}^{n} w_j(\hat{\xi}, \hat{c}, \alpha)(y_j - \hat{\xi})^2 - n\hat{c}^2 = 0 \tag{4.23}$$

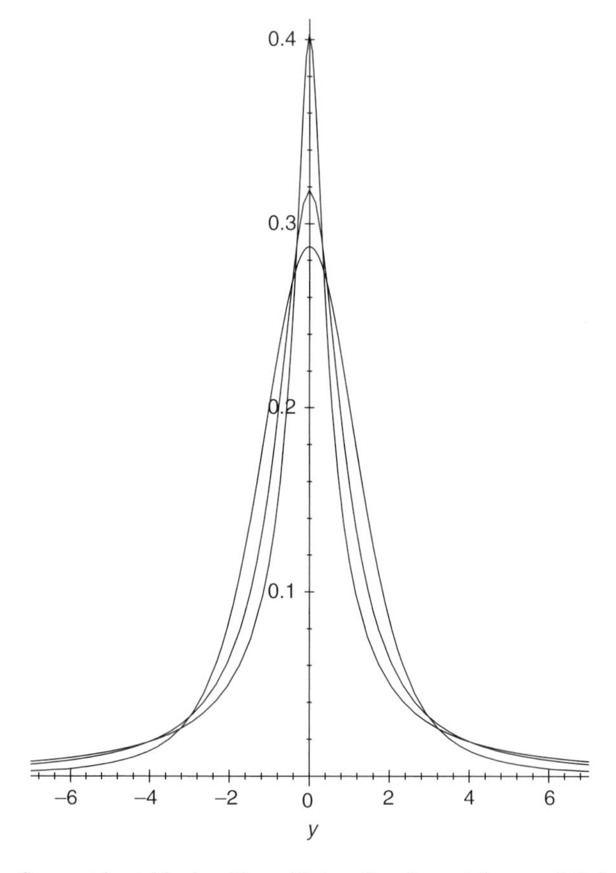

Fig. 4.10 Symmetric stable densities with $\xi = 0$ and $c = 1$ for $\alpha = 0.7$, 1.0 and 1.5.

where

$$w_j(\xi, c, \alpha) = \frac{-c\psi\{(y - \xi)/c\}}{y_j - \xi}.$$

The extension of these equations to regression models with stable errors can be found in McCulloch (1998a).

An alternative approach to calculating the stable densities is based upon the *inverse Fourier transform*. A stable density can be represented as the inverse Fourier transform of the characteristic function. However, such representations generally do not have closed forms. This requires that the integrals be evaluated numerically. Nolan (1997) has described a fast algorithm for the approximation of densities. The current version of this algorithm uses spline approximations to the densities in order to avoid the laborious task of density evaluation directly. A quasi-Newton hill climbing routine is applied to maximise the likelihood. Spline interpolations of the density function have also been developed by McCulloch

(1998*b*). DuMouchel (1973, 1975) has shown that the maximum likelihood estimator has the usual asymptotic properties by being asymptotically efficient and normally distributed.

Other methods for estimating parameters include the class of quantile or distribution fitting procedures. For example, it is possible to use a class of unbiased estimating functions of the form

$$g(\xi, c, \alpha) = \int_{-\infty}^{+\infty} \{\tilde{F}(t) - F(t; \xi, c, \alpha)\}\, \mathrm{d}H(t; \xi, c, \alpha) \tag{4.24}$$

where $\tilde{F}$ is the empirical CDF, and H is some appropriately chosen vector-valued integrating function whose dimension matches the dimension of the parameter space. To avoid too much analytical difficulty, the integrating function $H = (H_1, H_2, H_3)^{\mathrm{t}}$ can be set to some simple choice motivated by the desire to match quantiles. For example, we can use

$$H_1(t) = \begin{cases} 0, & t \le \xi \\ 1, & t > \xi \end{cases}$$

$$H_2(t) = \begin{cases} 0 & |t - \xi| \le k_1(\alpha)c \\ 1 & |t - \xi| > k_1(\alpha)c \end{cases}$$

$$H_3(t) = \begin{cases} 0 & |t - \xi| \le k_2(\alpha)c \\ 1 & |t - \xi| > k_2(\alpha)c \end{cases}$$

where $k_1(\alpha), k_2(\alpha) > 0$. If

$$k_j(\alpha) = F^{-1}(1 - \epsilon_j; 0, 1, \alpha)$$

then the method reduces to the fitting of the median and two quantiles. As α controls the behaviour in the extreme tails, it is helpful to fit quantiles far out from the median. For example, fitting both $\epsilon_1 = 0.05$ and $\epsilon_2 = 0.01$ gives a useful measure of the decay of the tail. On the other hand, if the shape of the distribution and its 'peakedness' in the centre of the distribution is of greater interest than the tail behaviour, it is reasonable to fit $\epsilon_1 = 0.25$ and $\epsilon_2 = 0.1$, say. Fama and Roll (1968) noticed that the 72nd and 28th percentiles of the stable laws are fairly insensitive to variation in the exponent α. This suggests that the location and scale parameters can be roughly estimated from the empirical quantiles by

$$\hat{\xi} = \tilde{F}^{-1}(0.5) \tag{4.25}$$

$$\hat{c} = \frac{1}{0.827}[\tilde{F}^{-1}(0.72) - \tilde{F}^{-1}(0.28)]. \tag{4.26}$$

These simple quantile estimates can be used in their own right or to initialise algorithms that converge to the maximum likelihood estimates or other estimators.

Quantile methods and maximum likelihood estimation both involve approximations to the distributions of the stable variates. So attention has turned to characteristic function fitting methods where the characteristic function can be expressed in closed form. The characteristic function is

$$\phi(t; \xi, c, \alpha) = \exp(\mathrm{i}\xi t - |ct|^{\alpha}).$$

Paulson *et al.* (1975) have proposed that the parameters be estimated by minimising

$$\int_{-\infty}^{+\infty} |\tilde{\phi}(t) - \phi(t; \xi, c, \alpha)|^2 \mathrm{e}^{-t^2}\, \mathrm{d}t \tag{4.27}$$

where

$$\tilde{\phi}(t) = n^{-1} \sum_j \mathrm{e}^{\mathrm{i}ty_j}$$

is the empirical characteristic function. The choice of $\exp(-t^2)$ as a weighting function is motivated by the need to ensure that this integral converges. Upon differentiating this objective function with respect to the three parameters, we see that this method is equivalent to solving the unbiased estimating equations

$$\int_{-\infty}^{+\infty} \{\tilde{\phi}(t) - \phi(t)\}\dot{\phi}(t)\mathrm{e}^{-t^2}\, \mathrm{d}t = 0 \tag{4.28}$$

where $\dot{\phi}(t)$ is the vector of derivatives of $\phi(t)$ with respect to the three parameters $\theta = (\xi, c, \alpha)$. Note that the imaginary components of the integral on the left-hand side are zero, because $\phi(-t) = \overline{\phi(t)}$ and $\tilde{\phi}(-t) = \overline{\tilde{\phi}(t)}$. Such equations are special cases of the family of equations of the form

$$\int_{-\infty}^{+\infty} \{\tilde{\phi}(t) - \phi(t)\} w(t)\, \mathrm{d}t = 0 \tag{4.29}$$

where $w(t)$ is an appropriately chosen vector weighting function such that $w(-t) = \overline{w(t)}$. This class of estimating functions was considered by Feuerverger and McDunnough (1981), who showed that the consistent root of this equation is asymptotically normal with asymptotic variance

$$n \operatorname{var}(\hat{\theta}) = \left[\iint w(s)w(t)\{\phi(s+t) - \phi(s)\phi(t)\}\, \mathrm{d}s\, \mathrm{d}t\right] \Big/ \left\{\int w(s)\psi(s)\, \mathrm{d}s\right\}^2.$$

for a one-parameter model. The extension to multiparameter models is straightforward. This result, and the other results of Feuerverger and McDunnough (1981) apply to general models beyond the stable family of distributions. The choice of $w(t)$ which minimises the asymptotic variance—that is, maximises the Godambe efficiency—can be found. This turns out to be the inverse Fourier transform of the score function. The resulting estimation procedure is then maximum likelihood. As such, there seems to be little point in using the empirical characteristic function equation with this optimal weight because maximum likelihood estimation can be accomplished more directly. A more practical choice of weighting function is one which has support on a finite set of values $t_1, \ldots, t_K$.

Another characteristic function technique proposed by Koutrouvelis (1980) has the additional advantage that the estimates for the parameters can be expressed as the solution to a linear regression problem. Koutrouvelis noticed that for stable laws, the characteristic function satisfies the equation

$$\log(-\log|\phi(t)|^2) = \log(2c^\alpha) + \alpha \log|t|. \tag{4.30}$$

So $v = \log(-\log|\phi(t)|^2)$ is a linear function of $u = \log|t|$. Based upon selected values $t_1, \ldots, t_K$ we can compute the 'variates' $v_1, \ldots, v_K$ and the 'covariates' $u_1, \ldots, u_K$ and perform a regression. The slope of the fitted regression line is an estimate of α. In turn, from this estimate of α and the v-intercept of the regression line, an estimate for c can be determined.

4.5 The relative likelihood principle

In the previous section, we noted that the problem of multiple solutions for the Cauchy location likelihood equation did not disappear asymptotically. Nevertheless, we could order the roots of the score function as estimators by calculating the likelihood at each root. The likelihood function is often interpreted as a *support function* that partially orders parameters by the value of the likelihood. The most appropriate root under this criterion would be that which globally maximises the likelihood. The rule which states that parameters can be partially ordered by the likelihood can be called the *relative likelihood principle*. It is not to be confused with the *weak likelihood principle* or to the *strong likelihood principle*. The relative likelihood principle can be stated as follows.

Definition 4.8 *Let θ_a and θ_b be two parameter values and E an observed event whose probability is controlled by the choice of θ_a or θ_b. The relative likelihood principle (RLP) states that, based upon E, the parameter value which better explains the event—in the sense, say, that $P(E; \theta_a) > P(E; \theta_b)$—is to be preferred. Equivalently, if $\ell(\theta_a; E) > \ell(\theta_b; E)$, then θ_a is to be preferred over θ_b.*

Should we believe the RLP? It seems reasonable, but we must proceed with care. The RLP is based upon the probability of the data given values of the parameter. What we wish to know is the plausibility of values of the parameter given the data. There would seem to be a link between these concepts. But they are not immediately equivalent.

It is often argued that maximum likelihood estimation is justified by the fact that the global MLE is asymptotically efficient. However, the usual Cramér conditions that are imposed for the asymptotic efficiency of the MLE only ensure that the consistent root of the likelihood equations is efficient; there is no guarantee that the global maximum of the likelihood corresponds to a consistent root. So, in the absence of any regularity on the model, it is possible for this strategy to come undone for some parametric models. Examples due to Kraft and LeCam (1956), LeCam (1979), Bahadur (1958) and Ferguson (1982) illustrate that the global maximum of the likelihood can correspond to an inconsistent root of the likelihood equation, while at the same time, some other root of the likelihood equation is a consistent estimator for the parameter. See also LeCam (1990) and Example 3.1 in Chapter 6 of Lehmann (1983). While inconsistent maximum likelihood estimates are well known from the examples of Neyman and Scott (1948), the examples due to Kraft and LeCam and others are more problematic for likelihood methodology. This is because they do not involve the use of nuisance parameters, and can be made to satisfy the regularity conditions of Cramér (1946), while, at the same time, the global MLE is inconsistent. While it is possible to invoke regularity conditions such as those of Wald (1949) to ensure that the global maximum likelihood estimate is consistent, the conditions are difficult to check for models involving multiple roots. As the examples above show, the Wald conditions can fail for models which are, in other respects, regular.

Example 4.9 An inconsistent global MLE

The model given by Ferguson (1982) is particularly compelling (see Figure 4.11). It has the following properties:

- The distributions are mixtures of Beta densities, which are well motivated by common statistical practice.
- The Cramér conditions are satisfied. So there is a consistent root of the likelihood equation.
- With probability one, as $n \to \infty$, the global MLE converges to 1, whatever the true value $\theta_0 \in [1/2, 1]$. So the global MLE is not consistent. Nevertheless, at the global MLE, the likelihood is continuous and differentiable. This last condition ensures that there is no simple way to dismiss the global MLE by a local analysis of the likelihood function.

To define Ferguson's family of densities, we set

$$\mathrm{d}(\theta) = (1-\theta)^{-1} \exp\{(1-\theta)^{-2}\}$$

for $\theta \in [1/2, 1]$. Then

$$f(y;\theta) = \theta b(y;1,1) + (1-\theta) b\{y; \theta \mathrm{d}(\theta), (1-\theta)\,\mathrm{d}(\theta)\} \tag{4.31}$$

where $b(y, \alpha, \beta)$ is a Beta density of the form

$$b(y;\alpha,\beta) = \frac{\Gamma(\alpha+\beta)}{\Gamma(\alpha)\Gamma(\beta)} y^{\alpha-1}(1-y)^{\beta-1} \quad 0 \le y \le 1.$$

Ferguson's model fails to satisfy a strong condition due to Wald (1949) that the family of log-likelihood ratios be uniformly integrable: there is no function $K(y) \ge 0$ such that

$$E_{\theta_0}\{K(Y)\} < \infty$$

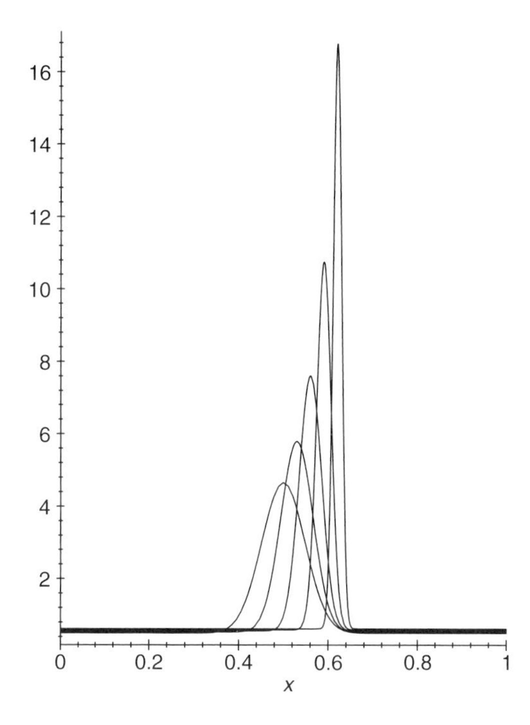

Fig. 4.11 A family of densities on the interval [0, 1] with an inconsistent global MLE, due to Ferguson (1982).

and

$$\ell(\theta; y) - \ell(\theta_0; y) < K(y)$$

for all y and θ. Here, $\ell(\theta; y) = \log f(y; \theta)$ is the log-likelihood.

Does Ferguson's model contradict the relative likelihood principle? It is difficult to argue that the global MLE is more plausible than a local MLE when the former is inconsistent and the latter is consistent.

4.6 Estimating the normal mean in stratified sampling

Suppose that observations of independent random variables $Y_{jk} \sim N(\xi, \sigma_j^2)$ are divided into m strata, the jth stratum consisting of a sample $y_{j1}, y_{j2}, \ldots, y_{jn_j}$. Let

$$\bar{y}_j = \frac{1}{n_j} \sum_{k=1}^{n_j} y_{jk} \quad \text{and} \quad s_j^2 = \frac{1}{n_j} \sum_{k=1}^{n_j} (y_{jk} - \bar{y}_j)^2$$

denote the sample mean and variance respectively in stratum j. Suppose we are interested in estimating the common mean ξ based on $y_{j1}, \ldots, y_{jn_j}$ for

$j = 1, \ldots, m$. The log-likelihood function is

$$\ell(\xi, \sigma_1, \ldots, \sigma_m) = \sum_{j=1}^{m} \left\{ -n_j \log \sigma_j - \sum_{k=1}^{n_j} \frac{(y_{jk} - \xi)^2}{2\sigma_j^2} \right\}. \tag{4.32}$$

Let $\hat{\sigma}_j(\xi)$ denote the MLE for σ_j in the restricted model obtained by fixing the value of ξ. Then

$$\begin{aligned} \hat{\sigma}_j^2(\xi) &= \frac{1}{n_j} \sum_{k=1}^{n_j} (y_{jk} - \xi)^2 \\ &= s_j^2 + (\bar{y}_j - \xi)^2. \end{aligned}$$

Then the *profile log-likelihood function* for ξ is

$$\ell_{\mathrm{PR}}(\xi) = \ell(\xi, \hat{\sigma}_1(\xi), \ldots, \hat{\sigma}_m(\xi)). \tag{4.33}$$

This can be maximised directly in ξ to obtain the MLE $\hat{\xi}$. However, Chaubey and Gabor (1981) have noted that the profile likelihood may well be multimodal. To see why this is the case, note first that we can also find the MLE by differentiating the log-likelihood in (4.32) with respect to ξ and plugging in $\hat{\sigma}_1(\xi), \ldots, \hat{\sigma}_m(\xi)$ for $\sigma_1, \ldots, \sigma_m$, respectively. The resulting equation has the form

$$\sum_{j=1}^{m} \frac{n_j(\bar{y}_j - \hat{\xi})}{s_j^2 + (\bar{y}_j - \hat{\xi})^2} = 0. \tag{4.34}$$

Equation (4.34) looks very similar in form to equation (4.7) for the Cauchy location parameter ξ, although the two models are quite different. This tells us that we can get multiple roots here as well, when the strata means $\bar{y}_1, \ldots, \bar{y}_m$ are widely dispersed compared to the strata standard deviations $s_1, \ldots, s_m$. Once again, the appearance of multiple roots seems to be a bellwether for the possible lack of good model fit. As the model assumes that the strata have a common mean, large variation between strata, as measured by the dispersion of the sample means $\bar{y}_1, \ldots, \bar{y}_m$, is an indicator that this model assumption could be incorrect.

Another problem with maximum likelihood estimation is that the MLE for ξ is consistent but not efficient. For instance, it is less efficient than the estimator derived from the estimating equation:

$$\sum_{j=1}^{m} w_j \frac{\bar{y}_j - \hat{\xi}}{s_j^2 + (\bar{y}_j - \hat{\xi})^2} = 0 \tag{4.35}$$

with $w_i = n_i - 2$, as advocated by Bartlett (1936) and Neyman and Scott (1948). Likelihood theory leads to the choice $w_j = n_j$. On the other hand, sufficiency and ancillarity arguments (Kalbfleisch and Sprott, 1970) lead to the same equation (4.35) but with $w_j = n_j - 1$. The class of estimating equations defined in (4.35) generally admit multiple roots.

4.7 Regression with measurement error

Stefanski and Carroll (1987) have considered generalised linear models in which the covariates cannot be observed directly, but can only be measured with a certain amount of measurement error. Suppose that a random variable Y has density

$$f_Y(y; u, \alpha, \beta, \phi) = \exp\left\{\frac{(\alpha + \beta u)y - b(\alpha + \beta u)}{a(\phi)} + c(y, \phi)\right\} \tag{4.36}$$

where α, β and ϕ are unknown real parameters, u is a real covariate, and $a(.)$, $b(.)$ and $c(.,.)$ are known real-valued functions. Now suppose that we cannot observe u, but that for each value of the u, we can take a measurement. We might model this measurement X as $X = u + \epsilon$, where ϵ has a normal distribution with mean zero and variance $\bar{\omega}$. We shall assume that $\bar{\omega} = a(\phi)\omega$, where ω is known. So the density of X is

$$f_X(x; u, \phi) = \frac{1}{\sqrt{2\pi a(\phi)\omega}} \exp\left\{-\frac{(x-u)^2}{2a(\phi)\omega}\right\}.$$

A sample (x_i, y_i), $i = 1, \ldots, n$, of independent observations is given. For convenience we represent the full vector of parameters (α, β, ϕ) by θ. The joint density of the data is

$$\prod_{j=1}^{n} f_{XY}(x_j, y_j; \theta, u_j) \tag{4.37}$$

where

$$f_{XY}(x, y; \theta, u) = f_X(x; u, \phi) f_Y(y; u, \alpha, \beta, \phi).$$

When written explicitly, the right-hand side becomes

$$\frac{1}{\sqrt{2\pi a(\phi)\omega}} \exp\left[\frac{1}{a(\phi)}\left\{y(\alpha + \beta u) - b(\alpha + \beta u) + c(y, \phi) - \frac{(x-u)^2}{2\omega}\right\}\right].$$

In this model, the covariates $u_1, \ldots, u_n$ act as nuisance parameters for the problem of estimating θ. Unfortunately, there are just too many of these nuisance parameters. The problem is reminiscent of the class of problems considered by Neyman and Scott (1948). Consider the submodel for the data in which we fix the value of θ and regard $u_1, \ldots, u_n$ as the parameters. Within this submodel, the joint density in (4.37) has exponential family form in the 'parameters' $u_1, \ldots, u_n$. A complete sufficient statistic for u_j is

$$\delta_j = x_j + y_j \omega \beta.$$

Stefanski and Carroll (1987) proposed that the nuisance parameters $u_1, \ldots, u_n$ could be eliminated from the model by *conditioning* on the complete sufficient statistics $\delta_1, \ldots, \delta_n$ for $u_1, \ldots, u_n$. There are two ways that we can consider to build estimating functions by conditioning on $\delta_1, \ldots, \delta_n$. One way is to calculate a gradient function for a conditional log-likelihood. We can write the joint log-likelihood for θ and u given y and $\delta = \delta(\theta)$ as

$$\ell_{\delta Y}(\theta, u; \delta, y) = \ell_{Y|\delta}(\theta; \delta, y) + \ell_{\delta}(\theta, u; \delta)$$

where

$$\ell_{Y|\delta}(\theta; \delta, y) = \log f_{Y|\delta}(y|\delta; \theta).$$

It seems natural to extract the conditional component of this log-likelihood and to take its gradient with respect to θ. For the full data set this becomes

$$g_S(\theta) = \sum_{j=1}^{n} \dot{\ell}_{Y|\delta}(\theta; \delta_j, y_j).$$

Calculating the gradient of $\ell_{Y|\delta}$ may look like a standard calculation of a score within a conditional model. However, this is not the case, because $\delta_1, \ldots, \delta_n$ depend upon the parameter of interest θ. It should be clearly understood that in calculating this derivative with respect to θ, the statistic δ is to be treated as a *constant* and not as a function of θ. This can be the source of some confusion. Note that $\delta_j = \delta_j(\theta)$ is a complete sufficient statistic for u_j in the *restricted model* where θ is given. For the purpose of calculating the gradient of the log-likelihood, δ_j must be treated as fixed. If we were to treat δ_j as functionally dependent upon the parameter θ in differentiating we would introduce a bias into the estimating function which could lead to inconsistent estimation. This point will be important for the example that follows.

Example 4.10 Normal regression with measurement error

Consider the special case where Y has a normal distribution with mean $\alpha + \beta u$ and variance σ^2. In the notation above, this makes $\phi = \sigma^2$, $a(\phi) = \phi$ and $\theta = (\alpha, \beta, \sigma^2)$. Once again, $\delta = x + y\omega\beta$. It can be checked that the conditional distribution of Y given δ is

$$Y|\delta \sim N\left(\frac{\alpha + \beta\delta}{1 + \omega\beta^2}, \frac{\sigma^2}{1 + \omega\beta^2}\right).$$

Therefore, based upon a sample $(x_1, y_1), \ldots, (x_n, y_n)$, the conditional log-likelihood is

$$\ell(\alpha, \beta, \sigma^2) = \frac{n}{2} \log\left(\frac{1 + \omega\beta^2}{2\pi\sigma^2}\right) - \frac{1}{2} \sum_{j=1}^{n} \frac{\{y_j(1 + \omega\beta^2) - (\alpha + \beta\delta_j)\}^2}{\sigma^2(1 + \omega\beta^2)}.$$

For convenience, let us set

$$\mu_j = \frac{\alpha + \beta\delta_j}{1 + \omega\beta^2}.$$

Differentiating with respect to α and equating to zero, we get the equation

$$\sum_{j=1}^{n}(y_j - \hat{\mu}_j) = 0.$$

Differentiating with respect to σ we get

$$\hat{\sigma}^2 = \sum_{j=1}^{n} \frac{(y_j - \hat{\mu}_j)^2(1 + \omega\hat{\beta}^2)}{n}.$$

So far so good, as far as α and σ are concerned. However, when we differentiate with respect to β we get

$$\frac{n\omega\hat{\beta}}{1 + \omega\hat{\beta}^2} - \sum_{j=1}^{n} \frac{(y_j - \hat{\mu}_j)^2\omega\hat{\beta} - (y_j - \hat{\mu}_j)(\delta_j - 2\omega\hat{\beta}\hat{\mu}_j)}{\hat{\sigma}^2} = 0.$$

The problem with this equation is that it can have multiple solutions for $\hat{\beta}$. This can be seen by writing the equation out in terms of the paired data (x_j, y_j) and simplifying. We obtain

$$-\omega\hat{\beta}^2 S_{xy}^2 + \hat{\beta}(\omega S_y^2 - S_x^2) + S_{xy}^2 = 0$$

where S_x^2, S_y^2 and S_{xy}^2 are the usual sample second moments. This equation is quadratic in $\hat{\beta}$. We are spared the indignity of having an equation with no solutions, because the discriminant of the quadratic is positive. However, the equation will have two solutions from which to choose. It is tempting to interpret the roots as corresponding to a local maximum and a local minimum of an objective function. However, this interpretation is not as well motivated as in the calculation of an MLE. In this case, we are not maximising a likelihood or a conditional likelihood.

Simple *ad hoc* methods can choose a root. For example, we certainly wish to recover the standard least squares solution when $\omega \to 0$. If we solve the equation for the roots, we get

$$\hat{\beta} = \frac{(\omega S_y^2 - S_x^2) \pm \sqrt{(\omega S_y^2 - S_x^2)^2 + 4\omega S_{xy}^4}}{2\omega S_{xy}^2}.$$

If we want the usual least squares estimator in the limit as $\omega \to 0$, we should take the upper root. *Ad hoc* considerations such as these are usually applied when multiple root problems arise. However, such methods are no solution to the general problem of choosing a root, because such choices will not fit within a general theory of estimation. Nor does this selection method extend to the case of multiple regression with measurement error that is discussed below.

Stefanski and Carroll also reported that a similar problem of multiple roots arises in logistic regression with errors in covariates. In this case, Y is assumed to be a binary random variable with mean p, which relates to (α, β) through the canonical link

$$\log\left(\frac{p}{1-p}\right) = \alpha + \beta u.$$

Once again, we assume the additive error model $Y = u + \epsilon$. Further analysis of this model can be found in Hanfelt and Liang (1995, 1997), where an objective function is constructed by a path-dependent integration approach.

For reasons of simplicity, we have only considered simple linear regression with measurement error up to this point. Stefanski and Carroll's general model is a multiple linear regression where β is a vector of coefficients, and u is a vector of covariates of the same dimension. The response Y is a scalar random variable whose error distribution is governed by the density in equation (4.36) as above. The parameters α and ϕ remain scalars, and the functions $a(\cdot)$ and $b(\cdot)$ remain real-valued. It is assumed that u cannot be measured exactly, and that X has a multivariate normal distribution with mean u and covariance matrix $a(\phi)\Omega$. Much of the analysis given above generalises easily to regression with multiple covariates. As was true for simple linear regression, error models lead to estimating equations with more than one solution. However, in higher dimensions, there is no 'quick fix'. Stefanski and Carroll proposed picking the root which is closest to the naive estimator obtained by ignoring measurement error. The concept of distance between two estimators can be treated flexibly here. It can either be defined operationally by iterating from the naive estimator using Newton–Raphson or some other iterative method, or it can be defined by Euclidean distance. However, Stefanski and Carroll have noted that such methods only have clear justification when the measurement error is small. Heyde and Morton (1998) have proposed three methods for selecting a root. We shall not consider these methods in detail at this point, as that will be discussed in detail in Sections 5.8 and 5.9.

Stefanski and Carroll (1987) also proposed another class of estimating functions for consistently estimating the regression parameters of generalised linear measurement error models. Following Lindsay (1980, 1982), they proposed estimating θ with a *conditional score* function $g_C(\theta)$. This conditional score has similar problems of multiple roots that are found with $g_S(\theta)$.

4.8 Weighted likelihood equations

Markatou *et al.* (1998) proposed a modification of the likelihood method for data in which there is some reason to believe that certain observations are not correctly modelled. They introduced a weighting function to the likelihood equation which adaptively downweights those observations which appear to be inconsistent with the model. Their proposed *weighted likelihood equations* take the form

$$\sum_{j=1}^{n} w(y_j, \theta, \hat{F}) g(\theta, y_j) = 0. \tag{4.38}$$

The data $y_1, \ldots, y_n$ are assumed to be a random sample from the distribution $F = F_\theta$, and $\hat{F}$ denotes the empirical distribution function. The function g can be any appropriate estimating function for a single observation from F_θ. However, the authors restrict their attention to the case where g is the vector-valued score function for θ based upon a single observation from F. We have already seen that weighted equations for parameters can arise in various contexts. In mixture models, e.g., the weights represent the probabilities for selection from each model component. These weights appear naturally when the *EM-algorithm* is invoked to solve the likelihood equations. *Weighted least squares* methods appear in symmetric stable law equations for location and scale parameters. In this case, the weights provide an adjustment to least squares methods for the heaviness of the tails of the distribution. However, in the context of Markatou *et al.* (1998) the weights serve a different purpose, namely to reflect the relative plausibility of the model assumptions for each observation.

The weight function w is assumed to take values in the closed interval $[0, 1]$, and is chosen to be larger or smaller according to model fit. In particular, $w(y_j, \theta, \hat{F})$ will be close to 1 provided that in a neighbourhood of the variable y_j the empirical distribution $\hat{F}$ is clearly *concordant* with the model distribution associated with the given value θ of the parameter. The value of w is close to 0 when there is a clear *discordance* between $\hat{F}$ and the model in a neighbourhood of y_j.

When the data come from a discrete distribution, such a measure of concordance and discordance is easy to define, because each observation has a natural neighbourhood in this case: Markatou *et al.* (1998) defined the 'neighbourhood about y_j' for discrete data to be the point y_j itself. A degree of concordance between the model with parameter θ and the observation y_j can be constructed from a *Pearson residual*, defined for a random sample of size n as $\psi(y_j, \theta, \hat{F})$, where

$$\psi(t, \theta, \hat{F}) = \frac{n^{-1} \#\{y_k : y_k = t\}}{P_\theta(Y_k = t)} - 1. \tag{4.39}$$

The Pearson residuals are unbiased estimating functions when regarded as functions of θ because

$$\begin{aligned} E_\theta\{\psi(t, \theta, \hat{F})\} &= \frac{n^{-1} E_\theta(\#\{Y_k : Y_k = t\})}{P_\theta(Y_k = t)} - 1 \\ &= \frac{P_\theta(Y_k = t)}{P_\theta(Y_k = t)} - 1 \\ &= 0. \end{aligned}$$

Markatou *et al.* (1998) proposed the weight function $w = w(y_j, \theta, \hat{F})$ given by

$$w = 1 - \frac{\psi^2}{(\psi + 2)^2}.$$

This may look a bit mysterious. An alternative representation is that the weight at y_j is given by

$$w(y_j, \theta, \hat{F}) = \frac{4\hat{p}_j p_j}{(\hat{p}_j + p_j)^2}$$

where $p_j = P_\theta(Y = y_j)$ and $\hat{p}_j$ is the proportion of the data at y_j. This function is symmetrical in p_j and $\hat{p}_j$, achieves its maximum at one when $p_j = \hat{p}_j$.

In the continuous case, definition (4.39) is replaced by

$$\psi(t, \theta, \hat{F}) = \frac{\int k(y; t)\, \mathrm{d}\hat{F}(y)}{\int k(y; t)\, \mathrm{d}F_\theta(y)} - 1$$

where k is some smooth kernel appropriate for kernel density estimation.

Markatou *et al.* (1998) noted that such weighted equations can have multiple solutions for θ. For example, they considered data of Lubischew (1962) describing bivariate measurements of two species of beetles. The data consisted of 21 bivariate observations for the species *Chaetocnema concinna* and 22 bivariate observations for the species *Chaetocnema heptapotamica*. To test the method, Markatou *et al.* (1998) artificially pooled the two species, and a weighted likelihood estimate for the location of a bivariate normal distribution was found. The results were in agreement with the data, because the weighting successfully separated the data by providing two roots as location estimates, one for each species.

Even when the data agree well with the model, the weighted likelihood approach will tend to downweight certain observations which, visually at least, are not in perfect accord with the model. A measure of this phenomenon is the *mean weighting statistic*, which is the average of the weights at the parameter estimate, namely

$$\tilde{w}^* = n^{-1} \sum_{j=1}^{n} w(y_j, \hat{\theta}, \hat{F}).$$

This statistic gives an indication of the proportion of the data that has been used in estimating the parameter. The statistic $n(1 - \tilde{w}^*)$ is therefore the amount of data discounted in estimation. A model-based version of this statistic is the expected downweighting parameter, defined as

$$n[1 - E_\theta\{w(Y, \theta, \hat{F})\}].$$

For models of discrete data, the expected downweighting cannot be tuned. However, models for continuous data have weights that are dependent upon the choice of kernel $k(y; t)$. As with kernel smoothing in other applications, the precise shape of the kernel is less important than the choice of bandwidth, that is, the spread or dispersion of the kernel. The expected downweighting parameter can be used much like a decision to trim a certain proportion of the data with *trimmed* or *Winsorised means*. The effect is to routinely discount the most suspect portion of the data. Unlike trimming, however, the exact proportion of the data to be discounted is not completely under the researcher's control. This need not be a disadvantage: the method can adaptively downweight more observations if the degree of discordancy is great.

Like other weighting methods such as *weighted least squares*, there is a simple algorithm for finding a solution to the weighted likelihood equations. Given an approximation $\theta^{(k)}$, we choose $\theta^{(k+1)}$ to solve

$$\sum_{j=1}^{n} w(y_j, \theta^{(k)}, \hat{F}) g(\theta^{(k+1)}, y_j) = 0.$$

The iteration can be started at the maximum likelihood estimate if this can be found in simple closed form. Markatou *et al.* (1998) have shown that when the data distribution and the model distribution agree, the convergence of the algorithm is faster than linear. The reader is cautioned that this is an asymptotic result. For fixed sample size, the discrepancy between the model and the data is such as to ensure that the convergence is linear.

4.9 Detecting multiple roots

In this and the following sections, we shall consider a number of questions about multiple roots.

- When should the researcher be suspicious that an estimating equation has multiple solutions?
- Suppose that we regard the number of roots of an estimating function as a random variable. Can we find its distribution? A complete answer to this question would also answer the previous question.
- When an estimating equation has multiple roots, how can we be sure that we have found all of them? Note that this was a problem in the mixture model analysis of the haemophilia data considered in Section 4.2. In other words, having found more than one solution, when do we stop searching?

This section shall be devoted to considering the first of these questions.

Turning the first question on its head, we begin by asking whether there are properties of likelihoods and estimating functions which are sufficient to ensure that any solution will be unique. In a likelihood analysis, the Hessian provides an important tool to investigate the unimodality of the likelihood. The likelihood equations have a unique root if the *Hessian* of the log-likelihood

$$\ddot{l}(\theta) = \left(\frac{\partial^2 l(\theta)}{\partial\theta_j \partial\theta_k} \right) \tag{4.40}$$

is *negative definite* for all values of θ.

General as this condition is, it is more than needs to be shown. To prove unimodality of the likelihood it is only necessary to show that the Hessian matrix is negative definite at the *stationary points* of the log-likelihood. Some mild regularity is necessary for this result.

Proposition 4.11 *Suppose that the following conditions hold:*

- *The parameter space Θ is an open, connected subset of $\mathbb{R}^k$.*
- *Let the log-likelihood $\ell : \Theta \to \mathbb{R}$ be twice continuously differentiable on Θ.*
- *Suppose that the global maximum of $\ell(\theta)$ is achieved at some point in Θ.*
- *Assume that $\ell(\theta) \to -\infty$ as θ goes to the boundary of Θ. (Note that the boundary may include the points at infinity if Θ is unbounded.)*
- *Finally, suppose that for all $\hat{\theta} \in \Theta$ satisfying $\dot{\ell}(\hat{\theta}) = 0$, the Hessian matrix $\ddot{\ell}(\hat{\theta})$ is negative definite.*

Then the equation $\dot{l}(\hat{\theta}) = 0$ will have a unique solution in Θ.

We shall not concern ourselves with a rigorous proof here. However, an illustration of the argument can be found in Figure 4.12. If a log-likelihood surface has two distinct local maxima then it must also have a saddle point somewhere between them. But a saddle point of the likelihood surface is a stationary point: that is, the likelihood equations will be satisfied there, without the Hessian of the log-likelihood being negative definite. It is important to note that the concavity of

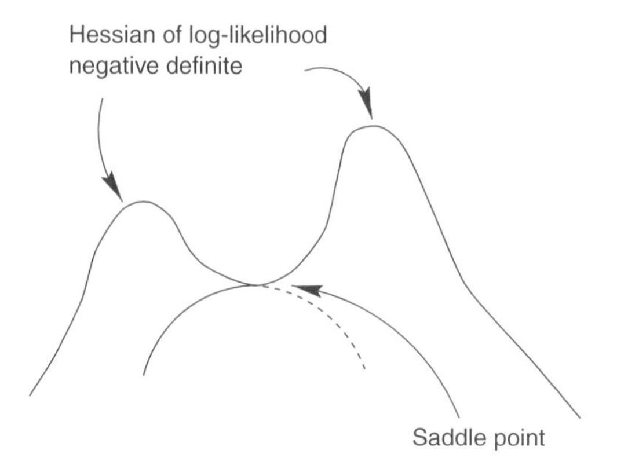

Fig. 4.12 A log-likelihood surface with two local maxima and a saddle point.

the log-likelihood is not invariant under a reparametrisation of the parameter space. To see that this is the case, let us consider a smooth reparametrisation $\tau = \tau(\theta)$. In the new parametrisation, the Hessian becomes

$$\ddot{\ell}(\tau) = \frac{\partial \theta}{\partial \tau} \ddot{\ell}(\theta) \left(\frac{\partial \theta}{\partial \tau} \right)^{\mathrm{t}} + \{\dot{\ell}(\theta)\}^{\mathrm{t}} \frac{\partial^2 \theta}{\partial \tau^2}. \tag{4.41}$$

If $\ddot{\ell}(\theta)$ is negative definite, the first term on the right-hand side will also be negative definite. However, the full expression may not be negative definite. Conversely, of course, the Hessian of the log-likelihood in the original model may not be negative definite while the Hessian after reparametrisation may be negative definite.

However, at any solution to the likelihood equations the second term will vanish because $\dot{\ell}(\theta) = 0$. So if the Hessian is negative definite at any solution to the likelihood equations, it will also be negative definite at a solution in a reparametrised model. This means that the assumptions of Proposition 4.11 are invariant under smooth reparametrisations of the model.

An argument based upon $\ddot{\ell}(\hat{\theta})$ was used by Copas (1975) to show that the Cauchy location-scale model has a unique MLE. In another example, Huzurbazar (1948) showed that linear exponential families have unique solutions to their likelihood equations under mild regularity. This follows from a special identity that is particular to exponential families, namely

$$-\ddot{\ell}(\hat{\theta}) = E_\theta\{\dot{\ell}(\theta)\dot{\ell}(\theta)^{\mathrm{t}}\}|_{\theta=\hat{\theta}}.$$

The right-hand side is the information matrix $I(\hat{\theta})$, which is positive definite under the usual regularity assumptions. So Proposition 4.11 can be invoked to conclude that the root of the likelihood equation for an exponential family is unique. Note that if θ is the natural parameter of the exponential family, then the identity $\ddot{l}(\theta) = -I(\theta)$ holds for all θ. However, the equation $\ddot{l}(\hat{\theta}) = -I(\hat{\theta})$ holds even when θ is not the natural parameter, because the condition is invariant under smooth reparametrisations of the parameter space.

In models where the Hessian matrix fails to be negative definite, multiple root problems need investigation. However, the researcher should avoid the trap of presuming that the negative definiteness of the Hessian matrix is necessary for the uniqueness of the root. An interesting case in point is provided by the *Tobit model*, where

$$Y_j^* = \beta x_j^{\mathrm{t}} + \epsilon_j, \quad Y_j = \max\{0, Y_j^*\}$$

for $j = 1, \ldots, n$. In this model, x_j is a vector of covariates, β is a coefficient vector, and $\epsilon_1, \ldots, \epsilon_n$ are independent $\mathcal{N}(0, \sigma^2)$ error terms. The random variable Y_j^* is not observed directly. Instead, we observe Y_j. Amemiya (1973) noticed that the Hessian matrix for the parameter vector $\theta = (\beta, \sigma^2)$ is not negative definite. Thus the question of multiple roots arose. However, Olsen (1978) showed that by letting

$$\zeta = \beta/\sigma \quad \text{and} \quad \xi = 1/\sigma$$

the Hessian is negative definite in the new parametrisation with $\theta = (\zeta, \xi)$. So multiple roots cannot occur. See Amemiya (1973), Greene (1990), Olsen (1978), Orme (1990) and Iwata (1993) for discussion of this model. Burridge (1981) discusses the concavity of the log-likelihood function in the case of regression with grouped data. See also Pratt (1981).

The task of detecting multiple roots for general estimating equations is more problematic than that for the likelihood equations, as there may not exist a statistically meaningful objective function whose stationary points correspond to roots of the estimating equation. Some geometrical insight into the nature of an estimating function can be obtained by interpreting a vector-valued estimating function $g(\theta)$ as a vector field on the parameter space Θ. There are two possibilities:

The first possibility is that the matrix $\dot{g}(\theta)$ is *symmetric* for all θ and all samples $y_1, \ldots, y_n$. In this case the vector field is conservative so that there exists a real valued function $\lambda(\theta)$ such that $g(\theta) = \dot{\lambda}(\theta)$. The function λ could be a log-likelihood or, in the case where the estimating function g is both unbiased as in (1.3) and information unbiased as in (1.4), may share some of the properties that are typically associated with log-likelihoods. In the case where Θ is one-dimensional, the symmetry condition is trivially satisfied. As $\dot{g}$ is symmetric, its eigenvalues will all be real. Those points $\hat{\theta} \in \Theta$ at which λ has a local maximum will correspond to points where the vector field g vanishes and the eigenvalues of $\dot{g}$ are all negative. Similarly, points at which λ is locally minimised will correspond to $\hat{\theta} \in \Theta$ where g vanishes and the eigenvalues will all be positive. Saddle points of λ will occur where g vanishes and the eigenvalues are mixtures of positive and negative quantities.

The second possibility is that the matrix $\dot{g}(\theta)$ is *not symmetric* in general. In this case, there will be no objective function whose gradient is $g(\theta)$. The points in Θ where the vector field vanishes will correspond to the zeros of $g(\theta)$. Despite the absence of an objective function, we can nevertheless determine those zeros of $g(\theta)$ which are analogs of local maxima and other zeros which are analogs of local minima. To do this we investigate the eigenvalues of $\dot{g}(\theta)$. Let $\kappa_1(\theta), \kappa_2(\theta), \ldots, \kappa_k(\theta)$ be the eigenvalues of $\dot{g}(\theta)$, in arbitrary order, where $k = \dim(\Theta)$. As $\dot{g}$ is not symmetric, these eigenvalues will generally be complex-valued. Therefore, we can write each eigenvalue $\kappa_j(\theta)$ in terms of its real and imaginary parts as

$$\kappa_j(\theta) = \Re\{\kappa_j(\theta)\} + \sqrt{-1}\Im\{\kappa_j(\theta)\}, \quad j = 1, \ldots, k.$$

A point $\hat{\theta} \in \Theta$ where $g(\hat{\theta}) = 0$, and where

$$\Re\{\kappa_1(\hat{\theta})\}, \Re\{\kappa_2(\hat{\theta})\}, \ldots, \Re\{\kappa_k(\hat{\theta})\} \tag{4.42}$$

are all negative is a *sink* for the flow of the vector field determined by $g(\theta)$. That is, we imagine that at each point θ of the parameter space, we place a vector of length and direction corresponding to $g(\theta)$. The resulting vector field will locally 'flow in' towards any sink and 'flow out' from any *source*. The latter is characterised as

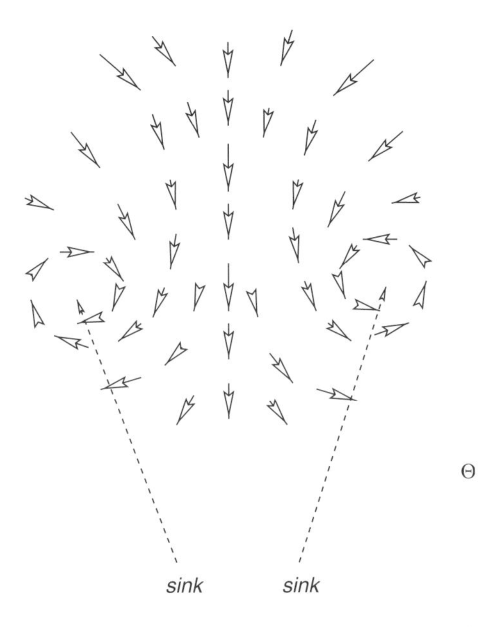

Fig. 4.13 A non-conservative vector field which vanishes at exactly two distinct points, both of which are sinks.

satisfying $g(\theta) = 0$ and having eigenvalues $\kappa_j(\theta)$ whose real parts are all positive (Figure 4.13).

Although an objective function $\lambda(\theta)$ such that

$$g(\theta) = \dot{\lambda}(\theta)$$

does not exist when $\dot{g}(\theta)$ is not symmetric, the sink and source of the vector field $g(\theta)$ can be regarded as analogs of maxima and minima. A sink of a vector field corresponds to a local maximum while a source corresponds to a local minimum. So it is natural to consider whether there is an immediate generalisation of the uniqueness result of Proposition 4.11 that is illustrated in Figure 4.12. In other words, if a vector field is such that all point at which it vanishes are sinks, is it true that there can be at most one such point? Unfortunately, this generalisation is false as Figure 4.13 demonstrates. Because there is a rotational component to the vector field, i.e., a non-zero component to the curl of the vector field, there does not exist an analog of a saddle point at which the vector field vanishes in the parameter space.

4.10 Finding all the roots

Next, we turn our attention to the problem of detecting all roots for estimating equations which admit the possibility of multiple roots. In principle, a careful search in the parameter space should uncover all the roots of any given estimating equation. However, in practice, this may be far too time-consuming, especially if the parameter space is of high dimension. We encountered an example of this problem early in this chapter when we considered the haemophilia data studied by Basford and McLachlan. For that data set it was far too time-consuming to search for roots by initialising the root search algorithm over the points of a grid covering the parameter space. It was expedient to be more selective in the choices of starting points. In that data set, we adapted the method of Markatou *et al.* (1998), which involved choosing as starting points those parameter estimates obtained from small subsamples of data points.

As this method seems applicable to a wide variety of estimating functions, we now give a brief description of the method in a more general context. We begin by noting that the roots of an estimating equation can often be divided into

- *reasonable roots*, which, upon examination, can be considered as candidates for estimation, and
- *unreasonable roots*, which arise in the estimating function for incidental reasons that have little to do with estimation.

Obviously, such a classification is not meant to be understood too formally. Nevertheless, it is possible to argue that some roots of an estimating equation arise for incidental reasons that have nothing to do with their being supported by the data. For example, roots corresponding to saddle points or local minima of likelihoods or quasi-likelihoods fall into this category. As we have seen, topological considerations often tell us that such roots must exist if two or more local maxima exist. Such *topological roots* will only be as stable as the local maxima which determine them. In other words, we expect that perturbations of the data which eliminate extraneous local maxima will usually eliminate these roots as well. On the other hand, in a neighbourhood of the reasonable roots, an estimating function will often be quite regular in appearance. The local properties of the function in a neighbourhood of the root will usually be in agreement with the known ergodic properties of the estimating function at the true value of the parameter. For example, the non-negativity of the *Kullback–Leibler* distance between densities is the ergodic counterpart of the principle that the maximum likelihood estimator is the global maximum of the log-likelihood.

Next, let us note that, in many cases, all reasonable roots of an estimating equation are those which are supported by some subset of the data. For example, in the case of the Cauchy location model, each relative maximum is either 'caused' by a visual outlier (i.e., supported by an outlying observation) or is the consistent root that is supported by the majority of the observations which lie in the center of the Cauchy distribution. The unreasonable local minima are not supported by observations or subsets of observations in this sense. To illustrate this idea, consider Figure 4.14. Here we see a score function produced by four variates, of which three are in reasonable accord and one is a visual outlier. The score function for the entire data set (above) is found by summing the scores for the individual observations (below). The result is an estimating equation with three roots. Among

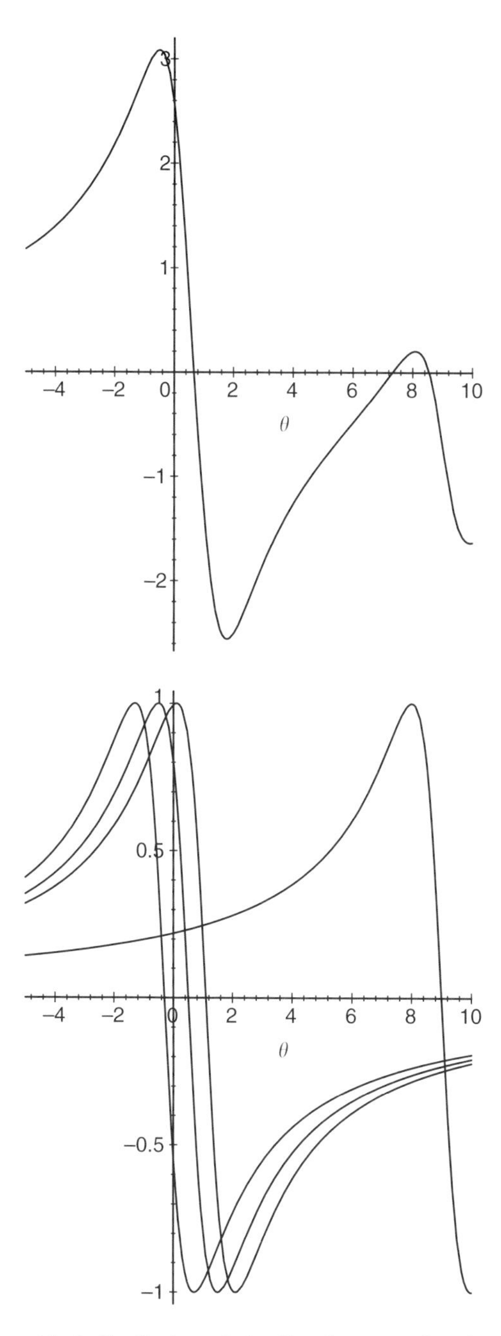

Fig. 4.14 The empirical distribution of the Cauchy score function from Figure 4.7. The extraneous local maximum arises from a single visual outlier.

these three roots, the middle root corresponds to a local minimum of the likelihood. Therefore, it is not a reasonable contender as a point estimate. Neither of the two other roots can be dismissed. The smallest root is obviously more credible as an estimate, but with a small data set the largest root cannot be ruled out. However, both reasonable roots are supported by the data: in a certain sense, the smallest root appears because of the three smallest variates, while the largest root arises in the likelihood equation because of the presence of the visual outlier on the right.

Suppose that $y_1, \ldots, y_n$ are n independent observations from some distribution with parameter θ. Let $m \leq n$ be the minimum number needed for the equation

$$g(\hat{\theta}, y_{i_1}, y_{i_2}, \ldots, y_{i_m}) = 0$$

to have a solution for all subsets $y_{i_1}, y_{i_2}, \ldots, y_{i_m}$ of size m with probability one. Typically, m will be the dimension of the parameter space, although counterexamples to this can be found.

Markatou *et al.* (1998) proposed that bootstrap samples $y_1^*, \ldots, y_m^*$ of size m be constructed by sampling m distinct elements of the data set $y_1, \ldots, y_n$. For each such bootstrap sample, the root θ^* found by solving

$$g(\theta^*, y_1^*, \ldots y_m^*) = 0$$

is to be used as a starting point for an appropriate algorithm which iterates to a root of the equation $g(\theta, y_1, \ldots, y_n) = 0$.

So for estimating the location parameter θ of the Cauchy location model using the likelihood equations, we will have $m = 1$. As the MLE for θ based on a sample of size one is the observation $\theta^* = y_j$, itself, the set of roots obtained would be those found by using iteration from the n original sample observations. In those estimation problems where $\binom{n}{m} \leq 100$ it is possible to do an exhaustive systematic search of all such starting points by using all subsets. For $\binom{n}{m} > 100$, Markatou *et al.* (1998) reported that randomisation with 100 bootstrap samples is sufficient, in the cases they considered, to ensure that all reasonable roots are detected.

An approach to root detection by placing a probability distribution on the parameter space has been proposed by Finch *et al.* (1989). Their method provides a way to estimate the probability that an iterative search from a *random starting point* (RSP) will find a root not observed in previous searches from RSPs. This is not the same as being able to determine the proportion of undetected roots. However, it is a useful surrogate for this proportion, as it helps the researcher to determine the efficacy of running additional iterative searches of a similar kind looking for extra roots.

Suppose that some probability distribution π is placed upon the parameter space. We begin by generating a random sample of size r from the distribution π, and use each of these RSPs as the r initial values of an algorithm, such as Newton–Raphson, which searches for roots. In general, these r iterative trials will converge to a number of distinct roots of the estimating equation which we can write as $\hat{\theta}_1, \ldots, \hat{\theta}_K$ where $K \leq r$ is a random variable. For each $j = 1, \ldots, K$, let D_j be the domain of convergence of the algorithm to $\hat{\theta}_j$. In most examples, we do not know much about D_j. We would ideally like to know something about the region

$$\Theta - \bigcup_{j=1}^{K} D_j$$

which is where the undetected roots reside. Since we cannot determine this set, perhaps the next best thing would be to try to find the probability of this set under the sampling scheme from the distribution π. Let us write this probability as

$$U_r = 1 - \sum_{j=1}^{K} \pi(D_j). \tag{4.43}$$

As D_j is a random set and K is random, U_r is itself a random variable. The quantity U_r is also unknown. However it can be *estimated*. Based upon a suggestion of Good (1953), Finch *et al.* (1989) suggested that U_r be estimated by

$$V_1 = \frac{S}{r}$$

where S is the number of observed $\hat{\theta}_j$ to which only one of the r RSP's converged. Strictly speaking, V_1 does not estimate U_r but rather U_{r-1}, the corresponding probability with one fewer trials. This is because $E(V_1) = E(U_{r-1})$, which was proved by Robbins (1968). This estimate was generalised by Starr (1979) to

$$V_t = \sum_{i=1}^{t} \left\{ \frac{\binom{t-1}{i-1}}{\binom{r}{i}} \right\} Q_i \tag{4.44}$$

where Q_i is the number of roots among the k discovered to which exactly i RSP's converged. It can be shown that $E(V_t) = E(U_{r-t})$. We refer the reader to Starr (1979) for the proof.

Through the use of statistics such as V_t, we can estimate the probability of detecting new roots with such a random search. So we can use such a measure to determine whether to continue searching further from additional RSPs. However, this does not tell us whether there are roots which are extremely unlikely to be detected because the choice of distribution π puts low probability on the domain of convergence of some root. The major hope for solving this problem may lie in bootstrap searches such as that of Markatou *et al.* (1998) mentioned above.

4.11 Root functionals and measures

One way to study the zeros of an estimating function is to examine the *moment measures* of the *point process* of zeros. Suppose that Θ is some closed subset of $\mathbb{R}^k$ that has non-empty interior. The roots of an $\mathbb{R}^k$-valued estimating function $g(\theta)$

define a point process N_g on Θ. That is, for each set B in Θ, we can define the counting random variable

$$N_g(B) = \#\{\theta \in B: \ g(\theta) = 0\}.$$

That is, $N_g(B)$ is the number of zeros of $g(\theta)$ that lie in the set B. In all practical examples, $N_g(B)$ is finite. The *mean measure* for this root process is

$$\Psi_g(B) = E\{N_g(B)\}. \tag{4.45}$$

The mean measure of a point process is roughly analogous to the scalar mean of a random variable or the vector mean of a random vector. Since a point process is a *locally finite counting measure* on Θ, its mean or expected value is also a measure on Θ. Of interest to us here will be the behaviour of the mean measure locally at a point $\theta \in \Theta$. This can be defined as

$$\psi_g(\theta) = \lim_{\epsilon \to 0} \frac{\Psi_g(B_\epsilon)}{\epsilon^k} \tag{4.46}$$

where

$$B_\epsilon = \left(\theta - \frac{\epsilon}{2}, \theta + \frac{\epsilon}{2}\right) \otimes \cdots \otimes \left(\theta - \frac{\epsilon}{2}, \theta + \frac{\epsilon}{2}\right)$$

for each $\epsilon > 0$, the displayed product being a k-fold Cartesian product of intervals. Note that there are really two values of the parameter θ being used to calculate $\psi_g(\theta)$. The other value is the one that is assumed in calculating the expectation that is found in formula (4.45). The local intensity function ψ, called the *mean intensity* is defined when the limit exists. Under standard regularity, we get

$$\Psi_g(B) = \int_B \psi_g(\theta)\, \mathrm{d}\theta.$$

The function Ψ_g is not without interest for the special case where the estimating function has a unique root with probability one. For this case, it reduces to the probability density function of the root. More generally, with estimating functions that admit multiple roots, the value of $\Psi_g(\Theta) - 1$ is of interest, as this is the expected number of extraneous roots to the estimating equation.

While the mean measure and the associated mean intensity are useful ways of describing the distributions of roots, they have the disadvantage that they do not characterise the point process of roots. The next tool that we shall consider overcomes this difficulty. We define the *zero-probability functional* of the roots to be Z_g, where $Z_g(B)$ is the probability that there are no roots in the set B. Unlike the mean measure, the zero-probability functional completely determines the distribution of the point process of roots. However, it does not have a simple representation in the same way as the mean measure through the mean intensity function.

Another functional that is of some interest in the study of the root process is the *probability generating functional*

$$A_g : \Xi \to \mathbb{R}$$

defined on the class Ξ of measurable functions on Θ taking values in the unit interval [0, 1]. This functional is defined by

$$A_g(\xi) = E\left\{ \prod_{\theta : g(\theta)=0} \xi(\theta) \right\}. \tag{4.47}$$

Like the zero-probability functional, the functional $A_g(\xi)$ characterises the dis-1tribution of the roots. However, it also has no simple representation, as the class Ξ is too large for a simple determination of the values of $A_g(\xi)$. Nevetheless, the probability generating functional allows for considerable flexibility in studying the distribution of the roots. It should be noted that $A_g(\xi)$ can be extended to the case where the function ξ is unbounded provided that the expectation in formula (4.47) remains finite. For example, when $g(\theta)$ has a unique root $\hat{\theta}$ and ξ is the identity function on Θ, the functional $A_g(\xi)$ reduces to $E(\hat{\theta})$.

The probability generating functional can be used to calculate moments of the point process of roots. For example, suppose we let I_B be the indicator function which is one on the set B and zero elsewhere. Then

$$\Psi_g(B) = -\left\{ \frac{\partial}{\partial \alpha} A_g(1 - \alpha I_B) \right\}_{\alpha=0}. \tag{4.48}$$

For additional properties of the probability generating functional, the reader is referred to Cressie (1993).

Example 4.12 Mean intensity function for the Cauchy location-model

In the following example we shall consider how to compute the mean intensity $\psi_g(\theta)$ for a one-parameter model. In this case, the set B_ϵ used in equation (4.46) is an interval of real values. So it is helpful to be able to compute the number of roots that fall in an interval without having to conduct an exhaustive search for the values of the roots. *Sturm's Theorem* for polynomial estimating functions was discussed in Chapter 3, and provides us with one such tool. If a simulation is to be performed for more than 10,000 trials, it is usually too slow to investigate the exact locations of all roots in each trial. An alternative is to count the number of roots in successive intervals by constructing the *Sturm chain* and tabulating the results in a histogram. Software designed for symbolic computation with polynomials is particularly convenient for such simulations because the symbolic algebra required to construct the Sturm chain can be called as a subroutine. Figure 4.15 shows the results of a simulation study using MAPLE V for 5000 trials of sample size $n = 5$ for the Cauchy location-model. Here the true value of θ was chosen to be zero.

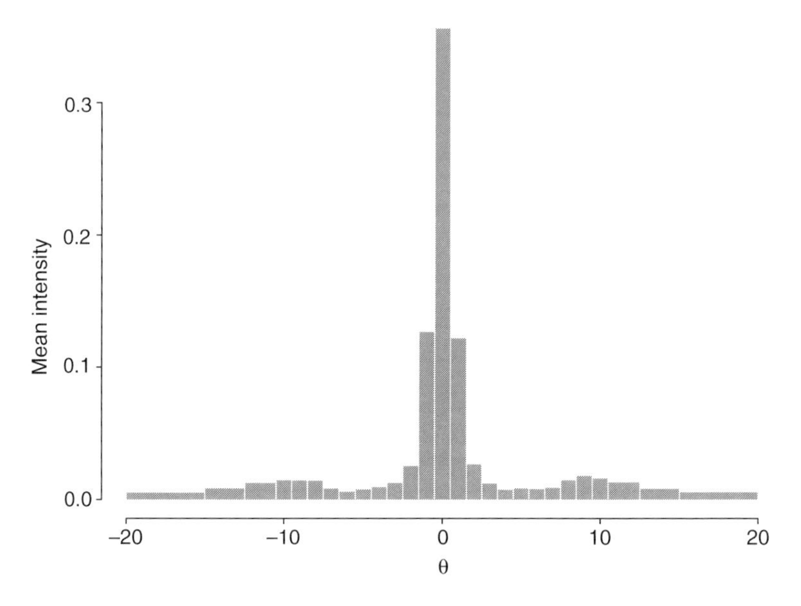

Fig. 4.15 Empirical histogram of the mean intensity for the Cauchy location-model.

The area of each histogram bar represents the average number of roots found in each interval among the trials. Certain features are evident from the histogram. First of all, the presence of the consistent root close to zero is clear from the large mode at zero. The curious secondary modes on each side of the primary mode can be explained by the fact that when the likelihood equation has multiple solutions, it will have local minima between the maxima. A more detailed investigation of the roots shows that the upcrossings of the score function are the principal cause of the secondary modes.

The histogram in Figure 4.15 provides us with an approximation to the mean intensity function $\psi_g(\theta)$. However, there are two sources of error in the histogram as an approximation to the mean intensity. The first of these arises because the roots have been grouped into bins. To eliminate this source of error, the bin widths must go to zero. The second source of error is due to sampling variability: the empirical averages over 5000 trials have been used rather than their expected values. To eliminate this source of variation we must sample more. Obviously, there must be a balance between the two. In practice, to approximate $\psi_g(\theta)$ we must let the bin widths go to zero and the number of trials go to infinity so that the expected number of roots in a bin also goes to infinity.

The mean intensity function is similar to the intensity function of the point process of local maxima examined by Skovgaard (1990). The only difference here is that we do not restrict ourselves to roots that are associated with local maxima. For many estimating functions, it is not necessary to find the roots explicitly in order to compute the root intensity function. Under certain regularity conditions, the root intensity ψ_g can be computed using the random vector

$$z = \dot{g}^{-1} g.$$

Suppose that the vector z has density function $f_g(z)$, where $z \in \mathbb{R}^k$. Then it can be shown that

$$\psi_g(\theta) = f_g(0).$$

Note that dependence of f_g and z upon the parameter θ is suppressed in this notation. The regularity conditions necessary to validate this formula were given by Skovgaard (1990). See Small and Yang (1999) for more discussion in connection with the problem of multiple roots. The regularity conditions required for Skovgaard's formula were generalised by Jensen and Wood (1998).

4.12 Smoothing the likelihood function

Multiple root problems can be regarded as examples of excess variation in estimating functions. In several areas of statistics, a standard tool to reduce variation is smoothing through the use of a moving average for a function, be it discrete or continuous. Daniels (1960) proposed the use of such a moving average to 'reduce the chance of selecting one of the erratic cusps of the likelihood function'. (Daniels, 1960, p. 162.) Daniels' smoothed likelihood was applied to the Cauchy location model by Barnett (1966, p. 164).

Suppose that θ is a real-valued parameter. Let $K_n : \mathbb{R} \to \mathbb{R}$ be a non-negative function such that

$$\int_{-\infty}^{+\infty} K_n(\eta)\, d\eta = 1$$

and

$$\lim_{n \to \infty} \int_{-\infty}^{+\infty} \eta^2 K_n(\eta)\, dy = 0.$$

If we regard K as a density function, then the second of these two conditions is sufficient to ensure that the weight of the density is close to zero when n is large. In the discussion that follows, the variable n will also denote the sample size. The smoothed likelihood, with kernel K_n for sample size n, is defined to be

$$\bar{\ell}_n(\theta) = \int_{-\infty}^{+\infty} \ell_n(\theta - \eta) K_n(\eta)\, d\eta \tag{4.49}$$

where as usual $\ell_n(\theta) = \log L_n(\theta)$. There exist obvious extensions of (4.49) for multiparameter models. The parameter value $\bar{\theta}_n$ which maximises $\bar{\ell}_n(\theta)$ is called a *smoothed maximum likelihood estimator*. We can also write $\bar{\theta}_n$ as the root of the smoothed score

$$\bar{g}_n(\theta) = \int_{-\infty}^{+\infty} \dot{\ell}_n(\theta - \eta) K_n(\eta)\, d\eta \tag{4.50}$$

provided we can interchange derivatives and integrals. Equation (4.50) also suggests an additional restriction on the choice of K_n, namely that it be chosen so that $\bar{g}_n(\theta)$ is an unbiased estimating function. However, such a restriction tends to make the choice of K_n analytically cumbersome in the general case. Fortunately, there are many models in which the smoothed score function will be unbiased.

If a smoothed likelihood is used in a k-parameter model, then the score function becomes vector-valued, and the weight function K_n becomes a real-valued function of k variables. However, in k dimensions, there is even greater variety in the choice of smoothing function than in dimension one. For example, when K_n is a uniform density function on some region, the region chosen could be spherical or rectangular with respect to the parametrisation. Of course, there are many other possibilities.

We shall now consider two examples. In the first case, the smoothed likelihood is used to eliminate multiple roots in the score function. In the second example, the pathologies in the likelihood will be more severe.

Example 4.13 Smoothed likelihood with uniform kernel

We shall consider a one-parameter model. A particularly simple choice of K_n is

$$K_n(\eta) = \begin{cases} \dfrac{1}{2\epsilon_n} & \text{for } |\eta| \le \epsilon_n \\ 0 & \text{for } |\eta| > \epsilon_n \end{cases}$$

for some $\epsilon_n > 0$. Then $\bar{\theta}_n$ will be a solution to the equation

$$L(\bar{\theta}_n - \epsilon_n) = L(\bar{\theta}_n + \epsilon_n).$$

A simulation study by Barnett (1966) found that for the Cauchy location-model it is possible to obtain efficiencies for $\bar{\theta}_n$ which exceed the efficiency of the MLE. In particular, a best improvement on the MLE was obtained by choosing $\epsilon = 2.0$ for a sample of size $n = 5$. In this case the improvement in efficiency was found to be approximately 10%.

The value of ϵ_n needs to be chosen sufficiently large so that the equation $L(\theta + \epsilon_n) = L(\theta - \epsilon_n)$ has only one solution. However, we must also have $\epsilon_n \to 0$ so that $\bar{\theta}_n$ is asymptotically efficient. Making ϵ_n large for a continuous likelihood function with finitely many relative extrema will ensure at most one solution, as Figure 4.16 illustrates.

Ensuring that the equation has only one solution in this manner will require that ϵ be data-dependent. On the other hand, to ensure efficiency we must examine

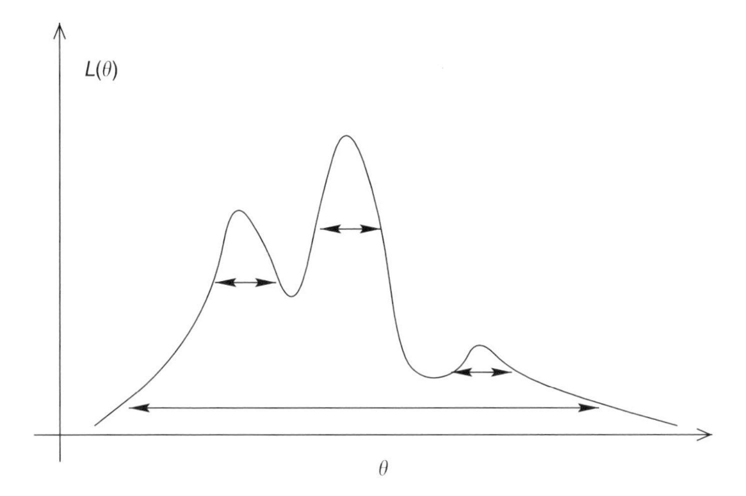

Fig. 4.16 Solving the equation $L(\theta + \epsilon) = L(\theta - \epsilon)$ for two values of ϵ.

the asymptotics. We can write

$$\ell_n(\theta \pm \epsilon) = \ell_n(\theta) \pm \epsilon \dot{\ell}_n(\theta) + \frac{\epsilon^2}{2}\ddot{\ell}_n(\theta) \pm \frac{\epsilon^3}{6} \dddot{\ell}_n\ [\theta + \xi(\pm\epsilon)]$$

under smoothness conditions, where $0 < \xi(\epsilon) < \epsilon$ and $-\epsilon < \xi(-\epsilon) < 0$. So we have

$$\frac{\ell_n(\theta + \epsilon) - \ell_n(\theta - \epsilon)}{2\epsilon} = \dot{\ell}_n(\theta) + \frac{\epsilon^2}{12}[\dddot{\ell}_n\ \{\theta + \xi(\epsilon)\} + \dddot{\ell}_n\ \{\theta + \xi(-\epsilon)\}]. \tag{4.51}$$

The left-hand side of (4.51) can be regarded as an estimating function with $\bar{\theta}_n$ as a root. As such, it is generally a biased estimating function. However, if we focus our primary attention on those models for which

$$E_\theta\{\ell_n(\theta - \epsilon)\} = E_\theta\{\ell_n(\theta + \epsilon)\}$$

then the estimating function is unbiased. A location-model with a symmetric density (such as the Cauchy) will satisfy this property. On the right-hand side, the first term is the score function, which is also unbiased under standard regularity conditions. Under the unbiasedness assumption, the term in brackets in (4.51) will be unbiased, and therefore will be of order $O_p(\sqrt{n})$. However, the asymptotic efficiency of a consistent root of (4.51) will be the asymptotic correlation between the left-hand side of (4.51) and the score function. So, if the consistent root of the smoothed likelihood equation is to be asymptotically efficient, then the second term on the right-hand side of (4.51) must be of smaller order than the first term. As the first term is $O_p(\sqrt{n})$, we must have $\epsilon = o_p(1)$.

Choosing $\epsilon_n = o_p(1)$ runs counter to the requirement that ϵ_n should be large enough to guarantee a unique root. A possible compromise is to use $\bar{\theta}_n$ as a starting point for Newton–Raphson iteration to a root of the score function or to define a one-step estimator. For this to work, the estimator $\bar{\theta}_n$ need only lie in a $\sqrt{n}$-neighbourhood of the true parameter value, and need not be efficient.

In the next example, the likelihood will have a singularity in the form of an infinite "spike" at each data point. We shall consider how to smooth such a likelihood.

Example 4.14 Smoothing a normal model with nuisance parameters

To illustrate the idea of smoothing, consider a sequence of n independent observations y_j, where $y_j \sim N(\mu, \sigma_j^2)$. Suppose we wish to estimate μ when the variances σ_j^2 are unknown. The sample mean can be used as a point estimate for μ, but it does not take into account the differing amounts of information about μ in the variables due to differing variance parameters. Clearly, an estimate which downweights the influence of outlying values is to be preferred. The joint likelihood is

$$L(\mu, \sigma_1, \ldots, \sigma_n) = \prod_{j=1}^{n} \frac{1}{\sigma_j} \exp\left\{-\frac{(y_j - \mu)^2}{2\sigma_j^2}\right\}$$

up to a constant of proportionality. For given μ, we can estimate σ_j by $|y_j - \mu|$. So a *profile log-likelihood* for μ is

$$\ell(\mu) = -\sum_{j=1}^{n} \log|y_j - \mu|$$

up to an additive constant.

Figure 4.17 shows the plot of the profile log-likelihood obtained from a sample of $n = 10$ observations, where $y_j \sim N(0, j)$. It can be seen that the log-likelihood

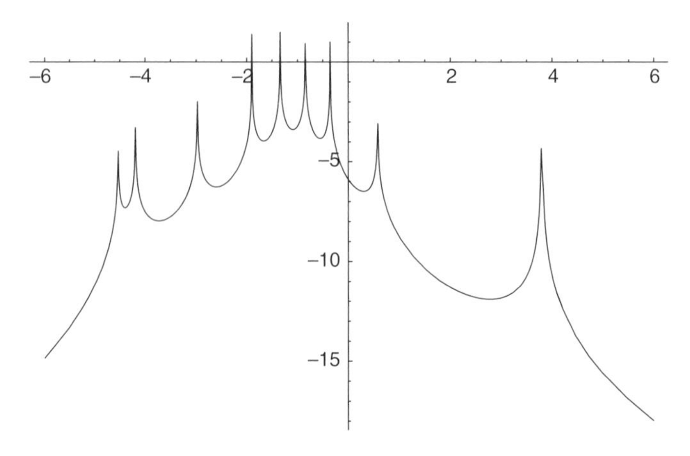

Fig. 4.17 The profile log-likelihood for $y_j \sim N(0, j)$, $j = 1, \ldots, 10$.

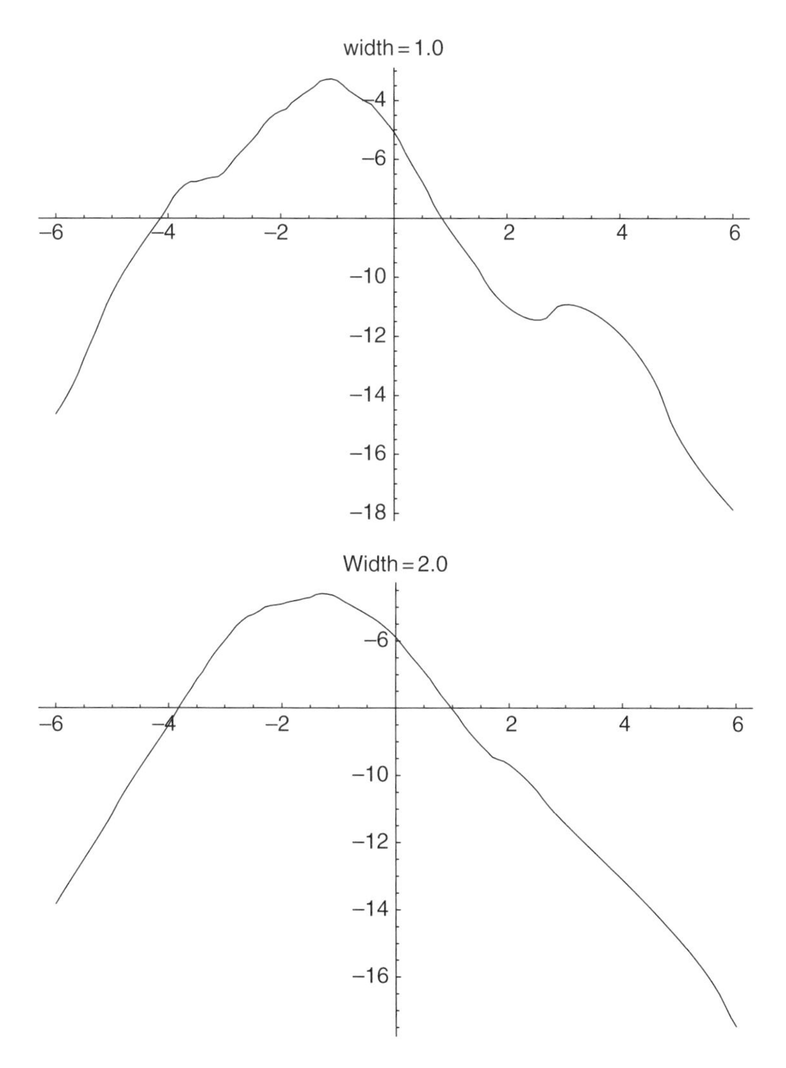

Fig. 4.18 Smoothed versions of the profile log-likelihood with two bandwidths.

goes to infinity at each value in the data. (The rightmost singularity in the graph is actually two almost identical singularities superimposed.)

As the log-likelihood function at these singularities is integrable, we can smooth the function using a kernel $K_n : \mathbb{R} \to \mathbb{R}$ as described above. Figure 4.18 shows the effect of smoothing the log-likelihood using a kernel which is a uniform density on $[-\epsilon, \epsilon]$ with $\epsilon = 1.0$ and $\epsilon = 2.0$. Both smoothed likelihoods are maximised close to the sample median at -1.09277. The sample mean, at -0.79685, is further to the right, and more heavily influenced by outlying values on the right side of the data set.

While standard likelihood asymptotics are not available for this model, it is nevertheless possible to construct exact confidence intervals for the parameter μ. Let $y_{1:n}, y_{2:n}, \ldots, y_{n:n}$ be the order statistics for the data. As all variables are symmetrically distributed about μ, it follows that

$$P_\mu\{y_{k:n} < \mu < y_{n+1-k:n}\} = 1 - 2\sum_{j=0}^{k-1} \binom{n}{j} \frac{1}{2^n}.$$

So $[y_{3:10}, y_{8:10}] = [-2.98, 0.58]$ is approximately a 90% confidence interval for μ.

5
Methodologies for root selection

5.1 Introduction

We have seen in Chapter 4 that many estimating equations have more than one root for practically occurring sample configurations. In certain cases multiple roots occur because the model under consideration is unnecessarily too narrow. In such cases the existence of multiple roots often indicates a discrepancy between the data at hand and the model under consideration. One approach to solving this problem is by embedding the current model into an appropriately chosen larger model. Three examples of this sort are discussed in Section 5.12. In other applications, the existence of multiple roots may reflect the structure of the data; the mixture model provides a case in point (Markatou *et al.*, 1988).

The problem of multiple roots is a wide-ranging phenomenon. With the availability of increasingly cheap and fast computational power one tends to use more and more flexible and realistic models to analyse even moderate sized data sets. The problem of multiple roots frequently occurs in these and other situations. It is not uncommon that the multiple roots issue can arise even in relatively simple parametric models. In this chapter, we examine various techniques for selecting a root as an estimate for the parameter of interest when an estimating equation has more than one solution.

In Section 5.2 we begin by discussing the general problem of solving an estimating equation. We emphasise the point that estimating a parameter is different from solving an estimating equation. The subsequent sections relate to this view in various forms. Note that estimating equations in the class $a(\theta)g(\theta, Y) = 0$, where $a(\theta)$ is a full-rank matrix not depending on data y, all have the same roots. We shall see however that a particular choice of $a(\theta)$ is sometimes preferred to other weighting matrices. When we are interested in an iterative algorithm for obtaining a point estimator using estimating function g, an appropriate choice of $a(\theta)$ may be crucial to ensure convergence properties of the algorithm. When confidence regions are desired, on the other hand, a proper choice of $a(\theta)$ may depend on distributional properties of the transformed estimating function.

In Section 5.3 we give a brief discussion of a class of irregular estimating functions. The estimating equations have multiple roots due to the irregularity of the corresponding estimating functions. In Section 5.4 we study the problem of solving an estimating equation by considering various iterative algorithms using some consistent estimators. In typical parametric models using the score estimating function, a one-step estimator gives an efficient estimator of the parameter. This efficient estimator is usually close in value to the efficient likelihood estimator. In Section 5.4.3, the concept of an efficient likelihood estimator is extended to semiparametric models using the Godambe efficiency of an estimating equation. The resulting Godambe efficient estimator reduces to the efficient likelihood estimator when the estimating function is a score function.

In Section 5.5 we consider a modification to the usual Newton–Raphson type iterative algorithm so that the algorithm will not converge to local minima and saddle points for a typical starting point. For a vector estimating function, the concepts of local minima and saddle points for scalar objective functions generalise to sources and saddle points of the vector field induced by the estimating function on the parameter space. Local maxima generalise to sinks of the vector field. The algorithm that we propose avoids convergence to sources and saddle points in the general case. On the other hand, a sink is always a fixed point of the algorithm. We show that a consistent root of the estimating equation, under mild regularity conditions, is a sink of the vector field, and therefore is guaranteed to be found by the algorithm.

In Section 5.6 a different type of modification to the Newton–Raphson algorithm is considered. The algorithm uses the property that an estimating function is information unbiased. This algorithm shares the same good property with the iterative algorithm studied in Section 5.5, namely, the algorithm only converges to sinks of the vector field of an estimating function. The algorithm considered in this section however improves the algorithm of Section 5.5 in that only sinks satisfying further conditions will be fixed points of the algorithm.

In Section 5.7 Muller's method is modified to take the advantage of a statistical property that pertains only to an estimating function. Namely, we shall incorporate into Muller's method the fact that an estimating function is information unbiased or can be made so by properly rescaling the original unbiased estimating function. A quite different strategy is discussed in Section 5.8, where the problem of multiple roots of an estimating equation is examined by considering the asymptotic form of the estimating equation. Along the same line, we will see that, when explicit formulas are available for the multiple roots, asymptotic inspections of the formula often shed light on the choice of the roots.

In Section 5.9 we examine the properties of a root to an estimating equation by studying some hypotheses at the root. Some specific hypotheses are considered in this section to examine the consistency of the root. In Section 5.10, the approach of testing the consistency of a root is studied more formally by considering a semiparametric analogue of the likelihood ratio test, a criterion studied in more detail in Chapter 6. The tests for root consistency use bootstrap resampling techniques. A root $\hat{\theta}$ at which the distribution of the bootstrap artificial likelihood ratio is closest to the chi-squared distribution with p degrees of freedom, where p is the dimension of the parameter of interest, is suggested to be the estimate for the parameter. Similar kind of techniques are introduced and discussed in Chapter 6.

In Section 5.11 we introduce the concept of shifted information, $I(\theta_1, \theta_2)$, which reduces to the Godambe efficiency when $\theta_1 = \theta_2$. The method based on the shifted information suggests choosing the root $\hat{\theta}$ as an estimator of the parameter θ_0 such that $\hat{I}_n(\hat{\theta})$ attains the maximum value among all roots of the estimating equation, where $\hat{I}_n(\hat{\theta})$ is an appropriate empirical version of $I(\theta_0, \hat{\theta})$.

In Section 5.12 we will restrict our attention to parametric models, where the modelling aspect of the multiple roots problem is discussed. We show, by the analyses of several examples, that model embedding may be a useful technique to overcome the problem of non-uniqueness of roots to a likelihood equation. Finally, Section 5.13 is devoted to a brief discussion of the situation where no root exists for a given estimating equation. Non-existence of a solution is usually due to data degeneracy. Many techniques may be considered for getting rid of data degeneracy. We will discuss an example involving the logistic regression model, where the maximum likelihood estimator lies on the boundary of the parameter space. We shall consider a Bayesian method by combining the likelihood function with the conjugate beta prior for the means of binary responses.

5.2 The problem of solving an estimating equation

Before describing the various methods for choosing an appropriate root of an estimating equation, in this section, we discuss the basic *problem of estimation* in the semiparametric framework based on estimating functions. Let $Y = (Y_1, \ldots, Y_n)$ be a random vector having joint distribution function $F(y) = F(\theta_0; y)$, where θ_0 is the true value of the parameter of interest θ. Since a semiparametric framework is assumed, knowing the value of θ_0 does not specify $F(y)$ completely. In practice, often the first two moments of $F(\theta; y)$ are assumed to be known functions of θ. Suppose that $g(\theta, y)$ is an unbiased estimating function, that is, the equality $E_\theta\{g(\theta, Y)\} = 0$ holds for any $\theta \in \Theta$. In particular, the following identity holds:

$$\int g(\theta_0, y)\, \mathrm{d}F(y) = 0. \tag{5.1}$$

Equation (5.1) gives a necessary condition for the value of θ_0 we seek to estimate. By replacing the unknown distribution function $F(y)$ with its empirical version using a sample $y = (y_1, \ldots, y_n)$, we arrive at the data version

$$g(\theta, y) = 0. \tag{5.2}$$

Now since y is known we can solve (5.2) for an estimate of θ_0. Suppose that (5.2) has a unique solution, say $\hat{\theta}_n$.

Such a root is a reasonable estimate of θ_0 under additional mild regularity conditions. For instance, $\hat{\theta}_n$ is $\sqrt{n}$-consistent provided the following asymptotics are valid:

$$\begin{aligned}\sqrt{n}(\hat{\theta}_n - \theta_0) &= -\sqrt{n}\,\{\dot{g}(\theta_0, Y)\}^{-1}\, g(\theta_0, Y) + O_p(1) \\ &= \left\{-\frac{1}{n}\dot{g}(\theta_0, Y)\right\}^{-1}\left\{\frac{1}{\sqrt{n}}g(\theta_0, Y)\right\} + O_p(1) \\ &= O_p(1).\end{aligned}$$

This will be the case for usual estimating functions. An example of this kind is the *unbiased additive estimating function** of the form

$$g(\theta, Y) = \sum_{j=1}^{n} g_j(\theta, Y_j).$$

The validity of the above asymptotics in this case is ensured by virtue of the law of large numbers and the central limit theorem.

A problem arises when (5.2) has multiple roots. For example, in the parametric case, (5.2) is simply the likelihood equation and all local extrama are possible solutions. For parametric models, fortunately, we have theories available to solve this problem. First, under the Cramér conditions, we know that there is a consistent root among the solutions to the likelihood equation. Next, if regularity conditions such as those of Wald (1949) are satisfied by the model, then the root which maximises the likelihood function globally is the consistent root (but see Example 4.9 for a counterexample). Therefore, a procedure for finding a desirable estimate involving the likelihood estimating equation may be summarised as follows:

1. Solve $g(\theta, Y) = \dot{\ell}(\theta; Y) = 0$ to get all possible roots, $\hat{\theta}_1, \ldots, \hat{\theta}_k$, where $\ell(\theta; Y)$ is the log-likelihood function.
2. Choose as the estimate the root $\hat{\theta} = \hat{\theta}_i$ at which the log likelihood attains its maximum value among values $\ell(\hat{\theta}_j; Y)$ for $j = 1, \ldots, k$.

The procedure shown above says that the problem of constructing a point estimate from a parametric model is not equivalent to the problem of solving the likelihood equation. The likelihood function also plays an essential role in this procedure. Similarly, extracting a point estimate from some moment assumptions is not equivalent to the problem of solving an optimal estimating equation. Moreover, in the semiparametric framework of estimating functions, the second component of the above procedure is not feasible because an objective function playing the role of a log likelihood in general does not exist.

When the plausibility of a root to an estimating equation is in doubt, we have to examine other constraints concerning the true value θ_0, in addition to the

* Some writers use the terminology *unbiased linear estimating function*, where linearity may be interpretated as the manner of dependence of the estimating function on a set of basis functions, such as a set of elementary estimating functions, but not on the parameter of interest nor on the observations.

unbiasedness condition (5.1). These constraints, in the present context, are naturally expressed in terms of the concerned estimating function $g(\theta, Y)$. Usually, the true value of θ_0 is subject to the following types of conditions:

$$E_{\theta_0}\{f_j[g(\theta_0, Y)]\} = \int f_j[g(\theta_0, y)]\,\mathrm{d}F(y) = 0, \quad j = 0, 1, \ldots, L \tag{5.3}$$

where $f_0(x) = x$ corresponds to the unbiasedness condition, and $f_j(\cdot)$ are known functions for $j = 1, \ldots, L$. To see what the other functions f_j will look like, suppose that the following matrices:

$$\Sigma = E_\theta\{g(\theta, Y)\, g^{\mathrm{t}}(\theta, Y)\} \quad \text{and} \quad \Gamma = E_\theta\{\dot{g}(\theta, Y)\}$$

exist and are known by assumption. It follows then the transformed estimating function:

$$-\Gamma^{\mathrm{t}}\, \Sigma^{-1} g(\theta, Y)$$

is information unbiased. In most cases, therefore, the information unbiasedness is a readily available property of an unbiased estimating function $g(\theta, Y)$. In this case, (5.3) takes the form

$$\int g(\theta_0, y)\,\mathrm{d}F(y) = 0$$

$$\int \{g(\theta_0, y)\, g^{\mathrm{t}}(\theta_0, y) + \dot{g}(\theta_0, y)\}\,\mathrm{d}F(y) = 0.$$

When these theoretical properties are available we may then search an estimate of the parameter of by solving the following problem:

$$\begin{aligned} &\text{solve} \quad g(\theta, y) = 0 \\ &\text{subject to} \quad \rho\{g^{\mathrm{t}}(\theta, y)\, g(\theta, y) + \dot{g}(\theta, y)\} \le \epsilon \end{aligned} \tag{5.4}$$

where ϵ is a preassigned tolerance limit, and $\rho(\cdot)$ is an appropriate norm. In the scalar case $\rho(x)$ can be simply the absolute value of x. When g is a vector, $\rho(x)$ can be the largest norm of eigenvalues of matrix x.

A root solving the first equation of (5.4) can fail to satisfy the second constraint. Asymptotically, when the sample size grows large enough, a consistent root will also satisfy the second constraint by virtue of the law of large numbers. Consequently, a root satisfying the second constraint of (5.4) may be more desirable than a root which does not. Some of the iterative algorithms discussed in the sequel are based on the idea (5.4).

5.3 A class of irregular estimating functions

Let $g(\theta, Y)$ be an unbiased estimating function and θ_0 be the true value of θ. Sometimes it may happen that the following relation holds:

$$E_{\theta_1}\{g(\theta_2, Y)\} = 0, \quad \forall\, \theta_2 \in \Theta_{\theta_1} \tag{5.5}$$

where Θ_{θ_1} is some subset of Θ depending on θ_1 such that $\Theta_{\theta_1} - \{\theta_1\}$ is non-empty. Equation (5.5) also defines the subset Θ_{θ_0} of the parameter space Θ corresponding to θ_0. The fact that Θ_{θ_0} contains other values besides θ_0 causes the problem when we solve the estimating equation $g(y, \theta) = 0$. We further suppose that the model under consideration is *identifiable*, that is, for $\theta_1, \theta_2 \in \Theta_{\theta_0}$ and $\theta_1 \neq \theta_2$, Y has different moment structure. We shall refer to an estimating function satisfying these conditions as an *irregular estimating function*.

Example 5.1 Circular model

Consider a bivariate random vector $\boldsymbol{X} = (Y, Z)^{\mathrm{t}}$ having mean vector $\mu = (\cos\theta, \sin\theta)^{\mathrm{t}}$ and the identity covariance matrix, where $\theta \in \Theta = [-\pi, \pi)$. Let θ_0 be the true value. Based on an independent sample, $x_j = (y_j, z_j)^{\mathrm{t}}$, $j = 1, \ldots, n$, the quasi-score takes the form

$$g(\theta, x) = -n(\bar{y}\sin\theta - \bar{z}\cos\theta) \tag{5.6}$$

where $\bar{y} = n^{-1}\sum_{j=1}^{n} y_j$, $\bar{z} = n^{-1}\sum_{j=1}^{n} z_j$. For any $\theta_1 \in [-\pi, \pi)$, let

$$\theta_2 = f(\theta_1) = \begin{cases} \theta_1 + \pi & \text{if } \theta_1 \in [-\pi, 0) \\ \theta_1 - \pi & \text{if } \theta_1 \in [0, \pi). \end{cases} \tag{5.7}$$

Then (5.6) satisfies (5.5) with $\Theta_{\theta_1} = \{\theta_1, f(\theta_1)\}$. That the circular model is identifiable follows from the following relation

$$\mu(\theta_0) = (\cos\theta_0, \sin\theta_0)^{\mathrm{t}} = -\mu(f(\theta_0)).$$

The quasi-score for the circular model (5.6) is therefore an irregular estimating function.

To see the particular feature when solving $g(\theta, y) = 0$ using an estimating function g satisfying (5.5), suppose that $\Theta_\theta = \{\theta, \theta - a\}$ consists of two points for each $\theta \in \Theta$, where a is a known constant. Now (5.5) corresponds to the following set of equations:

$$E_{\theta_0}\{g(\theta_0, Y)\} = 0 \tag{5.8}$$

$$E_{\theta_0}\{g(\theta_0 - a, Y)\} = 0. \tag{5.9}$$

In Section 5.2, we see that an estimating equation is obtained as the data version of the first moment constraint posed on the estimating function. For the irregular estimating function satisfying (5.8) and (5.9) we are therefore required to solve the following set of equations:

$$g(\theta, y) = 0, \tag{5.10}$$

$$g(\theta - a, y) = 0. \tag{5.11}$$

If $\hat{\theta}_1$ solves (5.10), then $\hat{\theta}_2 = \hat{\theta}_1 + a$ is a solution to (5.11). We therefore obtain two solutions, one consistently estimating θ_0 and the other consistently estimating $\theta_0 + a$.

Example 5.2 Circular model (continued)

The estimating equations (5.10) and (5.11) for the quasi-score (5.6) yield the solutions

$$\hat{\theta}_1 = \tan^{-1}\frac{\bar{z}}{\bar{y}} \quad \text{and} \quad \hat{\theta}_2 = f(\hat{\theta}_1)$$

respectively, where $\hat{\theta}_1 \in (-\pi/2, \pi/2)$ and $f(\cdot)$ is defined by (5.7). Which root is a reasonable estimate for θ?

For this problem, fortunately, we have a simple solution. First, note that there are $2n$ conditions concerning the parameter of interest θ:

$$E_\theta(Y_j - \cos\theta) = 0, \quad E_\theta(Z_j - \sin\theta) = 0, \quad j = 1, \ldots, n.$$

So, if we let

$$h(\theta, X) = (Y_1 - \cos\theta, \ldots, Y_n - \cos\theta, Z_1 - \sin\theta, \ldots, Z_n - \sin\theta)^{\mathrm{t}}$$

then $h(\theta, X)$ is an unbiased estimating function of dimension $2n$. This function is called an *elementary estimating function*. An estimate of θ can then be obtained by minimising the following criterion:

$$\psi(\theta; X) = \{h(\theta, X)\}^{\mathrm{t}}\, h(\theta, X). \tag{5.12}$$

The solution to this procedure, using sample x, can be obtained by solving the equation

$$\begin{aligned}\frac{\partial}{\partial\theta}\{\psi(\theta; x)\} &= 2n\bar{y}\sin\theta - 2n\bar{z}\cos\theta \\ &= 0\end{aligned}$$

yielding the (unique) answer $\hat{\theta}_1 = \tan^{-1}(\bar{z}/\bar{y})$.

Criterion (5.12) may be regarded as an extension of the method of least squares. In Section 6.6, we will make more general and systematic discussions on use of criterions such as (5.12).

One way to remove the irregularity of estimating functions is to construct a new estimating function so that Θ_θ defined by (5.5) is a singleton containing only the point θ. Embedding the model under consideration into a larger model can sometimes be helpful for this purpose.

Example 5.3 Circular model (continued)

Now suppose that the mean vector of $\boldsymbol{X}$ is instead $\mu = (r\cos\theta, r\sin\theta)^{\mathrm{t}}$, which contains the parameter θ and an additional parameter $r \geq 0$. The covariance matrix is still assumed to be the identity matrix. Note that if $r_0 = 0$ then θ is unidentifiable. Therefore we impose the condition that $r > 0$. Now the quasi-score is a two-dimensional vector given by

$$g_\theta(\theta, r, x) = -nr(\bar{y}\sin\theta - \bar{z}\cos\theta), \tag{5.13}$$

$$g_r(\theta, r, x) = n(\bar{y}\cos\theta - \bar{z}\sin\theta) - nr. \tag{5.14}$$

The estimating function

$$g(\theta, r, x) = (g_\theta(\theta, r, x), g_r(\theta, r, x))^{\mathrm{t}}$$

is unbiased. Further, it can be checked that

$$E_\theta\{g_\theta(\theta + \pi, r, X)\} = 0, \quad E_\theta\{g_r(\theta + \pi, r, X)\} = -2nr \neq 0.$$

The estimating function $g(\theta, r, x)$ is now a regular one. Solving $g(\theta, r, x) = 0$ gives the following unique solution:

$$\begin{aligned} \hat{\theta} &= \tan^{-1}(\bar{z}/\bar{y}), & \hat{r} &= \bar{y}/\cos\hat{\theta} \ \text{ if } \bar{y} \geq 0 \\ \hat{\theta} &= f\{\tan^{-1}(\bar{z}/\bar{y})\}, & \hat{r} &= \bar{y}/\cos\hat{\theta} \ \text{ if } \bar{y} < 0. \end{aligned} \tag{5.15}$$

Model embedding is a useful idea not limited to correcting the irregularity of an estimating function as discussed above. In Section 5.12 we shall explore this idea further in some parametric models for which the score functions have multiple roots.

5.4 Iterating from consistent estimators

5.4.1 INTRODUCTION

For estimating functions satisfying standard regularity conditions there is a unique isolated consistent root. So an obvious strategy for selecting a zero of $g(\theta, y)$ is to construct a consistent, albeit inefficient, estimator $\tilde{\theta}$ and to choose the root that is closest to $\tilde{\theta}$. The concept of the *closest root* could be measured by Euclidean distance or other metric in Θ. However, such a strategy does not pick a root in a parameterisation-invariant manner. That is, for a monotonic function $\xi = \xi(\theta)$, the fact that $\hat{\theta}$ is closest to $\tilde{\theta}$ in θ-space does not imply that $\hat{\xi} = \xi(\hat{\theta})$ is closest to $\tilde{\xi} = \xi(\tilde{\theta})$ in ξ-space.

Finding an estimate by iterating from appropriate consistent estimates is probably the most used and the most useful method in solving estimating equations. The idea of approximating an efficient estimator using a consistent estimate can be useful when an estimating equation has a unique solution or when multiple roots exist. When it is costly to solve an estimating equation that has a unique solution, then one may apply a one-step iteration to obtain an efficient estimator. When multiple roots are present, iterative methods can help identify the estimator that is both *consistent* and *efficient*. An estimator $\hat{\theta}$ is said to be a (*weakly*) *consistent estimator* of θ if for all $\theta \in \Theta$ we have

$$\lim_{n\to\infty} \Pr\{|\hat{\theta} - \theta| > \epsilon\} = 0.$$

That is, the probability of estimator $\hat{\theta}$ deviating from the parameter of interest θ by a small amount ϵ, when sample size grows, can be made as small as we wish. When this is true, $\hat{\theta}$ is said to converge *stochastically*, or *in probability*, to θ. So weak consistency of $\hat{\theta}$ is equivalent to convergence in probability. We shall use the o_p notation† to indicate that $\hat{\theta}$ is aconsistent estimator of θ, that is, $\hat{\theta} = \theta + o_p(1)$. Similarly, we can define the concept of *strong consistency*, which is equivalent to convergence of $\hat{\theta}$ to θ *with probability one*, that is

$$\Pr\left\{\lim_{n\to\infty} \hat{\theta} = \theta\right\} = 1.$$

Unless otherwise stated, by consistency we shall mean weak consistency. Many estimators commonly met in practice are in fact $\sqrt{n}$-consistent, meaning that

$$\hat{\theta} = \theta + O_p(n^{-1/2}).$$

That is, $\sqrt{n}\,(\hat{\theta} - \theta)$ is bounded in probability for every $\theta \in \Theta$.

We now begin with the study of one-step estimators for a scalar parameter θ.

5.4.2 EFFICIENT LIKELIHOOD ESTIMATORS

Let $\tilde{\theta}$ be a consistent estimator of θ. A *one-step estimator* for θ using $\tilde{\theta}$ is a value defined by

$$\hat{\theta} = \tilde{\theta} - \{\dot{g}(\tilde{\theta}, y)\}^{-1} g(\tilde{\theta}, y) \tag{5.16}$$

where $g(\theta, y)$ is an unbiased estimating function. In a parametric model the estimating function $g(\theta, y)$ can be a score function $\dot{\ell}(\theta; y)$, where $\ell(\theta; y)$ is a log-likelihood function. The one-step estimator (5.16) has been widely used in various applications. The popularity of this estimator is partly due to the particularly simple form. Now we first state a result concerning the efficiency of estimator (5.16) in one-dimensional parametric case. The more general case is treated later.

Proposition 5.4 *Let Y_j be independent and identically distributed random variables having density function $f(\theta; \cdot)$ for $j = 1, \ldots, n$. Suppose that the following conditions hold:*

1. *The parameter space is an open interval (a, b).*
2. *The support S of $f(\theta; \cdot)$ is functionally independent of θ.*

† Recall that a random variable X is of order $o_p(n^{\alpha})$ if

$$\lim_{n\to\infty} \Pr\left\{\left|\frac{X}{n^{\alpha}}\right| > \epsilon\right\} = 0 \quad \text{for all } \epsilon > 0.$$

3. *The third derivative* $(\partial^3/\partial\theta^3)f(\theta; y)$ *is continuous and bounded in* S, *i.e.,*

$$\left|\frac{\partial^3}{\partial\theta^3}\log f(\theta; y)\right| \leq M(y) \quad for\ y \in S,\ \ \theta_0 - c < \theta < \theta_0 + c \tag{5.17}$$

where $c > 0$ *and* $E_{\theta_0}\{M(Y)\} < \infty$.

4. *The integral* $\int f(\theta; y)\,\mathrm{d}\mu(y)$ *is three times differentiable with respect to* θ *under the integral sign.*
5. *Finally, the Fisher information*

$$I(\theta) = -E_\theta\left\{\frac{\partial^2}{\partial\theta^2}\log f(\theta; y)\right\}$$

satisfies $0 < I(\theta) < \infty$.

If $\ell(\theta) = \sum_{j=1}^{n}\log f(\theta; y_j)$ *and* $\tilde{\theta}_n$ *is* $\sqrt{n}$*-consistent, then the one-step estimator*

$$\hat{\theta}_n = \tilde{\theta}_n - \frac{\dot{\ell}(\tilde{\theta}_n)}{\ddot{\ell}(\tilde{\theta}_n)} \tag{5.18}$$

is asymptotically efficient. That is, $\hat{\theta}_n$ *is consistent and the variance of* $\hat{\theta}_n$ *tents to the inverse of the Fisher information. Further, we have*

$$\sqrt{n}(\hat{\theta}_n - \theta) \xrightarrow{\mathcal{L}} N(0, I^{-1}(\theta)). \tag{5.19}$$

The reader may consult Lehmann and Casella (1998, p. 454) for a proof. Since the efficiency result in (5.19) is a first-order asymptotic property, the denominator of the one-step estimator (5.18) can be replaced by any first-order equivalent random variable without altering the conclusion. A particularly important estimator obtained along this line is by using the method of scoring due to Fisher (1925). This estimator is a variant of (5.18) with the denominator replaced by its expectation evaluated at the starting point. This method is particularly useful when repeated iterations are desired and the evaluation of the denominator of (5.18) is costly.

Corollary 5.5 *Suppose that the assumptions of Proposition 5.4 hold and in addition that the Fisher information* $I(\theta)$ *is continuous in* θ. *Then the one-step estimator using Fisher's method of scoring*

$$\hat{\theta}_n = \tilde{\theta}_n + \frac{\dot{\ell}(\tilde{\theta}_n)}{nI(\tilde{\theta}_n)} \tag{5.20}$$

is an asymptotically efficient estimator of θ.

The two estimators defined by (5.18) and (5.20) are usually different because the observed information is in general different from its expected counterpart. Note, however, if there exists a one-dimensional sufficient statistic for θ, then the expected information will be idential with the observed information evaluated at the maximum likelihood estimator. Stuart (1958) heuristically argues that estimator (5.20) may be closer to the *efficient likelihood estimator* than (5.18). Here an efficient likelihood estimator means an efficient estimator defined as a root to the likelihood equation, which need not be the maximum likelihood estimator. Efron and Hinkley (1978) discuss the difference between the observed and the expected information in estimating the variance of the maximum likelihood estimator; see also Lindsay and Yi (1996).

A result stronger than that stated in Corollary 5.5 is possible for certain special models. An estimator is said *Fisher-consistent* if knowledge about the whole population is used then the estimator is equal to the true value of the parameter. Eguchi (1983) shows that, for curved exponential family model, a two-step version of (5.20) is *second-order efficient* (see Section 3.4.1 of Kass and Vos, 1997 for a definition) provided that $\tilde{\theta}_n$ is Fisher-consistent.

5.4.3 GENERAL CASE

Now we turn to the case of general estimating functions. Let $g(\theta) = g(\theta, y)$ be an unbiased vector-valued estimating function. The Godambe efficiency, $\text{eff}_\theta(g)$, a matrix depending on the sample size n, is defined by (2.5). In the parametric case when $g(\theta)$ is a score function the Godambe efficiency reduces to the Fisher information. Since the Godambe efficiency $\text{eff}_\theta(g)$ depends on the sample size n, in the following discussion it will be convenient to assume that, when divided by n, the Godambe efficiency converges to a positive definite matrix as $n \to \infty$, i.e.,

$$\Sigma_\theta = \lim_{n\to\infty} \frac{1}{n}\{\text{eff}_\theta(g)\}. \tag{5.21}$$

To generalise Proposition 5.4, we need to extend the concept of efficient likelihood estimator to the case of estimating functions.

Definition 5.6 *An estimator $\hat{\theta}_n$ is Godambe efficient with respect to an optimum estimating function $g(\theta)$ if*

$$\sqrt{n}(\hat{\theta}_n - \theta) \xrightarrow{\mathcal{L}} N\left(0, \Sigma_\theta^{-1}\right)$$

where Σ_θ is the limiting average Godambe efficiency defined by (5.21).

Proposition 5.7 *Let Y_j be independent random variables for $j = 1, \ldots, n$. Consider estimating function $g(\theta, Y)$, where $Y = \{Y_1, \ldots, Y_n\}$. The parameter space Θ is assumed to be an open subset of $\mathbb{R}^p$. In addition, suppose that the following conditions hold:*

1. *The estimating function $g(\theta, y)$ can be decomposed as*

$$g(\theta, y) = \sum_{j=1}^{n} h_j(\theta, y_j) \tag{5.22}$$

where each summand $h_j = (h_{j1}, \ldots, h_{jp})^t$ is a p-dimensional vector having mean zero. That is, $E_\theta\left\{h_j(\theta, Y_j)\right\} = 0$.

2. *The estimating function $g(\theta, Y)$ is twice differentiable with respect to θ, i.e., each component $h_{jd}(\theta)$, $j = 1, \ldots, p$; $d = 1, \ldots, p$, is twice differentiable. Further, the second derivatives are assumed to be bounded in probability. That is,*

$$\left|\frac{\partial^2}{\partial\theta^a\partial\theta^b} h_{jd}(\theta, y_j)\right| < M(y_j), \quad j = 1, \ldots, n; \quad a, b, d = 1, \ldots, p$$

where y_j and θ are in an open subset containing the true value θ_0 and $E_\theta|M(Y)| < \infty$.

3. *Assume that $\tilde{\theta}_n$ be a $\sqrt{n}$-consistent estimator of the true parameter θ_0. That is,*

$$\sqrt{n}(\tilde{\theta}_n - \theta_0) = O_p(1). \tag{5.23}$$

4. *The Godambe efficiency $\mathit{eff}_\theta(g)$ of (5.21) exists and is positive definite.*

Then the following one-step estimator is Godambe efficient:

$$\hat{\theta} = \tilde{\theta}_n - \{\dot{g}(\tilde{\theta}_n)\}^{-1} g(\tilde{\theta}_n). \tag{5.24}$$

Proof. We shall denote the j-th component of θ, $g(\theta)$ by θ^j, $g_j(\theta)$, and so on. It is convenient to use the same symbols to denote the same vectors. This convention can be extended to matrices and higher-order arrays. For instance, we shall use the notation

$$g_{ij}(\theta) = \frac{\partial}{\partial\theta^j} g_i(\theta), \quad g_{ijk}(\theta) = \frac{\partial^2}{\partial\theta^j\theta^k} g_i(\theta)$$

to denote the matrix $\dot{g}(\theta)$ and the $p \times p \times p$ array $\ddot{g}(\theta)$, respectively.

By Taylor expansion of $g(\tilde{\theta}_n)$ at θ_0 we have

$$g_j(\tilde{\theta}_n) = g_j(\theta_0) + g_{ja}(\theta_0)(\tilde{\theta}_n^a - \theta_0^a) + \tfrac{1}{2} g_{jab}(\theta_n^\dagger)(\tilde{\theta}_n^a - \theta_0^a)(\tilde{\theta}_n^b - \theta_0^b) \tag{5.25}$$

where each component of $\theta_n^\dagger$ is between the corresponding components of θ_0 and $\tilde{\theta}_n$. In (5.25) we used the Einstein summation convention: summation is automatically taken with respect to an index if it appears both as a subscript and as a superscript.

Now rewrite (5.24) as

$$\hat{\theta}^a = \tilde{\theta}_n^a - g^{ja}(\tilde{\theta}_n) g_j(\tilde{\theta}_n) \tag{5.26}$$

where $g^{ja}(\theta)$ denotes the (j, a)-th component of the inverse of $\dot{g}(\theta)$. Putting (5.25) into (5.26) and simplifying, we get

$$\sqrt{n}(\hat{\theta}^a - \hat{\theta}_0^a) = -\{n g^{ai}(\tilde{\theta}_n)\}\left\{\frac{1}{n} g_i(\theta_0)\right\} + \left\{1_{al} - g^{aj}(\tilde{\theta}_n) g_{jl}(\theta_0) - \frac{1}{2} g^{aj}(\tilde{\theta}_n) g_{jkl}(\theta_n^\dagger)(\tilde{\theta}_n^k - \theta_0^k)\right\} \sqrt{n}(\tilde{\theta}_n^l - \theta_0^l) \tag{5.27}$$

where 1_{al} is the (a, l)-th element of the $p \times p$ identity matrix.

Expanding $g_{ij}(\tilde{\theta}_n)$ at θ_0 gives

$$\frac{1}{n}g_{ij}(\tilde{\theta}_n) = \frac{1}{n}g_{ij}(\theta_0) + \frac{1}{n}g_{ijk}(\tilde{\theta}_n^*)(\tilde{\theta}_n^k - \theta_0^k) \tag{5.28}$$

where $\tilde{\theta}_n^*$ is between $\tilde{\theta}$ and θ_0. By $\sqrt{n}$-consistency of $\tilde{\theta}_n$, Assumption 2 and (5.28), we then have

$$1_{al} - g^{aj}(\tilde{\theta}_n)g_{jk}(\theta_0) = o_p(1).$$

Combining this fact and the following one:

$$g^{aj}(\tilde{\theta}_n)g_{jkl}(\theta_n^{\dagger})(\tilde{\theta}_n^k - \theta_0^k) = O_p(n^{-1})O_p(n)O_p(n^{-1/2}) = o_p(1)$$

we then have that the second term of (5.27) has order $o_p(1)$. Therefore, we have

$$\sqrt{n}(\hat{\theta}^a - \theta_0^a) = -\{ng^{ai}(\tilde{\theta}_n)\}\left\{\frac{1}{\sqrt{n}}g_i(\theta_0)\right\} + o_p(1)$$

or in usual matrix notation

$$\sqrt{n}(\hat{\theta} - \theta_0) = -\left\{\frac{1}{n}\dot{g}(\tilde{\theta}_n)\right\}^{-1}\left\{\frac{1}{\sqrt{n}}g(\theta_0)\right\} + o_p(1). \tag{5.29}$$

The coefficient $\{\dot{g}(\tilde{\theta}_n)/n\}^{-1}$ on the right-hand side of (5.29), by virtue of the law of large numbers, will tend to $\{n^{-1}E_{\theta_0}\dot{g}(\theta_0)\}^{-1}$ as $n \to \infty$. Since $g(\theta)$ is unbiased, applying the central limit theorem to $g(\theta_0)/\sqrt{n}$, we have

$$\frac{1}{\sqrt{n}}g(\theta_0) \stackrel{\mathcal{L}}{\to} N(0, A), \quad \text{where } A = \lim_{n\to\infty}\frac{1}{n}E_{\theta_0}\{g(\theta_0)g^{\text{t}}(\theta_0)\}.$$

Slutsky's Theorem now implies that $\sqrt{n}(\hat{\theta} - \theta_0)$ of (5.29) has a limiting normal distribution with covariance matrix

$$\begin{aligned}
\Omega &= \lim_{n\to\infty}\{n^{-1}E_{\theta_0}\dot{g}^t(\theta_0)\}^{-1} A \lim_{n\to\infty}\{n^{-1}E_{\theta_0}\dot{g}(\theta_0)\}^{-1} \\
&= \lim_{n\to\infty}\{E_{\theta_0}\dot{g}^t(\theta_0)\}^{-1}[E_{\theta_0}\{g(\theta_0)g^t(\theta_0)\}]\{E_{\theta_0}\dot{g}(\theta_0)\}^{-1} \\
&= \lim_{n\to\infty}\left(\frac{1}{n}\{E_{\theta_0}\dot{g}^t(\theta_0)\}[E_{\theta_0}\{g(\theta_0)g^t(\theta_0)\}]^{-1}\{E_{\theta_0}\dot{g}(\theta_0)\}\right)^{-1} \\
&= \Sigma_\theta^{-1}
\end{aligned}$$

where Σ_θ is defined by (5.21). That $\hat{\theta}$ is Godambe efficient follows from the definition. This completes the proof. □

Note that the conditions of Proposition 5.7 to ensure Godambe efficiency of $\hat{\theta}$ are only sufficient conditions. When one or some of the conditions are violated, the conclusion may still be valid. For example, the conclusions of Proposition 5.7 can often hold even when Condition 1 fails to be satisfied. For specific problems, it may be less laborious to verify directly a minimum set of conditions rather than to verify a set of general sufficient conditions such as those given in Proposition 5.7. See Heyde (1997, Section 4.2) for a discussion on similar issues in the context of quasi-likelihood estimation.

Corollary 5.8 *Suppose that the conditions of Proposition 5.7 hold and in addition that $J_n(\theta) = -E_\theta\{\dot{g}(\theta, Y)\}$ is continuous in θ. Then the one-step estimator*

$$\hat{\theta}_n = \tilde{\theta}_n + J_n^{-1}(\tilde{\theta}_n)g(\tilde{\theta}_n) \tag{5.30}$$

is also a Godambe efficient estimator.

5.4.4 CHOOSING $\sqrt{n}$-CONSISTENT ESTIMATORS

The choice of an appropriate $\sqrt{n}$-consistent estimator is essential to the success of the method of one-step estimation for root selection. Wolfowitz (1957), Le Cam (1969, p. 103) and Bickel *et al.* (1993, Section 2.5) give some general discussions on how to construct $\sqrt{n}$-consistent estimators. Unfortunately, the class of $\sqrt{n}$-consistent estimators is a very large one. So the choice of $\tilde{\theta}_n$ within this class is critically important to the success of one-step iteration, as asymptotic considerations may not hold until n is very large. For example, if $\tilde{\theta}_n$ is $\sqrt{n}$-consistent, then so is $\tilde{\theta}_n + n^{-1}a$, no matter how large the constant a is. Clearly, in practice, if $a = 10^8$, we would not consider $\tilde{\theta}_n$ and $\tilde{\theta}_n + n^{-1}a$ both appropriate.

When there are available more than one $\sqrt{n}$-consistent estimators, a sensible criterion to choose a particular estimator is to select one which has high correlation with an ideal estimator, which can be the efficient likelihood estimator if the estimating equation under consideration is a likelihood equation. The correlation between an initial estimator $\tilde{\theta}_n$ and an ideal estimator $\hat{\theta}$, in some cases, can be shown to be exactly equal to the *efficiency* of the initial estimator $\tilde{\theta}_n$. Here the efficiency of an estimator $\tilde{\theta}_n$ relative to a 'best' estimator $\hat{\theta}$ is defined to be the ratio of the variances

$$e(\tilde{\theta}_n) = \frac{\text{Var}(\hat{\theta})}{\text{Var}(\tilde{\theta}_n)}.$$

If $\hat{\theta}$ is a minimum variance unbiased estimator of θ and $\tilde{\theta}_n$ is any unbiased estimator of θ, then it can be shown that

$$\rho(\tilde{\theta}_n, \hat{\theta}) = e(\tilde{\theta}_n) \tag{5.31}$$

where $\rho(\tilde{\theta}_n, \hat{\theta})$ is the Pearson product moment correlation coefficient between $\tilde{\theta}_n$ and $\hat{\theta}$. Suppose that we wish to approximate a Godambe efficient estimator $\hat{\theta}$ which solves $g(\hat{\theta}) = 0$. If $\tilde{\theta}_n$ is a $\sqrt{n}$-consistent estimator of θ, then equation (5.31) will not hold exactly in general. For many regular models, we may however expect the relation (5.31) to hold approximately.

According to the criterion of the preceding paragraph, an often practical way of choosing an initial estimate is to use the (unique) root of the same type of estimating equation based on some subsamples or bootstrap resamples of the original data. In the Cauchy location model, for instance, the likelihood equation is expected to have a unique root if the 'outliers' are excluded from the sample. An equally useful method is to use the same data but based on a different strategy of estimation. A typical example is to use the moment estimator as the initial estimate in solving, for instance, the likelihood equations, with possible exceptions such as the Cauchy location model.

5.5 A modified Newton's method

5.5.1 INTRODUCTION

In the previous section we have seen that a one-step Newton–Raphson iteration, or some variant of it, provides a general procedure for constructing an efficient estimator from a $\sqrt{n}$-consistent estimator. In the likelihood case, the one-step estimators are efficient and therefore are expected to be close to the efficient likelihood estimator. For a general Godambe efficient estimating function, the one-step estimators are Godambe efficient. When multiple roots exist, the root that is closest, in Euclidean distance, to this efficient estimator can then be chosen as our estimate for the parameter of interest.

However, as noted at the end of last section, the one-step estimators often depend sensitively on the choice of the initial consistent estimators. Nor is the one-step estimation invariant under reparameterisation. To see this, suppose that θ is a scalar parameter and we wish to consider the problem of estimating a new parameter $\xi = \gamma(\theta)$ for a smooth monotone function $\gamma(\cdot)$. Suppose that $\tilde{\theta}$ is a consistent estimator for θ. That is, $\tilde{\theta} = \theta + o_p(1)$. Then we have

$$
\begin{aligned}
\tilde{\xi} &\equiv \gamma(\tilde{\theta}) \\
&= \gamma(\theta) + \dot{\gamma}(\theta^{\dagger})(\tilde{\theta} - \theta) \\
&= \xi + o_p(1)
\end{aligned}
$$

where $\theta^{\dagger}$ is between $\tilde{\theta}$ and θ, $\dot{\gamma}(\theta) = (\mathrm{d}/\mathrm{d}\theta)\,\gamma(\theta)$, and $\dot{\gamma}(\theta)$ is assumed to be bounded. It follows that $\tilde{\xi}$ is a consistent estimator for ξ. The same arguments remain unaltered for $\sqrt{n}$-consistency. From the two consistent estimators $\tilde{\theta}$ and $\tilde{\xi}$ we can construct two efficient one-step estimators $\hat{\theta}$ and $\hat{\xi}$ for θ and ξ, respectively. It can be verified, however, that $\hat{\xi}$ is not equal to $\gamma(\hat{\theta})$ in general.

Nevertheless, we can show that $\xi^{\dagger} = \gamma(\hat{\theta})$ is also a Godambe efficient estimator for ξ. To see this, first note that, since $\hat{\theta}$, which is Godambe efficient by assumption, has a limiting normal distribution

$$
\sqrt{n}(\hat{\theta} - \theta) \xrightarrow{\mathcal{L}} N(0, 1/\mathrm{eff}_{\theta}(g))
$$

we can therefore apply the delta method to $\xi^\dagger = \gamma(\hat{\theta})$ to show that

$$\sqrt{n}(\xi^\dagger - \xi) = \sqrt{n}\{\gamma(\hat{\theta}) - \gamma(\theta)\}$$

also has a limiting normal distribution with mean zero and variance given by

$$\left\{\frac{\partial \xi}{\partial \theta}\right\}^2 \{\text{eff}_\theta(g)\}^{-1}. \tag{5.32}$$

On the other hand, the limiting average Godambe efficiency Σ_ξ for ξ based on estimating function

$$h(\xi) = \left\{\frac{\partial \theta}{\partial \xi}\right\} g(\gamma^{-1}(\xi))$$

for ξ is equal to

$$\left\{\frac{\partial \theta}{\partial \xi}\right\}^2 \text{eff}_{\gamma^{-1}(\xi)}(g)$$

which can be calculated directly as follows:

$$\begin{aligned}
\Sigma_\xi &= \lim_{n\to\infty} \frac{1}{n} \text{eff}_\xi(h) \\
&= \lim_{n\to\infty} \frac{1}{n} [E_\xi\{h^2(\xi)\}]^{-1} \left\{E_\xi \frac{\partial\, h(\xi)}{\partial \xi}\right\}^2 \\
&= \lim_{n\to\infty} \frac{1}{n} \left(\frac{\partial \theta}{\partial \xi}\right)^2 \frac{[E_\theta\{\dot{g}(\theta)\}]^2}{E_\theta\{g^2(\theta)\}} \\
&= \frac{1}{\{\dot{\gamma}(\theta)\}^2} \text{eff}_\theta(g).
\end{aligned} \tag{5.33}$$

But Σ_ξ given by (5.33) is simply the reciprocal of the asmptotic variance (5.32) given by the delta method. It follows from the definition that $\xi^\dagger$ is Godambe efficient for ξ.

5.5.2 SCALAR CASE

An alternative strategy for finding a desirable root when multiple roots exist is to devise iterative algorithms that are robust against initial starting points. In this section and in Sections 5.6 and 5.7, we shall discuss some of the algorithms having this desirable property. These algorithms are modifications to the Newton–Raphson method, utilising the statistical properties of the estimating equations we wish to solve.

For scalar estimating functions, we shall initially consider the following iterative algorithm

$$\theta^{(j+1)} = \theta^{(j)} - \left[1 - \exp\{\gamma\, \dot{g}(\theta^{(j)})\}\right] \frac{g(\theta^{(j)})}{\dot{g}(\theta^{(j)})} \tag{5.34}$$

where γ is a positive constant functionally independent of the data. Note that the iteration (5.34) differs from Newton's method only in the modification of the coefficient $1 - \exp\{\gamma \dot{g}(\theta^{(j)})\}$. When $\gamma = +\infty$ and $\dot{g} < 0$, the algorithm reduces to the usual Newton's method. We shall see that the iterative algorithm defined by (5.34) has some desirable properties over Newton's method.

A basic property concerning the iteration (5.34) is the relation between the roots to estimating equation $g(\hat{\theta}) = 0$ and all possible *fixed points* of the iterative algorithm. A point $\hat{\theta}$ in the parameter space Θ is said to be a fixed point of the algorithm (5.34) if

$$\theta^{(j+1)} = \theta^{(j)} = \hat{\theta}. \tag{5.35}$$

Now suppose that $\hat{\theta}$ solves $g(\hat{\theta}) = 0$. Let $\theta^{(j)} = \hat{\theta}$. Then the second term on the right-hand side of equation (5.34) vanishes regardless of the value of γ. So relation (5.35) is satisfied. In other words, every zero of the estimating fuction $g(\theta)$ is a fixed point of the iterative algorithm (5.34), for every pair (α, γ). On the other hand, suppose that $\hat{\theta} \in \Theta$ satisfy condition (5.35). Putting (5.35) into (5.34) we have

$$[1 - \exp\{\gamma \dot{g}(\hat{\theta})\}] \frac{g(\hat{\theta})}{\dot{g}(\hat{\theta})} = 0.$$

It follows that $g(\hat{\theta}) = 0$, because

$$1 - \exp\{\gamma \dot{g}(\hat{\theta})\} \neq 0.$$

In other words, each fixed point $\hat{\theta}$ of the algorithm (5.34) must be a solution to the estimating equation $g(\hat{\theta}) = 0$.

To distinguish the algorithm (5.34) from Newton's method, we now suppose, for simplicity, that $g(\theta)$ is a score estimating function. We further suppose that regularity conditions such as those given by Wald (1949) hold for the underlying parametric model. Under these ideal conditions we set out the program of looking for the maximum likelihood estimator by solving the likelihood equation $g(\theta) = 0$. Suppose that there are more than one roots to the likelihood equation. Since the log-likelihood function is available, we may then classify the roots into two classes:

- The first class, C_1, consists of all local maxima of the log likelihood function.
- The second class, C_2, consists of all local minima and saddle points of the log likelihood function.

Since we know that the root we are looking for, now the maximum likelihood estimator, is in C_1, we then naturally wish the iterative algorithm (5.34) to behave differently according to whether a fixed point $\hat{\theta}$ is in C_1 or in C_2. The two classes of fixed points C_1 and C_2 can be characterised as follows:

$$C_1 = \{\hat{\theta}: \dot{g}(\hat{\theta}) < 0\}, \quad C_2 = \{\hat{\theta}: \dot{g}(\hat{\theta}) \geq 0\} \tag{5.36}$$

the desirable root being in the first class C_1.

Note that the formula (5.36) now provides a general classification of roots to an estimating equation $g(\theta) = 0$, which is not necessarily a likelihood equation. The formula (5.36) can be generalised in a straightforward manner when $g(\theta)$ is vector-valued. We shall study the multivariate case in the next subsection, where we shall also prove that, under mild regularity conditions, a desriable root will necessarily be in the class C_1.

From the above consideration we see that an iterative algorithm having the following properties is preferred to Newton's method for solving an estimating equation.

1. The iterative algorithm (5.34) will converge to every $\hat{\theta} \in C_1$ starting from a point close enough in Euclidean distance to $\hat{\theta}$. That is, every fixed point in C_1 is an *attracting fixed point.*
2. The iterative algorithm (5.34) will typically diverge for a starting point $\theta^{(0)}$ however close to $\hat{\theta} \in C_2$. That is, every fixed point in C_2 is a *repelling fixed point.*

The following proposition says that these properties hold for the algorithm (5.34).

Proposition 5.9 *Let $g(\theta) = g(\theta, y)$ be a smooth one-dimensional estimating function. Define the following function:*

$$w(\theta) = \theta - [1 - \exp\{\gamma \dot{g}(\theta)\}]\frac{g(\theta)}{\dot{g}(\theta)} \tag{5.37}$$

on the set $\{\theta\colon\ \dot{g}(\theta) \neq 0, \theta \in \Theta\}$, where γ is a positive constant.

Then the iterative algorithm

$$\theta^{(j+1)} = w(\theta^{(j)})$$

will have the following properties for any estimating function $g(\theta)$:

1. *If $\hat{\theta}$ is a root of $g(\hat{\theta}) = 0$, which satisfies $\dot{g}(\hat{\theta}) < 0$, then there exists $\epsilon > 0$ so that*
$$\lim_{j\to\infty} \theta^{(j)} = \hat{\theta}, \quad \text{for any } \theta^{(0)} \in (\hat{\theta} - \epsilon, \hat{\theta} + \epsilon).$$
2. *If $\hat{\theta}$ is a root of $g(\hat{\theta}) = 0$, which satisfies $\dot{g}(\hat{\theta}) > 0$, then for any $\epsilon > 0$ there exists $\theta^{(0)} \in (\hat{\theta} - \epsilon,\ \hat{\theta} + \epsilon)$ so that $\theta^{(j)}$ diverges as $j \to \infty$.*

We shall give a proof in Section 7.4 when we study the dynamical properties of estimating functions. In fact, we will see that the iterative algorithm (5.34) is motivated by considering an appropriate dynamical system induced by an estimating function. Proposition 5.9 says that the following algorithm:

$$\theta^{(j+1)} = \theta^{(j)} - [1 - \exp\{\gamma \dot{g}(\theta^{(j)})\}]\frac{g(\theta^{(j)})}{\dot{g}(\theta^{(j)})}, \quad \gamma > 0 \tag{5.38}$$

has the advantage over the traditional Newton's method in that (5.38) will not converge to any undesirable root defined by the set C_2. Divergence of the algorithm now plays the important role for checking the appropriateness of a root.

The iteration (5.38) still leaves the value of $\gamma > 0$ unspecified. Clearly, if we choose a value of $\gamma > 0$ very close to 0, then the second term on the right-hand side of (5.38) will be close to zero as well, implying that (5.38) will converge to (diverge from) a fixed point $\hat{\theta} \in C_1$ (C_2), at a very slow rate. On the other hand, if $\gamma > 0$ is large enough, then $\exp\{\gamma \dot{g}(\theta^{(j)})\}$ will become negligible if $\theta^{(j)}$ is close to $\hat{\theta} \in C_1$. This implies that, for a large value of $\gamma > 0$, the iterative algorithm (5.38) will behave essentially like the usual Newton's method for any fixed point in C_1. In the meantime if $\theta^{(j)}$ is close to $\hat{\theta} \in C_2$, then the value of $\exp\{\gamma \dot{g}(\theta^{(j)})\}$ will become arbitrarily large if γ is large. This means that the iterative algorithm (5.38) will explode starting from a value at the vicinity of any fixed point in C_2. From these discussions it becomes clear that a large value of γ may be preferred in general to ensure not only the desirable properties stated in Proposition 5.9 but also to achieve the nearly quadratic convergence rate comparable with Newton's method for fixed points in C_1. Since it holds in general that $\dot{g}(\theta) = O_p(n)$, a 'large' value of γ may be set to $\gamma = 1$ if n is large.

Summarising, as an alternative method to the Newton–Raphson iteration, the following algorithm:

$$\theta^{(j+1)} = \theta^{(j)} - [1 - \exp\{\dot{g}(\theta^{(j)})\}] \frac{g(\theta^{(j)})}{\dot{g}(\theta^{(j)})} \tag{5.39}$$

may be considered for use when an estimating equation has more than one solutions. The algorithm (5.39) behaves differently according to whether a fixed point $\hat{\theta}$ is in C_1 or in C_2. Further, the algorithm (5.39) converges to any fixed point in C_1 at approximately quadratic rate when the sample size is large.

The careful reader may have already noticed a property, which holds for Newton's method, now does not hold for the algorithm (5.38) any longer. The Newton–Raphson iteration

$$\theta^{(j+1)} = \theta^{(j)} - \frac{g(\theta^{(j)})}{\dot{g}(\theta^{(j)})}$$

remains unaltered if we replace $g(\theta)$ by

$$h(\theta) = a\, g(\theta) \tag{5.40}$$

where $a \neq 0$ is a constant functionally independent of θ. Or, equivalently, that a root of $g(\hat{\theta}) = 0$ is a repelling or attracting fixed point of Newton's iteration is a property that is invariant under transformation (5.40). This property however no longer holds for the algorithm (5.38), as can be verified by replacing $g(\theta)$ in (5.38) with $h(\theta)$ given by (5.40). As a matter of fact, the iteration (5.38) has been designed deliberately in a way so that this property is destroyed. For instance, if we change the sign of the log-likelihood function, then the maximum likelihood estimator $\hat{\theta}_{\mathrm{ML}}$ will become the minimum of minus the log-likelihood function. Therefore, $\hat{\theta}_{\mathrm{ML}}$ will belong to the class C_2 defined by minus the score estimating function, say, $-s(\theta)$. So the algorithm (5.38) will have different convergence properties for the same fixed point $\hat{\theta}_{\mathrm{ML}}$ according to if we use $s(\theta)$ or $-s(\theta)$. Estimating function $s(\theta)$, now a parametric score, differs from $-s(\theta)$ in that the information identity

$$E_\theta\{g^2(\theta)\} = -E_\theta\{\dot{g}(\theta)\}$$

holds for $g(\theta) = s(\theta)$ but not for $g(\theta) = -s(\theta)$. This property, with other mild regularity conditions, rules out the possibility that a desirable root will belong to C_2. See Proposition 5.10 for a weaker condition than the information identity for general estimating functions to ensure that a desirable root will belong to C_1.

To conclude this section we now briefly discuss the Godambe efficiency of one-step estimators using algorithm (5.39). Suppose that $\tilde{\theta}_n$ is a $\sqrt{n}$ consistent estimator. In last section, we have shown that $\hat{\theta} = \tilde{\theta}_n - g(\tilde{\theta})/\dot{g}(\tilde{\theta}_n)$ is a Godambe efficient estimator. We now show that the one-step estimator

$$\hat{\theta}^* = \tilde{\theta}_n - \{1 - e^{\dot{g}(\tilde{\theta}_n)}\}\frac{g(\tilde{\theta}_n)}{\dot{g}(\tilde{\theta}_n)} \tag{5.41}$$

is also Godambe efficient. We shall prove this fact by comparing $\hat{\theta}^*$ with $\hat{\theta}$.

Suppose that the second derivative of $g(\theta)/n$ is bounded in probability, and that

$$\sigma^2 = -\lim_{n\to\infty} E_{\theta_0}\{n^{-1}\dot{g}(\theta_0)\}$$

exists and is a positive number. From the expansion

$$\begin{aligned}\dot{g}(\tilde{\theta}_n) &= \dot{g}(\theta_0) + \ddot{g}(\theta_n^\dagger)(\tilde{\theta}_n - \theta_0)\\ &= -n\sigma^2 + O_p(n^{1/2})\end{aligned}$$

we see that $\exp\{\dot{g}(\tilde{\theta}_n)\} = o_p(n^{-1})$, which implies that $\exp\{\dot{g}(\tilde{\theta}_n)\}/\dot{g}(\tilde{\theta}_n) = o_p(n^{-2})$. On the other hand, by definition of $\hat{\theta}$ and (5.41) we have

$$\hat{\theta}^* = \hat{\theta} + g(\tilde{\theta}_n)\exp\{\dot{g}(\tilde{\theta}_n)\}/\dot{g}(\tilde{\theta}_n).$$

That $\hat{\theta}^*$ is Godambe efficient follows from the following relation between the estimator $\hat{\theta}^*$ and $\hat{\theta}$:

$$\sqrt{n}(\hat{\theta}^* - \theta_0) = \sqrt{n}(\hat{\theta} - \theta_0) + \{\sqrt{n}g(\tilde{\theta}_n)\}\frac{\exp\dot{g}(\tilde{\theta}_n)}{\dot{g}(\tilde{\theta}_n)} \tag{5.42}$$

$$= \sqrt{n}(\hat{\theta} - \theta_0) + o_p(n^{-1}) \tag{5.43}$$

together with the fact that $\hat{\theta}$ is Godambe efficient. In passing from (5.42) to (5.43) we used the fact that $\sqrt{n}g(\tilde{\theta}_n) = O_p(n)$.

The reader is advised not to read too much into efficiency results such as the one given above. This is because an estimator such as $\hat{\theta}^*$ is usually useful only when one is not sure that an initial estimate $\tilde{\theta}_n$ is close to the target root. In the meantime, a one-step estimators are usually trustworthy only when the initial estimators are close to the right value. These remarks apply to one-step estimators in general.

5.5.3 MULTIVARIATE CASE

The class of algorithms defined by (5.34) can be extended to the multivariate case for a vector-valued estimating function involving a parameter θ of dimension p. The generalised forms, similar to their univariate versions, differ from the multivariate version of Newton's method in that these proposed algorithms favour one type of root over the other. In the univariate case, we defined two classes of roots by (5.36). The algorithm (5.34) with appropriately chosen values of α and γ, such as algorithm (5.39), converges to every root in C_1 with properly chosen starting points.

Now we generalise the classification given by (5.36) to vector-valued estimating functions and give regularity conditions such that a $\sqrt{n}$-consistent estimator will be necessarily contained in the generalised class of C_1. Let $\hat{\theta}$ be a root to an estimating equation $g(\hat{\theta}) = 0$. Let $\kappa_j(\hat{\theta})$ be the complex eigenvalues of $\dot{g}(\hat{\theta})$, where $j = 1, \ldots, p$. In Section 4.9, we introduced generalisations of the concepts of local maxima, local minima and saddle points of a scalar objective function. Recall that a root $\hat{\theta}$ is a sink if $\Re\{\kappa_j(\hat{\theta})\} < 0$ for $j = 1, \ldots, p$; $\hat{\theta}$ is a source if $\Re\{\kappa_j(\hat{\theta})\} \geq 0$ for $j = 1, \ldots, p$, the inequality strictly holding for at least one j; and, finally, $\hat{\theta}$ is a saddle point if there is at least one j and m so that $\Re\{\kappa_j(\hat{\theta})\} > 0$ and $\Re\{\kappa_m(\hat{\theta}\} < 0$. Now (5.36) can be generalised as follows:

$$\begin{aligned} D_1 &= \{\hat{\theta}\colon \quad \Re\{\kappa_j(\hat{\theta})\} < 0, \quad j = 1, \ldots, p\} \\ D_2 &= \{\hat{\theta}\colon \quad \exists j, \Re\{\kappa_j(\hat{\theta})\} > 0\} \end{aligned} \tag{5.44}$$

Proposition 5.10 establishes conditions under which a $\sqrt{n}$-consistent root must belong to D_1. In Section 7.4, under similar conditions, we shall give a different proof using the *Liapunov's method.* In Proposition 5.10 we shall use a weaker condition than the information identity for a score function. A p-dimensional unbiased estimating function $g(\theta) = g(\theta, y)$ is said to be *first-order information unbiased* if $g(\theta)$ satisfies the following condition

$$E_\theta\{g(\theta)g^{\mathrm{t}}(\theta)\} = -E_\theta\{\dot{g}(\theta)\}\{1 + O(n^{-1/2})\}. \tag{5.45}$$

In most cases condition (5.45) can be written in the form

$$E_\theta\{\dot{g}(\theta) + g(\theta)g^{t}(\theta)\} = O(n^{1/2}). \tag{5.46}$$

The equation (5.46) becomes the usual information unbiasedness condition if the right-hand side is exactly zero. In other words, all information unbiased estimating functions are automatically first-order information unbiased.

Proposition 5.10 *Suppose the following conditions are satisfied by a p-dimensional unbiased estimating function* $g(\theta) = g(\theta, y)$*:*

(i) *The estimating function* $g(\theta)$ *is first-order information unbiased.*

(ii) *There exists a root* $\hat{\theta}$ *of* $g(\hat{\theta}) = 0$, *which is* $\sqrt{n}$*-consistent.*

(iii) *The weak law of large numbers can be applied to the quasi-Hessian matrix* $\dot{g}(\theta)$. *That is, if* θ_0 *is the unknown true value of* θ *and*

$$H = -n^{-1} E_{\theta_0}\{\dot{g}(\theta_0)\}$$

then we have that

$$-n^{-1}\dot{g}(\hat{\theta}) = H + o_p(1).$$

(iv) *Finally, the limit*

$$\Sigma = \lim_{n\to\infty} n^{-1} E_{\theta_0}\{g(\theta_0)g^{\mathrm{t}}(\theta_0)\}$$

is assumed to exist and is positive definite.

Then $\hat{\theta}$ *is an asymptotic sink in the sense that there exist constants* a_j *so that* $0 < a_j = O(1)$ *and*

$$\kappa_j(\hat{\theta}) = -a_j n + o_p(n), \quad j = 1, \dots, p$$

where $\kappa_j(\hat{\theta})$ *are eigenvalues of* $\dot{g}(\hat{\theta})$.

Proof. By the first-order information unbiasedness, we have

$$-\frac{1}{n} E_\theta\{\dot{g}(\theta)\} = \frac{1}{n} E_\theta\{g(\theta)g^{\mathrm{t}}(\theta)\} + O(n^{-1/2})$$

which implies that

$$H = \Sigma + O(n^{-1/2}).$$

On the other hand, the weak law of large numbers leads to the following relation:

$$-\frac{1}{n}\dot{g}(\hat{\theta}) = \Sigma + o_p(1). \tag{5.47}$$

Now let a_j be the eigenvalues of the positive definite matrix Σ, then

$$0 < a_j = O(1).$$

Finally, equation (5.47) implies that

$$-n^{-1}\kappa_j(\hat{\theta}) = a_j + o_p(1)$$

or

$$\kappa_j(\hat{\theta}) = -a_j n + o_p(n)$$

which is what was required to prove. □

Now we are ready to extend (5.34). Although parallelling general results can be obtained using α and γ as indices for the family of algorithms, we shall only state the result for the case $(\alpha, \gamma) = (1, 1)$ for brevity's sake. This particularly chosen pair of indices has some desirable properties; see the previous subsection for detailed discussions in the univariate case. The generalised algorithm takes the form

$$\theta^{(j+1)} = \theta^{(j)} - [\mathrm{I} - \exp\{\dot{g}(\theta^{(j)})\}]\{\dot{g}(\theta^{(j)})\}^{-1} g(\theta^{(j)}) \tag{5.48}$$

where I is the $p \times p$ identity matrix and the exponential of a square matrix A is defined by

$$\exp A = \mathrm{I} + A + \frac{1}{2}A^2 + \frac{1}{3!}A^3 + \cdots .$$

Now we extend Proposition 5.9 to the case when $g(\theta)$ is a vector-valued estimating function. In this case, the convergence properties of algorithm (5.48) essentially depend upon the properties of the quasi-Hessian matrix $\dot{g}(\theta)$. The term quasi-Hessian is used here in a rather informal sense. When $g(\theta)$ is not conservative, $\dot{g}(\theta)$ is not a Hessian matrix. In many cases, however, $g(\theta)$ is approximately conservative so that $\dot{g}(\theta)$ has many properties resembling a Hessian matrix in these cases.

Proposition 5.11 *Let $g(\theta) = g(\theta, y)$ be an estimating function of dimension $p \geq 1$. Define*

$$\theta^{(j+1)} = w\{\theta^{(j)}\}$$

$$w(\theta) = \theta - [\mathrm{I} - \exp\{\dot{g}(\theta)\}]\{\dot{g}(\theta)\}^{-1} g(\theta)$$

on the set $\{\theta\colon \det\{\dot{g}(\theta)\} \neq 0, \theta \in \Theta\}$. Let $\hat{\theta}$ be a root of $g(\hat{\theta}) = 0$, and $\kappa_k(\hat{\theta})$, $k = 1, \ldots, p$, be the complex eigenvalues of $\dot{g}(\hat{\theta})$.

1. *If the real parts, $\Re\{\kappa_k(\hat{\theta})\}$, of $\kappa_k(\hat{\theta})$, for any k, are negative, then there is an ϵ-neighbourhood $\mathcal{N}_\epsilon(\hat{\theta})$ of $\hat{\theta}$ such that*

$$\lim_{j\to\infty} \theta^{(j)} = \hat{\theta}, \quad \textit{for any } \theta(0) \in \mathcal{N}_\epsilon(\hat{\theta}).$$

2. *If there exists at least one $\kappa_k(\hat{\theta})$ such that $\Re\{\kappa_k(\hat{\theta})\} > 0$, then $\theta^{(j)}$ will diverge as j tends to infinity for a typical starting point at the vicinity of $\hat{\theta}$.*

A proof will be given in Section 7.4. Proposition 5.11 states that while a sink is a attracting fixed point of (5.48), iterative algorithm (5.48) however will not converge to any source or saddle point for a *typical starting point*. The meaning of a typical starting point will become clear in Chapter 7, where we study dynamical properties of estimating functions. Since, as in the univariate case, two sinks are usually isolated, algorithm (5.48) is therefore robust against the choice of initial starting points.

5.5.4 AN EXAMPLE

To illustrate the algorithms introduced in this section, we now consider the Cauchy location model with the density function

$$f(\theta; y) = \frac{1}{\pi\{1 + (y - \theta)^2\}}.$$

This is a sufficiently regular model so that the maximum likelihood estimator is the efficient likelihood estimator.

Let $\theta_0 = 0$ be the true value of the location parameter θ. For a pseudo random sample of size 5

$$y = (5.7, 4.6, 0.4, 0.1, -1.8)$$

the likelihood equation $s(\theta) = 0$ has three roots

$$\hat{\theta}_1 = 0.2, \quad \hat{\theta}_2 = 3.2, \quad \hat{\theta}_3 = 4.5.$$

While the second root $\hat{\theta}_2$ is a local minimum, the other two roots $\hat{\theta}_1$ and $\hat{\theta}_3$ are local maxima (see Figure 5.1). The maximum likelihood estimator is $\hat{\theta}_1 = 0.2$. The second derivatives (Hessians) of the log likelihood evaluated at the roots are found to be

$$\dot{s}(\hat{\theta}_1) = -3.4, \quad \dot{s}(\hat{\theta}_3) = 0.8, \quad \dot{s}(\hat{\theta}_3) = -1.5.$$

From these values and (5.36), we see that

$$\hat{\theta}_1, \hat{\theta}_3 \in C_1, \quad \hat{\theta}_2 \in C_2.$$

By Proposition 5.9 we see that both $\hat{\theta}_1$ and $\hat{\theta}_3$ are attracting fixed points and $\hat{\theta}_2$ is a repelling fixed point of the iterative algorithm (5.39).

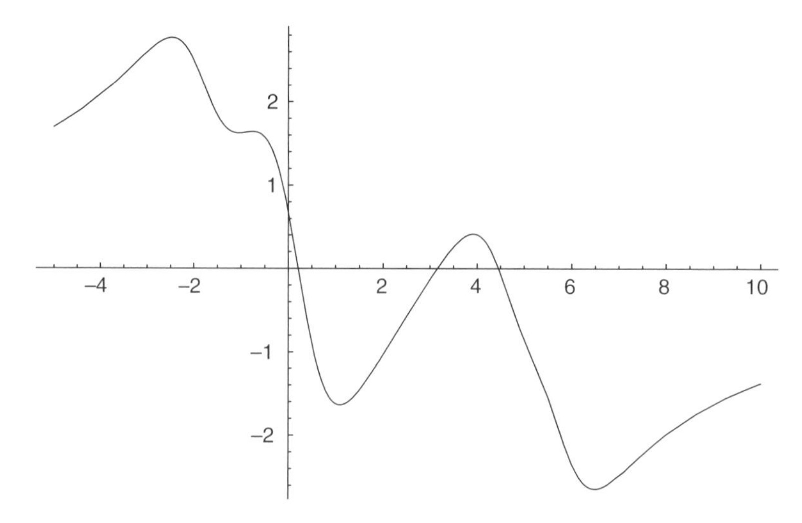

Fig. 5.1 The score function of the Cauchy location-model.

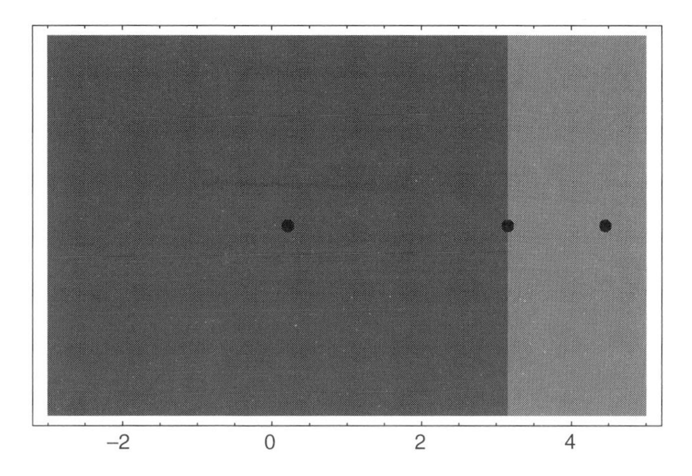

Fig. 5.2 Basins of attraction of the algorithm (5.39) for the Cauchy location model using a sample of size 5. The dots show the three roots to the likelihood equation, the smallest one being the maximum likelihood estimate.

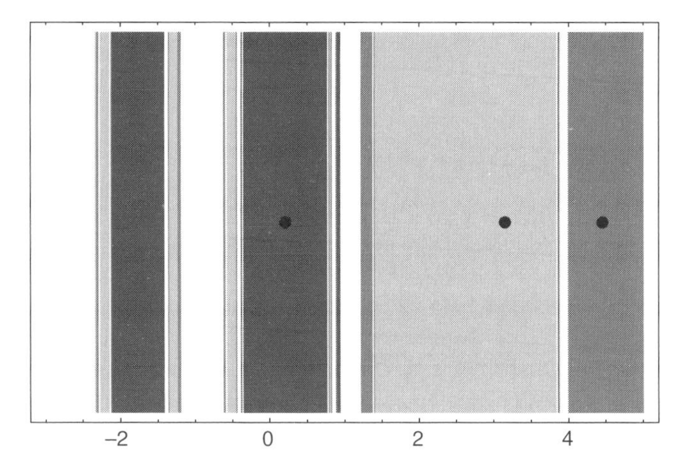

Fig. 5.3 Basins of attraction of Newton's method for the Cauchy location-model using the same data as in Figure 5.2. The dots show the three roots to the likelihood equation, the smallest one being the MLE.

Figures 5.2 and 5.3 show the results obatined by comparing the algorithm (5.39) with Newton's method. Shown in these figures are one-dimensional basins of attraction for the three roots. The basin of attraction of a fixed point $\hat{\theta}$ is the set of starting points which will lead to convergence to $\hat{\theta}$ under a particular iterative algorithm. Different algorithms will therefore have different basins of attraction for the same fixed point. For our present example, the basins of attraction for each algorithm are three disjoint sets in R. Each basin of attraction (connected or not) is represented in Figures 5.2 and 5.3 by a strip or several strips with a gray level different from the other two. We used the darkest gray level to colour strips corresponding to the basins of attraction of the maximum likelihood estimate $\hat{\theta}_1 = 0.2$, and the lightest gray level to colour strips corresponding to the basins of attraction of the second root, the local minimum $\hat{\theta}_2 = 3.2$. Initial points which lead to explosion of the algorithms are coloured white. Both Figures 5.2 and 5.3 used the same colouring convensions.

From Figure 5.2 we see that all initial starting points, in the shown interval, lead to convergence to either the first or the third root, which contrasts sharply with Newton's method shown in Figure 5.3. A notable feature in Figure 5.2 is that a starting point even very close to the local minimum $\hat{\theta}_2$ leads to convergence to either $\hat{\theta}_1$ or $\hat{\theta}_3$. The points to the left of $\hat{\theta}_2$ belong to the basin of attraction of $\hat{\theta}_1$, and the points to the right of $\hat{\theta}_2$ belong to the basin of attraction of $\hat{\theta}_3$. These properties are absent in Newton's method because $\hat{\theta}_2$ is also an attracting fixed point as can be seen from Figure 5.3. As a matter of fact, Figure 5.3 shows that many starting points (strips with the lightest gray level) converge to the local minimum of the log likelihood function. Figures 5.2 and 5.3 also seem to suggest that algorithm (5.39) has a much wider basin of attraction for the maximum likelihood estimator than that of the usual Newton's method. This implies that algorithm (5.39) may also have the desirable property of being more robust against the choice of initial starting points.

5.6 An algorithm based on the information identity

The modified Newton's methods discussed in the previous section are based on the idea that roots to an estimating equation $g(\hat{\theta}) = 0$ may be classified according to (5.44) and the fact that a desirable root $\hat{\theta}$, when the sample size gets alrge, must be contained in D_1 as defined by (5.44). At the same time, all roots belonging to D_2 are known to be undesirable, provided regularity conditions such as those given by Proposition 5.10 can be invoked. These methods will avoid undesirable convergence to sources and saddle points, which are inconsistent under mild conditions. However, convergence to sinks other than the consistent root remains a possibility. From Figure 5.2 we see that the modified iterative algorithm (5.39) converges to the undesirable local maximum (the black dot on the right) for starting points close enough to this root. To overcome this difficulty, in this section we shall consider another modification of Newton's method.

5.6.1 THE MODIFIED ALGORITHM

In Section 5.2, we discussed the general problem of estimation based on a regular estimating function. A regular estimating function is one which is subject to certain moment assumptions. There we emphasised the fact that the problem of solving an estimating equation, when difficulties arise, may necessitate the evaluation of other properties of the estimating function besides the unbiasedness condition. The iterative algorithm we shall consider in this section takes into account both the first and the second moment conditions of an estimating function. A general procedure along this line was given by (5.4).

We consider the scalar case first. Let $g(\theta)$ be an unbiased additive estimating functions. The following algorithm:

$$\theta^{(j+1)} = \theta^{(j)} + \frac{g(\theta^{(j)})}{v(\theta^{(j)})} \tag{5.49}$$

gives an iterative method, which is motivated by considerations such as the procedure formulated in (5.4). In (5.49), the data-dependent function $v(\theta)$ is assumed to satisfy the following condition:

$$v(\theta) = E_\theta \{g^2(\theta)\} + o_p(n). \tag{5.50}$$

The formula (5.50) says that $v(\theta)$ is a quantity which gives an estimator for the leading term of the variance of $g(\theta)$. Again the algorithm (5.49) is a modification to Newton's method. If we replace the coefficient of $g(\theta^{(j)})$ on the right-hand side of (5.49), i.e. $\{v(\theta^{(j)})\}^{-1}$, with the quantity $-\{\dot{g}(\theta^{(j)})\}^{-1}$, then we recover Newton's method. The major differences between algorithm (5.49) and Newton's method can thus be brought about by comparing the two quantities $v(\theta)$ and $-\dot{g}(\theta)$.

Now suppose that $\hat{\theta}$ is a solution to $g(\hat{\theta}) = 0$, which is a $\sqrt{n}$-consistent estimator of θ. We know that $\hat{\theta}$ is an attracting fixed point of Newton's method. From Proposition 5.9 we also know that $\hat{\theta}$ is an attracting fixed point of (5.38). We also wish the new algorithm (5.49) to converge to $\hat{\theta}$ with initial starting points close enough to $\hat{\theta}$. This will be true if we have

$$\lim_{n\to\infty} \frac{\dot{g}(\hat{\theta})}{v(\hat{\theta})} = -1 \tag{5.51}$$

so that, when the sample size is large, the iterations defined by (5.49) will become indistinguishable from the Newton–Raphson iterations with starting points in a proper neighbourhood of $\hat{\theta}$.

To see when (5.51) will be true, we first note that, by the law of large numbers, we have

$$\lim_{n\to\infty} \frac{\dot{g}(\hat{\theta})}{E_\theta\{\dot{g}(\theta)\}} = \lim_{n\to\infty} \frac{\dot{g}(\hat{\theta})/n}{E_\theta\{\dot{g}(\theta)/n\}} = 1 \tag{5.52}$$

provided that $\hat{\theta}$ is consistent and $g(\theta)$ satisfies some mild regularity conditions under which the law of large numbers may be applied in the above form. If we may further assume that $g(\theta)$ is approximately information unbiased so that $g(\theta)$ satisfies

$$E_\theta\{g^2(\theta)\} = -E_\theta\{\dot{g}(\theta)\}\{1 + O(n^{-1/2})\}$$

then we may deduce from (5.52) that

$$\lim_{n\to\infty} \frac{\dot{g}(\hat{\theta})}{E_\theta\{g^2(\theta)\}} = -1. \tag{5.53}$$

Finally, by (5.53) and (5.50), and noticing the fact that $\dot{g}(\theta)$ is usually of $O_p(n)$, we then conclude that (5.51) will be generally true when $\hat{\theta}$ is consistent and the mild regularity conditions imposed on $g(\theta)$ in the above discussions hold true.

The convergence to a desirable root is a basic requirement of an algorithm. How will the algorithm (5.49) behave for other roots of the estimating eqation? As in Section 5.5, here we also wish the iterative algorithm (5.49) to respond in a sensitive manner according to whether a root $\hat{\theta}$ is desirable or not. One thing, which is clear from the above discussions, is that iterations under (5.49) will be different from Newton's method for a root at which (5.51) is violated. Actually, we can say much more than this.

Proposition 5.12 *Let $g(\theta)$ be an unbiased and first-order information unbiased estimating function. Suppose that there exists a consistent estimator $\hat{\theta}$ of θ_0 so that $g(\hat{\theta}) = 0$. Let $v(\hat{\theta})$ be any consistent estimator of the variance of $g(\theta_0)$ such that the following is satisfied*

$$v(\hat{\theta}) = E_{\theta_0}\{g^2(\theta_0)\} + o_p(n)\,.$$

Then iterative algorithm (5.49) has the following properties.

A solution $\tilde{\theta}$ to estimating equation $g(\tilde{\theta}) = 0$ is an attracting fixed point of (5.49) if and only if the following condition:

$$-2 < \frac{\dot{g}(\tilde{\theta})}{v(\tilde{\theta})} < 0 \tag{5.54}$$

is satisfied.

We shall give a proof in Section 7.4. From Proposition 5.12 we immediately have the following results.

Corollary 5.13 *Suppose that the conditions of Proposition 5.12 are satisfied. Then we have*

(i) *A consistent root $\hat{\theta}$ of $g(\hat{\theta}) = 0$ is an attracting fixed point of (5.49) when n is large.*

(ii) *If $\tilde{\theta}$ is a root of $g(\tilde{\theta}) = 0$ and $\dot{g}(\tilde{\theta}) \geq 0$, then $\tilde{\theta}$ is a repelling fixed point of (5.49).*

Corollary 5.13 can be verified by directly checking the condition (5.54). Comparing Proposition 5.10 with Proposition 5.12 and Corollary 5.13, we see that algorithm (5.49) has the same desirable properties as algorithm (5.38) over Newton's method. That is, while any consistent root of an estimating equation is an attracting fixed point of algorithm (5.49), sources and saddle points are repelling fixed points of the algorithm. In addition, algorithm (5.49) has an important advantage over algorithm (5.38) in that iterations under (5.49) will not converge to any sink $\theta^\dagger$ which violates the condition (5.54).

To use algorithm (5.49) we need to specify $v(\theta)$. So far we have not discussed the issue of constructing an appropriate estimate of variance for $g(\theta)$. For this purpose, we shall now assume that $g(\theta, y)$ is an unbiased additive estimating function. That is, we assume that $g(\theta)$ takes the form

$$g(\theta, y) = \sum_{j=1}^{n} h_j(\theta, y_j)$$

where h_j satisfies

$$E_\theta \{h_j(\theta, Y_j) = 0\}, \quad j = 1, \ldots, n$$

and the Y_j's are independent. Now we can define $v(\theta)$ as follows:

$$v(\theta) = \sum_{j=1}^{n} h_j^2(\theta, y_j). \tag{5.55}$$

Another possibility is to let $v(\theta)$ to be

$$v(\theta) = \sum_{j=1}^{n} (h_j - \bar{h})^2, \quad \text{where } \bar{h} = n^{-1} \sum h_j(\theta, y_j). \tag{5.56}$$

Now we turn to consider the case when $g(\theta)$ is a vector-valued estimating function. Proposition 5.12 can be extended in this case as follows. Again see Section 7.4 for a brief derivation of the results.

Proposition 5.14 *Suppose that $g(\theta)$ is a p-dimensional unbiased estimating function. Further suppose that $g(\theta)$ is first-order information unbiased satisfying the condition (5.45). Let $\hat{\theta}$ be a root of $g(\hat{\theta}) = 0$, which is a consistent estimator of θ_0. Assume that $v(\hat{\theta})$ be any consistent estimator of the variance–covariance matrix of $g(\theta_0)$, i.e.,*

$$v(\hat{\theta}) = E_{\theta_0}\{g(\theta_0) g^{\mathrm{t}}(\theta_0)\} + o_p(n).$$

Then the iterative algorithm

$$\theta^{(j+1)} = \theta^{(j)} + \{v(\theta^{(j)})\}^{-1} \dot{g}(\theta^{(j)}) \tag{5.57}$$

has following properties:

(i) The consistent root $\hat{\theta}$ of $g(\theta)$ is an attracting fixed point of (5.57) when the sample size is large.

(ii) For any root $\tilde{\theta}$, let λ_j be the eigenvalues of

$$\{v(\tilde{\theta})\}^{-1} \dot{g}(\tilde{\theta})$$

where $j = 1, \ldots, p$. *If there exists a* k *such that* $|1 + \lambda_k| \geq 1$, *then* $\tilde{\theta}$ *is a repelling fixed point of (5.57).*

Again, as in the univariate case, when $g(\theta)$ is an unbiased additive estimating function

$$g(\theta, y) = \sum_{j=1}^{n} h_j(\theta, y_j)$$

we can choose $v(\theta)$ to be of the form

$$v(\theta, y) = \sum_{j=1}^{n} h_j(\theta, y_j) h_j^{\mathrm{t}}(\theta, y_j)$$

where each $h_j(\theta, y_j)$ is an unbiased p-dimensional function and the y_j's are independent.

5.6.2 AN EXAMPLE

Now we apply algorithm (5.49), with the choice of $v(\theta)$ given by (5.55), to the Cauchy location model with density function

$$f(\theta; y) = \frac{1}{\pi\{1 + (y - \theta)^2\}}.$$

We assume that the true value of the parameter is $\theta_0 = 0$. We generated 20 pseudo random numbers from this model. These values, arranged in ascending order, are

−393.16	−73.20	−10.85	−1.66	−1.64	−1.25	−1.08	−0.85	−0.53	−0.44
0.20	0.24	0.36	0.54	0.67	0.89	1.08	1.12	2.05	58.21

Using these pseudo observations, we found seven roots

$$-393.11, \quad -373.69, \quad -73.74, \quad -70.25, \quad 0.07, \quad 55.38, \quad 57.88$$

to the likelihood equation. Figure 5.4 shows the score function near these roots.

Evaluating the quantity

$$\{v(\tilde{\theta})\}^{-1}\dot{g}(\tilde{\theta})$$

at these roots gives the values

$$-199.05, \quad 0.50, \quad -6.91, \quad 0.47, \quad -0.69, \quad 0.44, \quad -3.74$$

respectively. Note that all values, except the value -0.69, which corresponds to the root $\hat{\theta} = 0.07$, are outside the stability interval $(-2, 0)$ given by (5.54). It follows therefore that $\hat{\theta} = 0.07$ is the only attracting fixed point of algorithm (5.49).

In contrast, the Newton–Raphson method is very sensitive to initial starting points. The poor convergence property of the Newton–Raphson algorithm to the

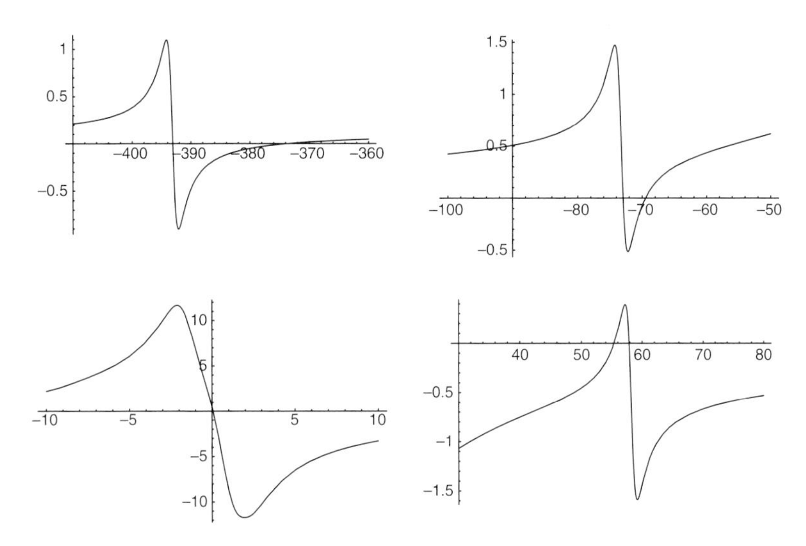

Fig. 5.4 The Cauchy score function near the roots for 20 pseudo random observations.

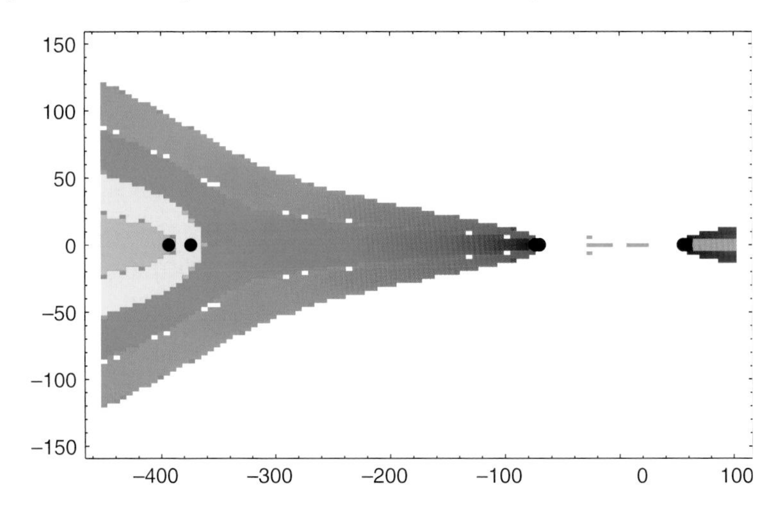

Fig. 5.5 Basins of attraction in the complex plane of Newton's method for the Cauchy location-model.

desirable root $\hat{\theta} = 0.07$ can be clearly seen from Figure 5.5. Since Newton's method is defined for complex values of $\theta^{(j)}$ as well, in Figure 5.5 we plotted the basins of attraction to each root using complex starting points in the square

$$\{a + b\sqrt{-1}: a \in [-400, 100], b \in [-150, 150]\}.$$

In Figure 5.5, the Newton–Raphson iterations were run 100 times for each starting point to determine its convergence property. The starting points converging to the desirable root, $\hat{\theta} = 0.07$, are plotted as red. From Figure 5.5 we see that the only region where the algorithm converges to the desirable root is the thin strip near the origin. All other shaded regions correspond to basins of attraction for other real roots. Basins of attraction for all complex roots are amalgamated as black, while white points denote non-convergence with respect to the stopping rule used. An equivalent plot for the algorithm (5.49), which is not shown here, shows that almost all complex initial starting points in the region $[-800, 800] \times [-800, 800]$ lead to convergence to the desirable root, $\hat{\theta} = 0.07$.

By plotting the domains of attraction in the complex plane, rather than solely along the real line, a much clearer picture of the behaviour of the various algorithms is obtained. So it seems useful sometimes to extend the study of root search algorithms into the complex plane.

5.7 A modified Muller's method

In this section, we discuss another iterative algorithm that is a modification of Muller's method introduced in Chapter 3. Muller's method does not take into account the statistical properties of the equation that is required to be solved. Similar to the algorithms discussed in the previous two sections, the method we shall discuss in this section is an alternative way to solve an estimating equation using the information available to us on the underlying estimating function. Since higher-order moment specifications on estimating functions are often impractical, the essential idea is again as formulated in (5.4).

Let $g(\theta, y) \in \mathbb{R}^p$ be an unbiased estimating function. Suppose that $g(\theta, y)$ is information unbiased, so that the quantity

$$\mathcal{D}(\theta_1, \theta_2) = E_{\theta_1}\{g(\theta_2, Y)g^{\mathrm{t}}(\theta_2, Y) + \dot{g}(\theta_2, Y)\} \tag{5.58}$$

is a zero matrix for any $\theta_1 = \theta_2 = \theta \in \Theta$, that is,

$$\mathcal{D}(\theta, \theta) = 0_{p\times p}, \quad \forall \theta \in \Theta$$

where $0_{p\times p}$ is a $p \times p$ matrix with all entries being zero. For $\theta_1 \neq \theta_2$, we generally have

$$\mathcal{D}(\theta_1, \theta_2) \neq 0_{p\times p}, \quad \text{if } \theta_1 \neq \theta_2.$$

It is this property which makes the quantity $\mathcal{D}(\theta_1, \theta_2)$ a useful diagnostic tool.

For our present purpose we shall set the value $\theta_1 = \theta_0$, where θ_0 be the true value of the parameter. When this is done, (5.58) is a function of θ_2 along. For convenience, we shall use the same symbol $\mathcal{D}$ to define

$$\mathcal{D}(\theta) = E_{\theta_0}\{g(\theta, Y)g^{\mathrm{t}}(\theta, Y) + \dot{g}(\theta, Y)\} \tag{5.59}$$

which will be referred to as the *information divergence* from now on. The information divergence (5.59) provides a measure of discrepancy between a value θ and the true value of the parameter. If θ is far away from θ_0 in Euclidean distance then we also expect that $\mathcal{D}(\theta)$ will be different from the zero matrix, and vice versa.

To use $\mathcal{D}(\theta)$ for diagnostic purpose, we need to estimate the right-hand side of (5.59) for a given value of θ using the available sample informations. If we may assume that g is an unbiased additive estimating function

$$g(\theta, y) = \sum_{j=1}^{n} h_j(\theta, y_j)$$

where each $h_j(\theta, y_j)$ is unbiased and the y_j's are independent, then the following quantity

$$\hat{\mathcal{D}}(\theta) = \sum_{j=1}^{n} \{h_j(\theta, y_j) h_j^{\mathrm{t}}(\theta, y_j) + \dot{h}_j(\theta, y_j)\} \tag{5.60}$$

gives a consistent estimator for $\mathcal{D}(\theta)$. Note that this will be true even if the y's are not identically distributed, provided mild regularity conditions on the variances of these variables hold true. We shall call (5.60) the *empirical information divergence*.

With aid of (5.60) we can compare the plausibility of roots to an estimating equation $g(\hat{\theta}) = 0$ by checking the deviation of $\hat{\mathcal{D}}(\hat{\theta})$ from the ideal zero matrix. For this criterion to work we require that $\hat{\mathcal{D}}(\hat{\theta})$ can be made as close to the zero matrix as we wish if $\hat{\theta}$ is consistent and the sample size can get arbitrarily large. That the random matrix $\hat{\mathcal{D}}(\hat{\theta})$ will tend to zero as the sample size gets large can be appreciated by applying the law of large numbers to the right-hand side of

$$\hat{\mathcal{D}}(\hat{\theta}) = \sum_{j=1}^{n} \{h_j(\hat{\theta}, y_j) h_j^{\mathrm{t}}(\hat{\theta}, y_j) + \dot{h}_j(\hat{\theta}, y_j)\}$$

and then using the assumption that $g(\theta)$ is information unbiased.

Note that an equivalent criterion to $\hat{\mathcal{D}}(\theta)$ of (5.60) is to measure how far the following matrix

$$\hat{\mathcal{G}}(\theta) = \left[\sum_{j=1}^{n} \{h_j(\theta, y_j) h_j^{\mathrm{t}}(\theta, y_j)\}\right]^{-1} \left[-\sum_{j=1}^{n} \{\dot{h}_j(\theta, y_j)\}\right] \tag{5.61}$$

is from the $p \times p$ identity matrix. The population counterpart of $\hat{\mathcal{G}}(\theta)$ is given by

$$\mathcal{G}(\theta) = [E_{\theta_0}\{g(\theta, Y) g^{\mathrm{t}}(\theta, Y)\}]^{-1} [-E_{\theta_0}\{\dot{g}(\theta, Y)\}]. \tag{5.62}$$

The assumption that $g(\theta)$ is information unbiased implies that $\mathcal{G}(\theta_0)$ is the identity matrix. Moreover, if $\hat{\theta}$ is consistent for θ_0, then $\hat{\mathcal{G}}(\hat{\theta})$ will tend to the identity matrix for the same reasons that $\hat{\mathcal{D}}(\hat{\theta})$ will tend to the zero matrix.

One way for checking the closeness of $\hat{\mathcal{G}}(\theta)$ to the identity matrix is by investigating the eigenvalues of $\hat{\mathcal{G}}(\theta)$. Suppose that $\hat{\theta}$ is a $\sqrt{n}$-consistent estimator of θ_0. Then we can show (Small *et al.*, 2000, Section 4.4) that

$$\lambda_j = 1 + O_p(n^{-1/2}) \tag{5.63}$$

where the λ_js are eigenvalues of

$$\hat{\mathcal{G}}(\hat{\theta}) = \hat{\Sigma}\hat{H}^{-1}$$

and $\hat{\Sigma}$ and $\hat{H}$ are defined as

$$\hat{\Sigma} = \sum_{j=1}^{n}\{h_j(\hat{\theta}, y_j)h_j^{\mathrm{t}}(\hat{\theta}, y_j)\}, \quad \hat{H} = \sum_{j=1}^{n}\{\dot{h}_j(\hat{\theta}, y_j)\}.$$

The above discussions suggest a modification to Muller's method for solving a non-linear estimating equation. Muller's method makes a quadratic fit at each step using the previous three values, $\hat{\theta}^{(j-k)}$, $k = 1, 2, 3$. The next value $\hat{\theta}^{(j)}$ is the one of the two solutions to a quadratic equation. The value chosen for $\hat{\theta}^{(j)}$ is usually the closer of the two solutions to $\hat{\theta}^{(j-1)}$. In our modification of Muller's method, *we propose choosing that solution for which the empirical information divergence (5.60) is closer to the zero matrix*. Or, equivalently, $\mathcal{G}(\theta)$ of (5.62) is closer to the identity matrix. In the scalar case, the closeness to zero may be measured by its absolute value. For higher dimensional estimating functions a criterion for selecting the value from each fit is no longer obvious. A possible answer is to use the result given by (5.63) as a guide to choose the root from each quadratic fit.

To compare this method with Muller's method, we give the results of a small numerical experiment, again using the Cauchy location-model with the true value of the location parameter set to be $\theta_0 = 0$. For 11 pseudo random observations

$$0.255,\ -13.276,\ 16.996,\ 0.285,\ -1.043,\ -1.680,\ 1.541,$$
$$-0.843,\ 4.684,\ 14.630,\ 0.166$$

there are three roots

$$\hat{\theta}_1 = -0.024,\ \hat{\theta}_2 = 13.026,\ \hat{\theta}_3 = 14.302$$

to the likelihood equation. While $\hat{\theta}_2$ is a local minimum, the other two roots are local maxima. The first root $\hat{\theta}_1 = -0.024$ is the maximum likelihood estimate. The corresponding values of the empirical information divergence are

$$\hat{\mathcal{D}}(\hat{\theta}_1) = -0.022,\ \hat{\mathcal{D}}(\hat{\theta}_2) = 18.876,\ \hat{\mathcal{D}}(\hat{\theta}_3) = -2.475$$

respectively. The MLE, as expected, has almost zero empirical information divergence.

Table 5.1 reports on the use of Muller's method and the modified method based on the empirical information divergence criterion for finding these roots. The results of first few iterations using the following two sets of random starting points are shown in Table 5.1:

$$\text{case } 1 = (7.837,\ -0.253,\ 7.267)$$
$$\text{case } 2 = (8.765,\ -0.644,\ 2.075).$$

Table 5.1 Iterations using Muller's method and the modified Muller's method based on the empirical information divergence. There are three roots −0.024, 3.026 and 14.302 to the Cauchy likelihood equation for an artificial sample of size 11. Two sets (case 1 and case 2) of random starting points are used. The true value is $\theta_0 = 0$.

Modified method			Muller's method		
Case 1					
7.837	−0.253	7.267	7.837	−0.253	7.267
0.541	7.837	−0.253	10.252	7.837	−0.253
−0.068	0.541	7.837	11.917	10.252	7.837
−0.030	−0.068	0.541	13.360	11.917	10.252
−0.024	−0.030	−0.068	13.044	13.360	11.917
−0.024	−0.024	−0.030	13.025	13.044	13.360
−0.024	−0.024	−0.024	13.026	13.025	13.044
—	—	—	13.026	13.026	13.025
—	—	—	13.026	13.026	13.026
Case 2					
8.765	−0.644	2.075	8.765	−0.644	2.075
0.082	8.765	−0.644	9.380	8.765	−0.644
−0.067	0.082	8.765	11.744	9.380	8.765
−0.027	−0.067	0.082	13.465	11.744	9.380
−0.024	−0.027	−0.067	13.052	13.465	11.744
−0.024	−0.024	−0.027	13.024	13.052	13.465
−0.024	−0.024	−0.024	13.026	13.024	13.052
—	—	—	13.026	13.026	13.024
—	—	—	13.026	13.026	13.026

For the cases shown there, the modified method not only converges to the desirable root but also seems to converge fast. We note however that one needs to be careful in using this method when complex numbers are encountered. Although the empirical information divergence is well defined for complex numbers as well, the interpretation of this measure is then less straightforward. A possible idea, when a complex value, say $a + bi$, arises, is to stop iterating and restart the algorithm with new initial values. Or, one may replace $a + bi$ by the real part a to continue iterating. The convergence properties of the latter procedure would, however, be quite sophisticated.

5.8 Asymptotic examinations

An estimating function $g(\theta, y)$ is a function of the data y and the parameter of interest θ. It is sometimes possible to write $g(\theta, y)$ in a form

$$g(\theta, y) = G(\hat{F}|\theta), \tag{5.64}$$

where $\hat{F}$ is the empirical distribution function and $G(\cdot|\theta)$ is a functional of $\hat{F}$ indexed by the parameter θ. Let F be the true distribution function, and θ_0 be the true value of θ. Applying the central limit theorem to the binomial variable $n\hat{F}$, we have that

$$\sqrt{n}\{\hat{F}(t) - F(t)\} \xrightarrow{\mathcal{L}} N(0,\ F(t)\{1 - F(t)\}).$$

It therefore follows that an estimating function possessing the structure (5.64) will asymptotically tend to a limit depending on F

$$g(\theta, Y) \longrightarrow G(F|\theta) \tag{5.65}$$

provided G is a continuous functional. Note that the right-hand side of (5.65) vanishes if we set $\theta = \theta_0$, that is

$$G(F|\theta_0) = 0$$

which follows from the definition of unbiasedness of $g(\theta, y)$.

When an estimating function $g(\theta, y)$ has multiple roots it often happens that the limit $G(F|\theta)$ will also have more than one root. When this is the case, as suggested by Heyde and Morton (1998), it may be helpful to examine the structure of the asymptotic equation

$$G(F|\theta) = 0 \tag{5.66}$$

in order to shed light on discriminating the multiple roots arising from finite sample configurations. This works provided of course the sample size is large enough.

To illustrate how this idea may be used we consider the problem of estimating the mean of a non-normal distribution. This example is discussed in detail by Heyde (1997) and Heyde and Morton (1998). See the next section for a different approach.

Example 5.15 Estimation of the mean of a non-normal distribution

Let $Y_1, \ldots, Y_n$ be *i.i.d.* with unknown mean θ and known variance σ^2. Suppose that we wish to use the following estimating function

$$g(\theta, Y) = \sum_{j=1}^{n} [a(Y_j - \theta) + \{(Y_j - \theta)^2 - \sigma^2\}] \tag{5.67}$$

to estimate θ, where a is a given constant.

Let θ_0 be the true value of θ. Rewrite (5.67) as follows:

$$g(\theta, Y) = n\{(\theta_0 - \theta)^2 + a(\theta_0 - \theta)\} + n(S_n^2 - \sigma^2) + n\{a + 2(\theta_0 - \theta)\}\bar{Y} \tag{5.68}$$

where

$$\bar{Y} = \frac{1}{n}\sum_{j=1}^{n}(Y_j - \theta_0) \quad S_n^2 = \frac{1}{n}\sum_{j=1}^{n}(Y_j - \theta_0)^2.$$

By the law of large numbers, $\bar{Y} \to 0$ and $S_n^2 \to \sigma^2$, so the second and the third terms of (5.68) vanish asymptotically. It follows then that estimating function (5.68), ignoring a factor of n, asymptotically will take the form

$$\begin{aligned} G(F|\theta) &= (\theta_0 - \theta)^2 + a(\theta_0 - \theta) \\ &= (\theta - \theta_0)[\theta - (\theta_0 + a)]. \end{aligned}$$

Setting $G(F|\theta) = 0$, we get two asymptotic roots,

$$\tilde{\theta}_1 = \theta_0, \quad \tilde{\theta}_2 = \theta_0 + a \tag{5.69}$$

the second one being the wrong root. Thus we may conclude that, when sample size is large enough, the solution that exceeds the other one by an amount close to a is the undesirable root.

Now solving $g(\theta, Y) = 0$ for a fixed sample size gives two roots

$$\hat{\theta}_1 = \bar{Y} + \tfrac{1}{2}a - \Delta, \quad \hat{\theta}_2 = \bar{Y} + \tfrac{1}{2}a + \Delta$$

where

$$\Delta = \sqrt{\bar{Y}^2 + \frac{1}{4}a^2 - \frac{1}{n}\sum_{j=1}^{n} Y_j^2 + \sigma^2}.$$

That $\hat{\theta}_1$ is preferred to $\hat{\theta}_2$ as an estimator of θ follows from the fact that:

$$\hat{\theta}_2 - \hat{\theta}_1 = 2\Delta \to a.$$

Note that in the above example the solutions are available analytically. So we could have examined the asymptotic behaviour of the roots directly in order to judge the consistency of the roots. The approach based on examining the asymptotic equation (5.66) is useful because equation (5.66) is often much simpler to analyse even when the finite sample version of the estimating equation is difficult to solve. When explicit analytic solutions are available, however, it may be more straightforward to have an analysis based on the algebraic formulae. The following example illustrates this point in the case of estimating the correlation coefficient. In the analyses of this example we will also point out that one needs to avoid potential pitfalls stemming from the algebraic formulae.

Example 5.16 Correlation coefficient of bivariate normal distribution

Explicit algebraic formulae exist if the estimating function can be reduced to a polynomial of degree $d \leq 4$. This is the case when we wish to estimate the correlation coefficient using the cubic estimating function of (4.1). As the estimating equation is of degree three, we can find explicit formulas for the three roots using

Cardano's formula:

$$\hat{\rho}_1 = \tfrac{1}{6}\, A_1 - 6\, A_2 + \tfrac{1}{3}\, S_1 \tag{5.70}$$

$$\hat{\rho}_2 = \frac{-1 + \sqrt{-3}}{12}\, A_1 + 3\,(1 + \sqrt{-3})\, A_2 + \frac{1}{3}\, S_1 \tag{5.71}$$

$$\hat{\rho}_3 = \frac{-1 - \sqrt{-3}}{12}\, A_1 + 3\,(1 - \sqrt{-3})\, A_2 + \frac{1}{3}\, S_1 \tag{5.72}$$

where

$$S_1 = \frac{1}{n} \sum_{j=1}^{n} x_j y_j$$

$$S_2 = \frac{1}{n} \sum_{j=1}^{n} (x_j^2 + y_j^2)$$

$$A_1 = 12\sqrt{B} + 144\, S_1 - 36\, S_1\, S_2 + 8\, {S_1}^3$$

$$A_2 = \frac{1}{9A_1}(-3 + 3S_2 - S_1^2)$$

and B is given by

$$B = -12 + 36S_2 + 132{S_1}^2 - 36{S_2}^2 - 48S_2{S_1}^2 + 12{S_1}^4 + 12{S_2}^3 - 3{S_2}^2{S_1}^2.$$

To investigate the asymptotics of the root associated with each formula we can apply the limits $S_1 \to \rho$, $S_2 \to 2$, as $n \to \infty$. After some tedious algebra, we find that $\hat{\rho}_1 \to \rho$. The roots $\hat{\rho}_2$ and $\hat{\rho}_3$ converge to $+\sqrt{-1}$ and $-\sqrt{-1}$, respectively, as $n \to \infty$. It follows that in most cases $\hat{\rho}_1$ is the root we want.

As every cubic has at least one real root, it is tempting to suppose that $\hat{\rho}_1$ is real. However, the situation is not that simple. For example, when $S_1 = -0.1$ and $S_2 = 1.0$ then we have

$$\hat{\rho}_1 \approx 0.2 - 0.4\sqrt{-1}.$$

An additional problem with $\hat{\rho}_1$ is that unlike the MLE, $\hat{\rho}_1$ is *not equivariant* with respect to reflections of the data through the x- and y-axes. To be equivariant, an estimator $\hat{\rho} = \hat{\rho}(S_1, S_2)$ should satisfy the equation

$$\hat{\rho}(-S_1, S_2) = -\hat{\rho}(S_1, S_2).$$

However this is not the case for $\hat{\rho}_1$. This problem arises because Cardano's method is not itself equivariant under these reflections (see Figure 5.6).

So even when it is possible to find an analytic formula for a consistent root of an estimating equation it is naive to assume that this formula is the obvious estimator. Since the uniqueness of the consistent root is an asymptotic result, two consistent root selection mechanisms may have radically different properties for a given sample size.

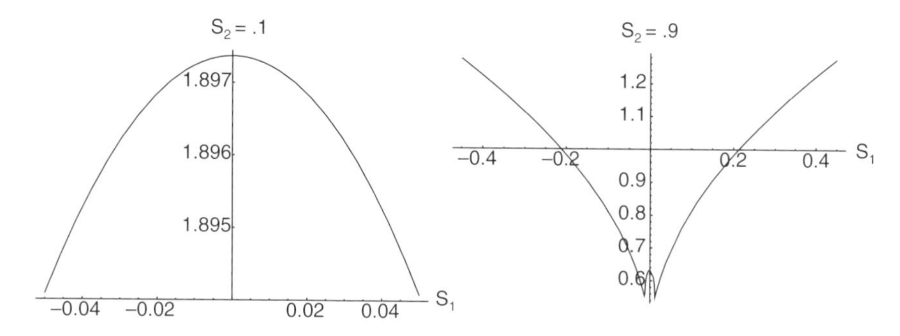

Fig. 5.6 Plots of $\Re[\hat{\rho}_1(S_1, S_2) + \hat{\rho}_1(-S_1, S_2)]$ for two values of S_2.

5.9 Testing the consistency of roots

Suppose that there are roots $\hat{\theta}_1, \ldots, \hat{\theta}_m$ to an estimating equation $g(\theta) = 0$. The problem of choosing a correct root may be formally studied as one of testing the consistency for each root $\hat{\theta}_i$. In other words, we wish to consider testing the null hypotheses

$$H_0^{(i)}: \; \hat{\theta}_i \text{ is } \sqrt{n}\text{-consistent}$$

for $i = 1, \ldots, m$. The alternative to $H_0^{(i)}$ is that $\hat{\theta}_i$ is not $\sqrt{n}$-consistent. In order to make this hypothesis meaningful, we assume that the root $\hat{\theta}_i$ is defined simultaneously for all sample sizes, so that its asymptotic properties can be examined. To select a root we would find that root $\hat{\theta}_i$ at which a test statistic is least significant.

Heyde (1997) and Heyde and Morton (1998) have proposed two methods for testing the consistency of roots. These methods are

1. picking the root $\hat{\theta}_i$ for which $\dot{g}(\theta)$ behaves asymptotically like $E_\theta\, \dot{g}(\theta)$ when evaluated at $\theta = \hat{\theta}_i$; and
2. using a least squares or goodness-of-fit criterion to select the best root.

Both of these methods can be interpreted as selecting a root using a test statistic at each root. In the first case, the hypothesis

$$H_1: \dot{g}(\theta)\{E_\theta \dot{g}(\theta)\}^{-1} \to \mathrm{I}_{k\times k}$$

where $\mathrm{I}_{k\times k}$ is the $k \times k$ identity matrix, is examined at $\theta = \hat{\theta}_1, \ldots, \hat{\theta}_m$ to determine at which root the test seems to be least significant. A test statistic for H_1 is based upon a comparison of $\dot{g}(\theta)$ and $E_\theta\{\dot{g}(\theta)\}$. The root that minimises the value of the test statistic is chosen as the desired root of the estimating function. This test is reasonable because, by the law of large numbers, $\dot{g}(\theta)/n$ will tend to $E_{\theta_0}\{\dot{g}(\theta)\}/n$, where θ_0 is the true value of the parameter.

In the second case, the test is constructed more directly from the data by examining the hypotheses

$$H_2: Y \sim P_\theta$$

at $\theta = \hat{\theta}_1, \ldots, \hat{\theta}_m$. Here Y represents the observed data set and P_θ its distribution under the assumption that θ is the true value of the parameter. The test statistic for H_2 can be constructed by *partitioning* the sample space into cells and constructing a *chi-square goodness-of-fit test*, or using a *weighted least squares* criterion such as

$$\sum_{i=1}^{n} w_i(\theta)[Y_i - E_\theta(Y_i)]^2 \tag{5.73}$$

using an appropriate weighting function $w(\theta)$.

In using such a test procedure, as noted by Singh and Mantel (1998), it is important to construct a criterion $Q(\theta; Y)$ such that the minimiser $\hat{\theta}$ of the objective function $Q(\theta; Y)$ will yield a consistent estimator of θ. As an illustration of this point let us consider a sequence of random variables, $Y_1, \ldots, Y_n$, independently distributed with the same distribution having mean and variance θ, where the parameter space $\Theta = \{\theta \colon \theta > 0\}$ is the positive part of the real axis. Singh and Mantel (1998) discussed a weighted least square type criterion

$$Q(\theta; Y) = \sum_{j=1}^{n} \frac{(Y_j - \theta)^2}{\theta}. \tag{5.74}$$

The minimiser of (5.74), under the constraint that $\theta > 0$, can be found analytically as

$$\tilde{\theta} = \sqrt{n^{-1} \sum_{j=1}^{n} Y_j^2}. \tag{5.75}$$

Using the law of large numbers we see that

$$\tilde{\theta} \to \sqrt{\theta^2 + \theta} \quad \text{as } n \to \infty.$$

This shows that the minimiser $\tilde{\theta}$ of the criterion (5.74) is not a consistent estimator. Such an objective function therefore would not help in distinguishing a consistent root from a wrong one.

To see how the criterion (5.74) behaves asymptotically, let $\hat{\theta}$ be any consistent estimator of the true value θ_0. Let $\theta^\dagger = \hat{\theta} + c$, where c is a positive constant. Again by the law of large numbers we have

$$\frac{1}{n} Q(\hat{\theta}; Y) \to 1 \tag{5.76}$$

$$\frac{1}{n} Q(\theta^\dagger; Y) \to \frac{\theta_0 + c^2}{\theta_0 + c}. \tag{5.77}$$

Therefore if $c < 1$, then the criterion (5.74) will choose the inconsistent root $\theta^\dagger$.

To appreciate how the multiple root problem may arise in this situation let us consider the following quadratic estimating function considered by Heyde and Morton (1998) and Singh and Mantel (1998)

$$g(\theta, Y) = \sum_{j=1}^{n} [a(Y_j - \theta) + \{(Y_j - \theta)^2 - \theta\}] \tag{5.78}$$

in estimating θ_0. Solving $g(\theta, Y) = 0$ we have two roots

$$\hat{\theta}_1 = \bar{Y} + \frac{1+a}{2} - \Delta \quad \hat{\theta}_2 = \bar{Y} + \frac{1+a}{2} + \Delta$$

where $\bar{Y} = \sum_{j=1}^{n} Y_j/n$ and

$$\Delta = \sqrt{\bar{Y} + \left(\frac{1+a}{2}\right)^2 - S_n^2} \quad S_n^2 = \frac{1}{n}\sum_{j=1}^{n}(Y_j - \bar{Y})^2.$$

Assuming that $1 + a > 0$, then asymptotically we have that

$$\hat{\theta}_1 \approx \bar{Y}$$

and

$$\hat{\theta}_2 - \hat{\theta}_1 = 2\Delta \to 1 + a\,.$$

So in this case $\hat{\theta}_1$ is consistent while $\hat{\theta}_2$ is inconsistent. Since

$$\hat{\theta}_2 \approx \hat{\theta}_1 + (1 + a)$$

by the above general discussions, the criterion (5.74) will be asymptotically smaller for any a satisfying the condition $-1 < a < 0$. In other words, the test statistic (5.74) will choose the wrong root when $a \in (-1, 0)$.

Singh and Mantel (1998) suggested using a criterion based on *minimum chi-square estimation* when the latter procedure gives a consistent estimator. These authors suggest that in some cases one may use a modified least squares test statistic of the form

$$\left[\sum_{i=1}^{n} w_i(\theta)\ \{Y_i - E_\theta(Y_i)\}\right]^2 . \tag{5.79}$$

This criterion applied to the above example suggests the use of the test statistic

$$Q(\theta; Y) = \left\{\sum_{j=1}^{n} \frac{Y_j - \theta}{\theta}\right\}^2 \tag{5.80}$$

as a modification to the least squares criterion (5.74). Note that minimising (5.80) now does give a consistent estimator of the parameter.

In Section 5.2 we discussed the general problem of estimation using an estimating function. We pointed out there that point estimation in this framework is not equivalent to the problem of solving the corresponding estimating equation. This is because an estimating function usually has to simultaneously satisfy several conditionas such as thoese given by (5.3). This idea leads to the consideration of several iterative algorithms discussed in the previously sections. In Section 5.7, for instance, we defined a criterion called the empirical information divergence, $\hat{\mathcal{D}}(\theta)$, and discussed how Muller's method may be improved using this criterion.

Restricted to parametric models, the first two equations corresponding to the general equations (5.3) can be written as

$$E_{\theta_0}\left\{\dot{\ell}(\theta; Y)\right\}\Big|_{\theta=\theta_0} = 0, \tag{5.81}$$

$$E_{\theta_0}\left[\left\{\dot{\ell}(\theta; Y)\right\}^2 + \ddot{\ell}(\theta; Y)\right]\Big|_{\theta=\theta_0} = 0 \tag{5.82}$$

where $\ell(\theta; Y)$ is the log-likelihood function. The data version of equation (5.81) is the likelihood equation. The empirical version of equation (5.82) is a special case of the empirical information divergence defined by(5.60), which now takes the form

$$\phi(\theta) = \frac{1}{n}\sum_{j=1}^{n}\left[\left\{\dot{\ell}(\theta; y_j)\right\}^2 + \ddot{\ell}(\theta; y_j)\right]. \tag{5.83}$$

Note that $\phi(\theta)$ relates to the empirical information divergence in the form

$$\phi(\theta) = \frac{1}{n}\hat{\mathcal{D}}(\theta).$$

For the same reasons as stated in Section 5.7, when the likelihood equation admits multiple solutions, the consistent one, $\hat{\theta}$ say, is expected to satisfy the property that $\phi(\hat{\theta}) \approx 0$. Gan and Jiang (1999) proposed a test for the consistency of each root based on a studentised version of $\phi(\hat{\theta})$. The idea of using the criterion $\phi(\theta)$ of (5.83) traces at least back to White (1982), who used a similar method for testing model misspecification.

Since a test based on (5.83) involves the second moment of the score function, the accuracy of such a test may not be very satisfactory unless the sample size is quite large. Using a random sample of size 20 from the Cauchy location-model with $\theta_0 = 0$ we found three roots

$$\hat{\theta}_1 = 0.07, \quad \hat{\theta}_2 = 55.38 \quad \text{and} \quad \hat{\theta}_3 = 57.88$$

to the likelihood equation. The corresponding values of $\phi(\hat{\theta})$ were found to be

$$\phi(\hat{\theta}_1) = 0.174, \quad \phi(\hat{\theta}_2) = 0.03 \quad \text{and} \quad \phi(\hat{\theta}_3) = -0.053.$$

For this data set, the criterion $\phi(\hat{\theta})$ deviates most from its ideal value of zero at the most favourable root $\hat{\theta}_1 = 0.07$.

5.10 Bootstrap quadratic likelihood ratio tests

In this section, we consider a particular type of test that is computer-oriented. These tests are simple to implement and are applicable to a wide variety of problems. Test procedures, such as those discussed in the preceding section, that require the calculation of probabilities or expectations, may be cumbersome and inappropriate for simulation studies involving thousands of trials. For this reason tests based upon resampling techniques become particularly appealing.

5.10.1 BOOTSTRAP QUADRATIC LIKELIHOOD

Let $y = \{y_1, \ldots, y_n\}$ be a random observation, where each y_j is independent of y_k for $j \neq k = 1, \ldots, n$. In this section, we shall focus on additive estimating functions of the form

$$g(\theta, y) = \sum_{j=1}^{n} h_j(\theta, y_j)$$

where, as before, each $h_j(\theta, y_j)$ is assumed to be unbiased. The quasi-score function, for instance, provides an example of this sort. Note that if $g(\theta, y)$ is an additive estimating function then any function of the form $a(\theta)g(\theta, y)$ is also an additive estimating function, where $a(\theta)$ is a full-rank matrix functionally independent of y. In this section, we shall consider a general vector-valued estimating function g, which has dimension $p = \dim\theta \geq 1$. Sometimes there may exist $k \geq \dim\theta$ conditions, each giving an unbiased estimating function for θ. So these conditions lead to a k-dimensional estimating function for θ. As will be discussed in Section 6.6, theories are available so that the k unbiased estimating functions can be optimally combined to give an estimating function of the same dimension as that of the parameter of interest. We shall assume that this has already been done for $g(\theta, y)$.

The bootstrap. Let $\hat{F}$ be the empirical distribution function, putting mass $1/n$ on each datum of y. Let Y_i^* be a random variable distributed according to $\hat{F}$ so that

$$\Pr\left\{Y_i^* = y_j\right\} = \frac{1}{n}, \qquad i, j = 1, \ldots, n. \tag{5.84}$$

The sampling scheme (5.84) defines the usual non-parametric bootstrap method. Now denote by

$$Y^* = \{Y_1^*, \ldots, Y_n^*\}$$

a bootstrap sample of size n. Note that the Y_i^*'s are independent and identically distributed even though the y_j's may not have the same distribution. Now if we define

$$j^* = i_j \qquad \text{if } Y_j^* = y_{i_j}, \quad j = 1, \ldots, n$$

then a bootstrap version of the additive estimating function $g(\theta, y)$ is given by

$$g(\theta, Y^*) = \sum_{j=1}^{n} h_{j^*}(\theta, Y_j^*) \,.$$

The generalised bootstrap. The bootstrap tests to be introduced later in this section can be generalised to estimating functions of the form

$$g(\theta, y) = \sum_{i_1 \neq \cdots \neq i_k} h_{i_1 \cdots i_k}(\theta, y_{i_1}, \ldots, y_{i_k})$$

where, as in the case of unbiased additive estimating functions, each summand

$$h_{i_1 \cdots i_k}(\theta, y_{i_1}, \ldots, y_{i_k})$$

is assumed to be unbiased. In this case, a bootstrap procedure more general than the one given above can be defined in a straightforward fashion. Let

$$m = \binom{n}{k} = \frac{n!}{k!(n-k)!} \,.$$

Let $Y_j^\dagger$ be a k-valued random vector

$$Y_j^\dagger = (Y_{j1}^\dagger, \ldots, Y_{jk}^\dagger), \quad j = 1, \ldots, m$$

drawn *without* replacement from $\hat{F}$. Now a bootstrap sample, $Y^\dagger$, is an $m \times k$ random matrix given by

$$Y^\dagger = (Y_1^\dagger, \ldots, Y_m^\dagger)^{\mathrm{t}} = \begin{pmatrix} Y_{11}^\dagger & \cdots & Y_{1k}^\dagger \\ & \vdots & \\ Y_{m1}^\dagger & \cdots & Y_{mk}^\dagger \end{pmatrix}.$$

The bootstrap random sample (now a matrix) is then repeatedly drawn an appriopriate number of B times to produce B bootstrap matrices

$$Y_1^\dagger, \ldots, Y_B^\dagger$$

which are used to approximate the distribution of a statistic, etc. Note that the generalised bootstrap reduces to the bootstrap procedure described in the previous paragraph when $k = 1$. In the rest of this section we shall focus on the case $k = 1$ to avoid notational complications.

The bootstrap quadratic likelihood. Let θ and ξ be two values in the parameter space. Usually, but by no means always, these values are assumed to be close to each other. The value of θ is usually the parameter of interest for which inference is desired, while the value of ξ may be a rough guess or a hypothesised value of θ. At the value of ξ, we define the following quantity associated with an unbiased estimating function $g(\theta, y)$

$$J(\xi, y) = -\tfrac{1}{2}\left[\dot{g}(\xi, y) + \{\dot{g}(\xi, y)\}^{\mathrm{t}}\right]$$

which is a symmetric matrix depending on ξ as well as the observations. Note that if $g(\theta, y)$ is a score function, then $J(\xi, y)$ is nothing but the (minus) Hessian matrix of the log likelihood at ξ. For a general estimating function, however, the matrix $J(\xi, y)$ cannot be interpretated as a Hessian matrix associated with the estimating function $g(\theta, y)$. This is because an estimating function $g(\theta, y)$ may not be the gradient of any scalar objective function.

Using $J(\xi, y)$, we now define a function for any unbiased estimating function $g(\theta, y)$ as follows:

$$\ell(\theta, \xi) = -\tfrac{1}{2}(\theta - \xi)^{\mathrm{t}}\, J(\xi, y)\, (\theta - \xi) + \{g(\xi, y)\}^{\mathrm{t}}\, (\theta - \xi)\,. \tag{5.85}$$

We deliberately used in (5.85) the symbol ℓ, which has been reserved for the log likelihood function up to now, because the scalar objective function $\ell(\theta, \xi)$ has many properties resembling a genuine log likelihood function. We shall refer to $\ell(\theta, \xi)$ as the *approximate quadratic likelihood function* associated with an estimating function $g(\theta, y)$. In (5.85), ξ is assumed to be known so that the approximate quadratic likelihood function $\ell(\theta, \xi)$ should be interpreted as a function of θ. In defining $\ell(\theta, \xi)$ we have implicitly assumed that $\dot{g}(\xi, y) \neq 0$ so that $J(\xi, y)$ does not vanish at ξ. A value of ξ satisfying this property is sometimes called a *non-singular point* in the parameter space. In Section 6.5, we shall see that $\ell(\theta, \xi)$ is motivated by a decomposition of the vector field in the parameter space induced by the estimating function $g(\theta, y)$. Likelihood-like properties of this function will also be studied in detail in Section 6.5.

To study properties of a root to an estimating equation using $\ell(\theta, \xi)$ of (5.85), we shall apply the bootstrap method introduced eariler to study distributional properties of $\ell(\theta, \xi)$. To define the bootstrap counterpart of $\ell(\theta, \xi)$, we replace in (5.85) the quantities $g(\xi, y)$ and $J(\xi, y)$ by $g_*(\xi) = g(\xi, Y^*)$ and $J_*(\xi) = J(\xi, Y^*)$, respectively, where Y^* is a bootstrap sample according to (5.84). The *bootstrap quadratic likelihood function* is defined as follows:

$$\ell_*(\theta, \xi) = -\tfrac{1}{2}(\theta - \xi)^{\mathrm{t}}\, J_*(\xi)\, (\theta - \xi) + \{g_*(\xi)\}^{\mathrm{t}}\, (\theta - \xi)\,. \tag{5.86}$$

As in (5.85), we shall also treat ξ in (5.86) as known and regard the bootstrap quadratic likelihood function $\ell_*(\theta, \xi)$ as a function of θ.

Since our primary interest in this section is to test formally if a root to an estimating equation is consistent, we are therefore interested in the *bootstrap quadratic likelihood ratio*, which is naturally defined by

$$\gamma_*(\xi) = 2\{\ell_*(\tilde{\theta}, \xi) - \ell_*(\tilde{\theta}_0, \xi)\} \tag{5.87}$$

where ξ is known, $\tilde{\theta}$ and $\tilde{\theta}_0$ are the unconstrained and constrained maximisers of $\ell_*(\theta, \xi)$ for a given hypothesis.

Now suppose that $\hat{\theta}$ is a solution to $g(\hat{\theta}) = 0$. If $\hat{\theta}$ is a $\sqrt{n}$-consistent estimator of θ, then, for a large sample size, we would expect that the true value of the parameter θ_0 differs only slightly from the value of $\hat{\theta}$. This then leads to the consideration of the natural hypothesis

$$H_0 \, : \, \theta_0 = \hat{\theta} \, .$$

Under this hypothesis, and taking $\xi = \hat{\theta}$ in (5.87), we then see that the bootstrap quadratic likelihood ratio can be written as

$$\gamma_*(\hat{\theta}) = \left\{ g(\hat{\theta}, Y^*) \right\}^{\mathrm{t}} \left\{ J(\hat{\theta}, Y^*) \right\}^{-1} g(\hat{\theta}, Y^*) \tag{5.88}$$

where the symmetrised quasi-Hessian matrix $J(\hat{\theta}, Y^*)$ is calculated according to

$$J(\hat{\theta}, Y^*) = -(1/2) \left[\dot{g}(\hat{\theta}, Y^*) + \left\{ \dot{g}(\hat{\theta}, Y^*) \right\}^{\mathrm{t}} \right] . \tag{5.89}$$

Sometimes it may be too costly to compute in (5.88) the inverse of $J(\hat{\theta}, Y^*)$ using formula (5.89) for each bootstrap replication. If this is the case, one may replace $J(\hat{\theta}, Y^*)$ by its bootstrap expectation. To compute the expectation of $J(\hat{\theta}, Y^*)$ we assume that $g(\theta, y)$ is an unbiased additive estimating function so that

$$\dot{g}(\hat{\theta}, Y^*) = \sum_{j=1}^{n} \dot{h}_{j^*}(\hat{\theta}, Y_j^*) \, .$$

Under the resampling scheme (5.84), we therefore have

$$\begin{aligned} E_* \left\{ \dot{g}(\hat{\theta}, Y^*) \right\} &= \sum_{j=1}^{n} E_* \left\{ \dot{h}_{j^*}(\hat{\theta}, Y_j^*) \right\} \\ &= \sum_{j=1}^{n} \left\{ \frac{1}{n} \sum_{k=1}^{n} \dot{h}_k(\hat{\theta}, y_k) \right\} \\ &= \sum_{j=1}^{n} \left\{ \frac{1}{n} \dot{g}(\hat{\theta}, y) \right\} \\ &= \dot{g}(\hat{\theta}, y) \, . \end{aligned}$$

It follows then that the bootstrap expectation of $J(\hat{\theta}, Y^*)$ is equal to $J(\hat{\theta}, y)$. That is,

$$E_* \left\{ J(\hat{\theta}, Y^*) \right\} = J(\hat{\theta}, y) \, .$$

The above considerations lead finally to the following criterion:

$$\gamma^{\dagger}(\hat{\theta}) = \left\{g(\hat{\theta}, Y^*)\right\}^{t} \left\{J(\hat{\theta}, y)\right\}^{-1} g(\hat{\theta}, Y^*) \tag{5.90}$$

for the problem of testing the consistency of a root to an estimating equation. The test statistic $\gamma^{\dagger}(\hat{\theta})$ is intuitively appealing because it has the following familar form:

$$\{g(\xi, Y)\}^{t} \ \Sigma^{-1}(\xi)\, g(\xi, Y)\,. \tag{5.91}$$

In (5.91), $g(\theta, Y)$ is an unbiased estimating function, ξ is a value usually determined from a concerned hypothesis, and $\Sigma(\theta)$ is an appropriate positive definite matrix, usually the covariance matrix of $g(\theta, Y)$. The quantity given by (5.91) is sometimes called a *generalised score statistic*. When g is a score function, ξ the value specified by a null hypothesis, and $\Sigma(\xi)$ the Fisher information, then (5.91) reduces to the usual score statistic. The fact that $\gamma^{\dagger}(\hat{\theta})$ of (5.90) gives a bootstrap version of a generalised score statistic partially justifies the use of $\ell(\theta, \xi)$ given by (5.85) as an approximate likelihood function. In Section 6.5.2, we shall formally define the generalised score and Wald statistics and study their relations with the approximate quadratic likelihood ratios.

5.10.2 BOOTSTRAP QUADRATIC LIKELIHOOD RATIO TEST

Let E_* denote the expectation under the bootstrap sampling (5.84). Let $\hat{\theta}$ be a root to $g(\hat{\theta}) = 0$. First, we note that the bootstrap estimating function $g(\hat{\theta}, Y^*)$ is unbiased given the observations. This is because

$$\begin{aligned}
E_*\left\{g(\hat{\theta}, Y^*)\right\} &= E_*\left\{\sum_{j=1}^{n} h_{j^*}(\hat{\theta}, Y_j^*)\right\} \\
&= \sum_{j=1}^{n} E_*\left\{h_{j^*}(\hat{\theta}, Y_j^*)\right\} \\
&= \sum_{j=1}^{n}\left\{\frac{1}{n}\sum_{k=1}^{n} h_k(\hat{\theta}, y_k)\right\} \\
&= g(\hat{\theta}, y) \\
&= 0
\end{aligned}$$

where the first equality is due to the assumption that g is additive, and the third equality follows from the definition of E_*. Similarly, the bootstrap variance

$\hat{\Sigma} = \text{var}_* \{g(\hat{\theta}, Y^*)\}$ can be calculated as follows:

$$\begin{aligned}
\hat{\Sigma} &= E_* \left\{ \sum_{i=1}^{n} h_{i*}(\hat{\theta}, Y_i^*) \sum_{j=1}^{n} h_{j*}^{t}(\hat{\theta}, Y_j^*) \right\} \\
&= E_* \left[\sum_{j=1}^{n} \left\{ h_{j*}(\hat{\theta}, Y_j^*)\, h_{j*}^{t}(\hat{\theta}, Y_j^*) \right\} + \sum_{j \neq k} \left\{ h_{j*}(\hat{\theta}, Y_j^*)\, h_{k*}^{t}(\hat{\theta}, Y_k^*) \right\} \right] \\
&= \sum_{j=1}^{n} E_* \left\{ h_{j*}(\hat{\theta}, Y_j^*)\, h_{j*}^{t}(\hat{\theta}, Y_j^*) \right\} \\
&= \sum_{j=1}^{n} \left[\sum_{k=1}^{n} \frac{1}{n} \left\{ h_k(\hat{\theta}, y_k)\, h_k^{t}(\hat{\theta}, y_k) \right\} \right] \\
&= \sum_{j=1}^{n} \left\{ h_j(\hat{\theta}, y_j)\, h_j^{t}(\hat{\theta}, y_j) \right\} \qquad (5.92)
\end{aligned}$$

where the third equality is due to independence of the Y^*'s and the unbiasedness of $h_{j*}(\hat{\theta}, Y_j^*)$, and the fourth equality follows directly from the definition of E_*.

Now we are in a position to state a result concerning the distribution of the bootstrap quadratic likelihood ratio $\gamma^{\dagger}(\hat{\theta})$ defined by (5.90). The following proposition says that the bootstrap distribution of $\gamma^{\dagger}(\hat{\theta})$, at any root $\hat{\theta}$ of g, has an approximate weighted chi-squared distribution as $n \to \infty$. The proof is similar to that of Proposition 6.8 and thus is omitted.

Proposition 5.17 *Assume that $g(\theta, y)$ is an unbiased additive estimating function of the form*

$$g(\theta, y) = \sum_{j=1}^{n} h_j(\theta, y_j)$$

where each $h_j(\theta, y_j)$ is also assumed unbiased. Suppose that $\hat{\theta}$ is a root to $g(\hat{\theta}, y) = 0$. Let $\hat{\Sigma} = var_* \{g(\hat{\theta}, Y^*)\}$ be the bootstrap variance of $g(\hat{\theta}, Y^*)$, which is given by (5.92). Let $\hat{J} = J(\hat{\theta}, y)$ be defined as before. Then, under mild regularity conditions, conditional on y, we have that

$$\gamma^{\dagger}(\hat{\theta}) \xrightarrow{\mathcal{L}} \sum_{j=1}^{p} \hat{\lambda}_j Z_j^2 \qquad (5.93)$$

as $n \to \infty$, where $\hat{\lambda}_j$ are eigenvalues of $\hat{\Sigma} \hat{J}^{-1}$.

Using the above results we now consider the appropriateness of a root $\hat{\theta}$ of $g(\hat{\theta})$ as an estimator of θ_0, the true value of the parameter. Consider testing the hypothesis

$$H_0 : \ \hat{\theta} \text{ is a } \sqrt{n}\text{-consistent estimator of } \theta_0 \qquad (5.94)$$

against the alternative that $\hat{\theta}$ is not consistent. We shall assume that $g(\theta)$ is first-order information unbiased.

Now define

$$\hat{\Sigma} = \sum_{j=1}^{n} h_j(\hat{\theta}, y_j)\, h_j^{\mathrm{t}}(\hat{\theta}, y_j), \qquad \hat{J} = -\frac{1}{2}\sum_{j=1}^{n}\left\{\dot{h}_j(\hat{\theta}, y_j) + \dot{h}_j^{\mathrm{t}}(\hat{\theta}, y_j)\right\}$$

where $\hat{\theta}$ is a root of $g(\hat{\theta}) = 0$ satisfying (5.94). Let

$$H = -\frac{1}{n}E_{\theta_0}\{\dot{g}(\theta_0)\}, \qquad \Sigma = \frac{1}{n}E_{\theta_0}\{g(\theta_0)\, g^{\mathrm{t}}(\theta_0)\}$$

both quantities being of $O(1)$. By the first-order information unbiasedness of g, we have

$$\Sigma = H + O(n^{-(1/2)}). \tag{5.95}$$

So H is symmetric to $O(n^{-1/2})$. On the other hand, by the weak law of large numbers and using (5.94), we have

$$\frac{1}{n}\hat{\Sigma} = \Sigma + o_p(1), \qquad \frac{1}{n}\hat{J} = (1/2)(H + H^t) + o_p(1) = \Sigma + o_p(1)$$

the last equality being due to (5.95).

Combing these results, we conclude that under (5.94) the empirical information matrices $\hat{\Sigma}$ and $\hat{H}$ are asymptotically equivalent, i.e.,

$$\hat{\Sigma}\hat{J}^{-1} = \mathrm{I}_{k\times k} + o_p(1) \tag{5.96}$$

where $\mathrm{I}_{k\times k}$ is a $p \times p$ identity matrix. Now let $\hat{\lambda}_j$ be the eigenvalues of $\hat{\Sigma}\hat{J}^{-1}$. By (5.96), $\hat{\lambda}_j - 1$ are eigenvalues of an $o_p(1)$ matrix, implying that $\hat{\lambda}_j = 1 + o_p(1)$. Summarising these results we then have

Proposition 5.18 *Suppose that $g(\theta, y)$ is first-order information-unbiased. Assume that the conditions of Proposition 5.17 hold. Under hypothesis (5.94), the bootstrap quadratic likelihood ratio $\gamma^{\dagger}(\hat{\theta})$ of (5.90) has an approximate chi-squared distribution with p degrees of freedom.*

This result suggests a method for root selection. *The root $\hat{\theta}$, at which the distribution of the bootstrap quadratic likelihood ratio is closest to $\chi^2(p)$, is a natural estimate for the parameter*. This method is applied in the following section to two cases. The first example concerns a scalar parameter from a parametric model. The second example deals with a vector parameter in a generalised linear model when measurement errors are present. The estimating function in the latter case is not conservative.

5.10.3 EXAMPLE: CAUCHY LOCATION-MODEL

We first briefly discuss the application of the bootstrap test to a relatively simple likelihood example with a scalar parameter. We consider the Cauchy location-model with density function $f(\theta; y) = 1/[\pi\{1+(y-\theta)^2\}]$. There are many works on multiple roots of this model (e.g., Barnett, 1966; Reeds, 1985; Small and Yang, 1999). We have also discussed this model from several different perspectives in the previous sections.

We take the score as our estimating function to test the performance of the root consistency test. If $\hat{\theta}$ is a root satisfying $\hat{\theta} = O_p(n)$, then we can show that the bootstrap quadratic likelihood ratio (5.90) at $\hat{\theta}$ tends to $-2n$ and the eigenvalue $\lambda = \hat{\Sigma}/\hat{J}$ tends to -2 as $n \to \infty$. So the power of the root consistency test tends to unity as $n \to \infty$. Note that the alternative hypothesis $\hat{\theta} = O_p(n)$ is reasonable since 'wrong' roots are of the same magnitude as 'outliers' for this example.

For $\theta_0 = 0$, we found five roots of the likelihood equation for a sample of size 20. Figure 5.7 shows the distributions of (5.90) at two roots 0.07 (left) and 55.38 (right). The bootstrap distribution at the first root is much closer to $\chi^2(1)$ than that at the second one. Bootstrap distributions at other three roots, 57.88, -73.74, -70.25, are similar to that at the second one. We thus conclude that the root $\hat{\theta} = 0.07$ is preferable to any other root we found.

5.10.4 EXAMPLE: LOGISTIC REGRESSION WITH MEASUREMENT ERRORS

In Section 4.7, we introduced the generalised linear model in which the covariates can be observed only indirectly. We discussed in some detail how the problem of multiple roots may arise in this setting, and illustrated the point using the normal regression with errors in variables.

Our second example in this section also falls into the category of generalised linear models with measurement errors. Instead of normal models, we now consider the logistic regression when the covariate is observed indirectly with

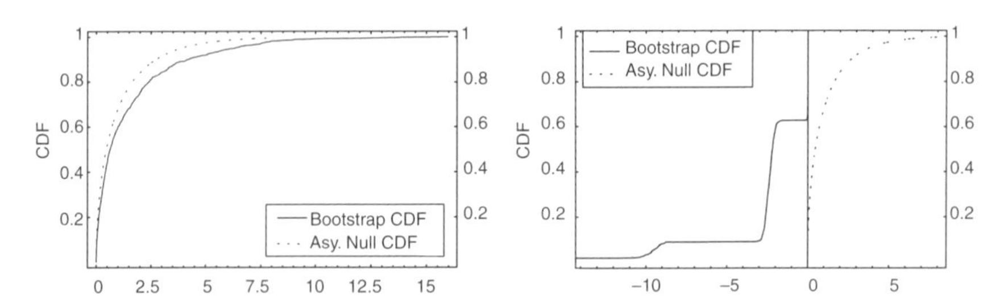

Fig. 5.7 Distributions of the quadratic bootstrap likelihood ratio at two roots, $\hat{\theta}_1 = 0.07$ (left) and $\hat{\theta}_2 = 55.38$ (right), of the Cauchy likelihood equation with $\theta_0 = 0$. The bootstrap distributions (solid curves) are compared with the asymptotic null distribution of $\chi^2(1)$ (broken curves).

additive error,

$$\text{logit}(\pi) = \alpha + \beta^{t}x, \quad z = x + \epsilon$$

where π is the mean of a binary response y. In this model measurements are made on z rather than on x directly. The error is assumed to be normal, $\epsilon \sim N_{p-1}(0, \ \Psi)$, with known covariance matrix Ψ, and is independent of y. This model is often referred to as semiparametric or *functional model* since we treat x as unknown but *fixed* variables. It has the advantage of avoiding the need to specify the distribution of x as is done in the so-called *structural model*. The latter approach can be sensitive to the choice of the distribution of x (Fuller, 1987, Section 3.3; Schafer, 1987; Carroll *et al.*, 1995).

Following Hanfelt and Liang (1995), we shall consider the following estimating function:

$$g(\theta, y) = \sum_{i=1}^{n} (1, \ d_i)^{t}(y_i - \mu_i^c) \tag{5.97}$$

for $\theta = (\alpha, \beta^{t})^{t}$, where d_i and μ_i^c are given respectively by

$$d_i = A_i + (\mu_i^c - 1)\Psi\beta$$
$$\mu_i^c = \{1 + \exp[-\{\alpha + (A_i - 0.5\Psi\beta)^{t}\beta\}]\}^{-1} \quad A_i = z_i + y_i\Psi\beta.$$

The function (5.97) is based on a conditional score of Stefanski and Carroll (1987),

$$\frac{\partial \log L}{\partial \theta} - E\left\{\frac{\partial \log L}{\partial \theta}\bigg| A\right\} = \sum_{i=1}^{n} (1, \ x_i)^{t}(y_i - \mu_i^c)$$

by eliminating the 'nuisance parameters' x_i. The estimating function (5.97) is not conservative since the Jacobian $\dot{g}(\theta)$ with respect to θ can be shown to be an asymmetric matrix. Estimating functions similar to (5.97) often have multiple roots, a problem which can be serious when measurement errors are large. Hanfelt and Liang (1995, 1997) suggested an approximate likelihood ratio test based on some empirically defined paths in the parameter space. Their approach is discussed in next chapter.

We shall now consider the case when both α and β are scalars. A Monte Carlo study has been carried out in three cases. In the first two cases we set $n = 100$ and $\Psi = 0.8$, and in the third case we set $n = 50$ and $\Psi = 0.4$. In all cases we let the true parameter $\theta_0 = (-1.4, 1.4)$, a value reported in a large cohort study by Stefanski and Carroll (1985). The x values were pseudo random numbers generated from $N(0, \ 0.1^2)$.

Figure 5.8 shows the distributions of the bootstrap quadratic likelihood ratios (5.90) for one data set in each of the three cases. Shown in each plot of Figure 5.8 are also the null distribution $\chi^2(2)$ for testing the consistency of the roots, and the distributions of $\hat{\lambda}_1 Z_1^2 + \hat{\lambda}_2 Z_1^2$, where $(\hat{\lambda}_1, \hat{\lambda}_2)$ are eigenvalues of $\hat{\Sigma} J^{-1}$. These values are positive at the roots shown on the left panel for cases (a), (b) and (c)

$$(\hat{\lambda}_{1a}, \hat{\lambda}_{2a}) = (1.6, 0.9), \quad (\hat{\lambda}_{1b}, \hat{\lambda}_{2b}) = (1.4, 1.0) \quad \text{and} \quad (\hat{\lambda}_{1c}, \hat{\lambda}_{2c}) = (1.1, 0.6)$$

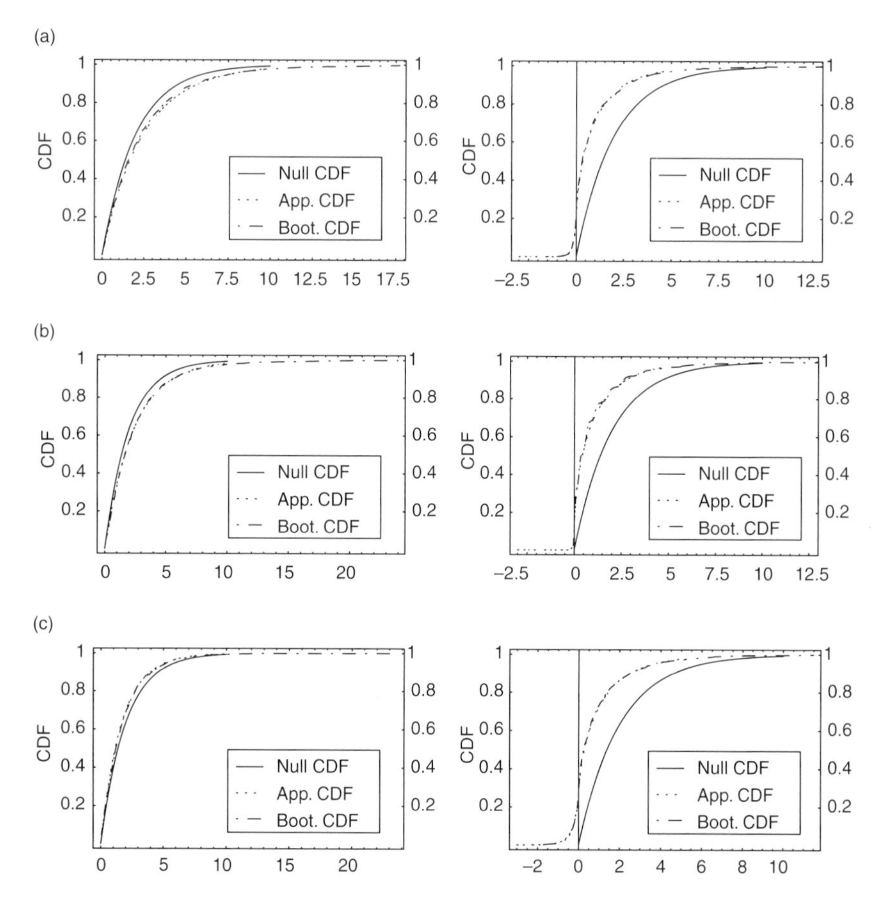

Fig. 5.8 Distributions of the quadratic bootstrap likelihood ratios using estimating function (5.97) arising from the errors-in-variables logistic regression. Two roots were found for each of the three artificial samples, marked a, b and c. The distribution of the bootstrap likelihood ratio at each root (broken curves) is compared with its asymptotic null distribution $\chi^2(2)$ (solid curves). Shown are also the approximate distributions of the test statistics using the weighted *chi*-squared distributions (dotted curves).

and (positive, negative) at the corresponding roots shown on the right panel

$$(\tilde{\lambda}_{1a}, \tilde{\lambda}_{2a}) = (1.0, -0.1), \quad (\tilde{\lambda}_{1b}, \tilde{\lambda}_{2b}) = (1.0, -0.0) \quad \text{and}$$
$$(\tilde{\lambda}_{1c}, \tilde{\lambda}_{2c}) = (1.0, -0.2).$$

In all cases the distributions of the bootstrap quadratic likelihood ratios (broken curves) are seen to be very close to their theoretical approximations given by $\hat{\lambda}_1 Z_1^2 + \hat{\lambda}_2 Z_1^2$ (dotted curves). The bootstrap distributions on the left panel are much closer to the null distribution $\chi^2(2)$ (solid curves). This suggests that the roots on the left panel for cases (a), (b) and (c)

$$\hat{\theta}_{1a} = (-1.1, 2.3), \quad \hat{\theta}_{1b} = (-2.2, 2.0) \quad \text{and} \quad \hat{\theta}_{1c} = (-2.2, 2.0)$$

are preferred to the corresponding roots on the right panel

$$\hat{\theta}_{2a} = (-1.0, 8.8), \quad \hat{\theta}_{2b} = (-7.9, 9.6) \quad \text{and} \quad \hat{\theta}_{2c} = (-5.5, 8.7)$$

as estimates for the value of

$$\text{`true parameter'} = (-1.4, 1.4)\,.$$

These roots selected by the bootstrap quadratic likelihood ratio tests are also closer to the true value in Euclidean distance.

5.11 An information theoretic approach

The concept of *information* plays an important role in various disciplines of science and technologies. The Fisher information is a key concept in statistical inference based on parametric models. We have seen that the Godambe efficiency, an extension of the Fisher information, plays a similar important role in the theory of estimating functions. For a location model, the Godambe efficiency may be regarded as a special case of a more general information measure, $I(\theta_1, \theta_2)$ say, which we shall define shortly. We shall call $I(\theta_1, \theta_2)$ the *shifted information*, which reduces to the Godambe efficiency $\text{eff}_\theta(g) = I(\theta, \theta)$ for $\theta_1 = \theta_2 = \theta$. Let θ_0 be the true value of the parameter. The shifted information $I(\theta_0, \theta)$ is then a function of θ alone. We shall see that for a number of problems the shifted information $I(\theta) = I(\theta_0, \theta)$ attains its maximum value at $\theta = \theta_0$. That is, the Godambe efficiency $\text{eff}_\theta(g)$ will crosscut the shifted information $I(\theta)$ at the true value $\theta = \theta_0$ where $I(\theta)$ has its maximum value. This fact suggests a method for detecting the appropriateness of a root to an estimating equation. We shall discuss this method for location models alone.

Let $Y_1, Y_2, \ldots, Y_n$, be independent and identically distributed random variables from a location model $f(y - \theta)$, where $f(\cdot)$ is a density function. Suppose for simplicity that θ is a scalar parameter. So the score estimating function has the form

$$g(\theta) = \sum h(y_i - \theta)$$

where

$$h(x) = -\frac{f'(x)}{f(x)} \quad \text{and} \quad f'(x) = \frac{\mathrm{d}}{\mathrm{d}x} f(x)\,.$$

Within the location model, we define the *shifted information* to be

$$I(\theta_1, \theta_2) = \frac{\left\{E_{\theta_1} h'(Y - \theta_2)\right\}^2}{E_{\theta_1} \{h(Y - \theta_2)\}^2} \tag{5.98}$$

where E_{θ_1} denotes the expectation with respect to the density $f(y-\theta_1)$ and $h'(x) = (\mathrm{d}/\mathrm{d}x)h(x)$. In particular, $I(\theta,\theta)$ is the Godambe efficiency at θ. See Godambe (1960). In the following discussion we shall consider the special case $\theta_1 = \theta_0$, where θ_0 is the true value of the parameter.

Now let us assume that for all $t \in R$,

$$\lim_{|y|\to\infty} \frac{f(y+t)\, f'(y)}{f(y)} = 0. \tag{5.99}$$

Note that this condition is satisfied by common distributions including the Cauchy and normal. Integrating by parts, we get

$$\begin{aligned}
E_{\theta_0}\left\{h'(Y-\theta)\right\} &= \int_{-\infty}^{\infty} h'(y-\theta)\, f(y-\theta_0)\,\mathrm{d}y \\
&= \int_{-\infty}^{\infty} f(y-\theta_0)\,\mathrm{d}\,h(y-\theta) \\
&= -\int_{-\infty}^{\infty} h(y-\theta)\, f'(y-\theta_0)\mathrm{d}y \\
&= -\int_{-\infty}^{\infty} \frac{f'(y)\, f'(y+t)}{f(y)}\mathrm{d}y
\end{aligned}$$

where $t = \theta - \theta_0$, and in passing from the second equality to the third one we used condition (5.99). Similarly, we have

$$E_{\theta_0}\left\{h^2(Y-\theta)\right\} = \int_{-\infty}^{+\infty} \left\{\frac{f'(y)}{f(y)}\right\}^2 f(y+t)\,\mathrm{d}y\,.$$

By the Cauchy–Schwarz inequality,

$$\begin{aligned}
&\left\{\int_{-\infty}^{\infty} \frac{f'(y)\, f'(y+t)}{f(y)}\mathrm{d}y\right\}^2 \\
&= \left\{\int_{-\infty}^{\infty} \frac{f'(y)\, f'(y+t)}{f(y)\, f(y+t)} f(y+t)\,\mathrm{d}y\right\}^2 \\
&\le \left[\int_{-\infty}^{+\infty} \left\{\frac{f'(y)}{f(y)}\right\}^2 f(y+t)\,\mathrm{d}y\right]\left[\int_{-\infty}^{\infty} \frac{\{f'(y)\}^2}{f(y)}\mathrm{d}y\right]
\end{aligned}$$

which is equivalent to

$$I(\theta_0,\theta) \le I(\theta_0,\theta_0)\,. \tag{5.100}$$

Note that the Fisher information for a location model is a constant irrespective of the location parameter. The inequality (5.100) states that the shifted information is always bounded above by the Fisher information for location models.

To use this inequality as a method for root selection, we need to estimate the function $I(\theta) = I(\theta_0, \theta)$. Let $\hat{\theta}$ solve the equation $g(\hat{\theta}) = 0$. We define the empirical version of $I(\theta)$ by

$$\hat{I}_n(\hat{\theta}) = \frac{\left\{n^{-1} \sum_{j=1}^{n} h'(y_j - \hat{\theta})\right\}^2}{n^{-1} \sum_{j=1}^{n} h^2(y_j - \hat{\theta})} = \frac{\left\{\sum_{j=1}^{n} h'(y_j - \hat{\theta})\right\}^2}{n \sum_{j=1}^{n} h^2(y_j - \hat{\theta})}. \tag{5.101}$$

If $\hat{\theta}$ is consistent, then, by the law of large numbers, $\hat{I}_n(\hat{\theta})$ of (5.101) will tend in probability to $I(\theta_0)$. By the information inequality (5.100), we therefore conclude that if $\hat{\theta}$ is consistent then the quantity $\hat{I}_n(\hat{\theta})$ will be larger with probability 1 than any other value $\hat{I}_n(\hat{\theta}')$ evaluated at $\hat{\theta}'$, as n tends to infinity. For heavy-tailed distribution, it is sometimes appropriate to replace the means in (5.101) by some trimmed means.

The inequality (5.100) is similar in some respects to the inequality for the Kullback–Leibler information,

$$K(\theta_0, \theta) = E_{\theta_0}\left[\log\left\{\frac{L(\theta)}{L(\theta_0)}\right\}\right] \leq \log\left[E_{\theta_0}\left\{\frac{L(\theta)}{L(\theta_0)}\right\}\right] = 0$$

where equality holds if $\theta = \theta_0$. This suggests that for certain models, the empirical version of the information for an estimating function can replace the likelihood as an objective function to be maximised. Therefore, we could choose the root $\hat{\theta}$ at which $\hat{I}_n$ or a trimmed analog of $\hat{I}_n$ is maximum. For example, a simulation of 2000 trials for samples of size $n = 10$ from the Cauchy location model found that the root which maximised the trimmed version of $\hat{I}_n$ was the global maximum of the likelihood approximately 97% of the time.

Maximising an information function may be most helpful when the estimating function is not the score function. For the method to be practicable, it is necessary that the inequality (5.100) hold for an additive estimating function, where we would define more generally

$$I(\theta_0, \theta) = \frac{\left\{E_{\theta_0} \dot{h}(\theta)\right\}^2}{E_{\theta_0} h^2(\theta)}. \tag{5.102}$$

Necessarily, in applications we need to replace (5.102) by an appropriate estimate of it.

5.12 Model enlargement

5.12.1 INTRODUCTION

A quite different view is taken in this section in dealing with the problem of multiple roots. In this section, we shall restrict our attention to parametric models. We assume that random variables $Y_1, \ldots, Y_n$ are independent and identically distributed from a population with density function $f(\theta; y)$, where $\theta \in \Theta$ is the parameter of interest. We assume that $f(\theta; \cdot)$ is of known form so that the value of the parameter θ completely specifies the model. That is, the probabilistic mechanism governing the data is thoroughly determined by the form of $f(\theta; \cdot)$ together with the value of θ. At preliminary stages of real data analyses, it is a usual practice to choose a parameterisation in such a way that the dimension θ is not very large so that the problem will be kept relatively simple at an initial stage. Accordingly, such a model contains certain amount of arbitrariness that needs to be corrected when the discrepancy between the observations and the mathematical model is too large. More formally, the selection of a good model may be studied by considering an appropriate criterion for *model selection*. Model selection is an important element of data analyses, the treatment of which however is beyond the scope of this book.

As we shall see from the examples discussed later, multiple roots to likelihood equations can occur as a result of the choice of too narrow a model so that the data can not be fitted satisfactorily using the specified model. Adding more flexibility to such a model by introducing additional parameters is often a useful way to solve the problem. Again how large a new model should be chosen is inevitably arbitrary to some extent. When such difficulty arises the *principle of parsimony* may provide a useful guide: enlarge a model with as few extra parameters as possible to ensure that the likelihood equation of the enlarged model will have a unique solution.

More formally, we start from a parametric model $f(\theta; y)$, specified by the known function f and the unknown parameter θ belonging to a parameter space Θ. There are several possibilities to expand this model. We shall consider the possibly simplest way so that the enlarged model will be specified by the same density function f but with a new set of parameters (θ, η). With some possible abuse of notation we shall use the same symbol f to denote the enlarged model by $f(\theta, \eta; y)$, where $\theta \in \Theta$ has dimension r and $\eta \in \Theta'$ has dimension s. For each given value of $\eta_0 \in \theta'$, the family of densities

$$\mathcal{M}_\theta = \{f(\theta, \eta;\ y) : \ \theta \in \theta,\ \eta = \eta_0\}$$

forms a subspace in the space

$$\mathcal{M} = \{f(\theta, \eta;\ y) : \theta \in \theta,\ \eta \in \Theta'\}.$$

We shall call $\mathcal{M}$ a *parent model* and $\mathcal{M}_\theta$ a *sub-model* associated with the value $\eta_0 \in \theta'$. Our focus here is to construct such a parent model $\mathcal{M}$ so that the likelihood equation for this model will have a unique solution.

5.12.2 MODEL EMBEDDING VIA MIXING

Now we discuss a general method for model embedding via the idea of mixing. Suppose that $f(\theta; y)$ is a density function with a real-valued parameter θ. Consider a transformation t from the positive real axis to $\mathbb{R}$,

$$t : \mathbb{R}^+ \to \mathbb{R}.$$

Suppose that t is a monotone function. Denote the inverse of t by t^{-1}. The logarithm, e.g., is a function satisfying these properties. A proper choice of t will depend on properties of the concerned density function $p(\theta;\ y)$. As will be seen shortly, for some density functions, considerations on relations between the parent model and the sub-model will lead to a natural choice of such a function.

Now let α be a mixing proportion, a real value between 0 and 1. We shall suppose that α is known. The *mixture embedding* is defined as a two-parameter family specified by density functions of the form

$$p(\theta, \eta;\ y) \equiv c(\alpha, \theta, \eta)\, t^{-1}[\alpha\, t\{f(\theta;\ y)\} + (1-\alpha)\, t\{f(\eta;\ y)\}] \tag{5.103}$$

where $c(\alpha, \theta_{r1}, \theta_{r2})$ is a normalising constant so that $p(\theta, \eta;\ y)$ is a density function. Note that when $\theta = \eta$ or $\alpha(1-\alpha) = 0$, the *mixing density* (5.103) will reduce to the original sub-model given by $f(\theta;\ y)$.

To see how the mixing density (5.103) will look like, we discuss the problem of enlarging the Cauchy location model. Now the sub-model is specified by the density function

$$f(\theta;\ y) = \frac{1}{\pi\{1 + (y-\theta)^2\}}\,. \tag{5.104}$$

Let the mixing proportion $\alpha = \frac{1}{2}$. Then the mixing density can be written as

$$p(\theta, \eta;\ y) \propto t^{-1}[t\{f(\theta;\ y)\} + t\{f(\eta;\ y)\}]\,. \tag{5.105}$$

Comparing (5.105) and (5.104), we see that if the mixing density $p(\theta, \eta;\ y)$ is to be in the Cauchy location-scale family, then the monotone function t will necessarily have the form

$$t(x) = \frac{1}{x}\,. \tag{5.106}$$

Putting (5.106) and (5.104) into (5.105), we have

$$\begin{aligned} p(\theta, \eta;\ y) &\propto \frac{1}{\{1 + (y-\theta)^2\} + \{1 + (y-\eta)^2\}} \\ &\propto \frac{1}{\sigma^2 + (y-\xi)^2} \end{aligned} \tag{5.107}$$

where

$$\sigma = \sqrt{1 + \left(\frac{\theta - \eta}{2}\right)^2} \quad \text{and} \quad \xi = \frac{\theta + \eta}{2}\,.$$

From (5.107) we conclude that $p(\theta, \eta;\ y)$ of (5.107), using new parameterisation (σ, ξ), must be of the form

$$p(\sigma, \xi;\ y) = \frac{\sigma}{\pi\left\{\sigma^2 + (y-\xi)^2\right\}} \tag{5.108}$$

because integration of $p(\theta, \eta; y)$ with respect to y over the real axis must be unity. The density functions $p(\sigma, \xi; y)$ define the Caucy location-scale family with location parameter ξ and scale parameter σ. Note that if $\theta = \eta$ then $\sigma = 1$, so (5.108) will reduce to the Cauchy location-model.

5.12.3 EXAMPLES

Now we shall demonstrate the technique of model enlargement as a general method to solving the problem of multiple roots by studying several examples. In each of the following examples there exists a parent model within the same family so that the sub-model can be embedded in this model in a natural way.

Example 5.19 Circular model

This is the example we have already studied in Section 5.3. In Example 5.2, we were interested in estimating the angle of the mean of a bivariate random variable assuming that the mean is on the unit circle. We found that there are multiple roots associated with the quasi-score and examined this problem through the irregularity of the underlying estimating function. In Example 5.3 we extended the unit circle model to a larger model assuming that the mean of the bivariate random variable is on a circle but with *unknown* radius r. The (regular) two-dimensional estimating function (5.13) and (5.14) jointly gives a unique solution to the problem.

Example 5.20 Cauchy location-scale model

The Cauchy location-model is known to be prone to visual outliers. Setting the scale parameter to a predetermined number such as unity and treating the problem in the restricted location model may be in serious conflict with the observed data. A natural generalisation is to treat the data $y_1, \ldots, y_n$ as iid observations from the Cauchy location-scale model having density function (5.108), which was derived as the mixing density using the function t given by (5.106).

Differentiating the total log likelihood

$$\ell = n \log \sigma - \sum_{i=1}^{n} \log\{\sigma^2 + (y_i - \theta)^2\} - n \log \pi$$

with respect to θ and σ, we get the score estimating functions

$$\frac{\partial \ell}{\partial \theta} = \sum_{i=1}^{n} \frac{2(y_i - \theta)}{\sigma^2 + (y_i - \theta)^2},$$

$$\frac{\partial \ell}{\partial \sigma} = \frac{n}{\sigma} - \sum_{i=1}^{n} \frac{2\sigma}{\sigma^2 + (y_i - \theta)^2}.$$

Solving the likelihood equations

$$\frac{\partial \ell}{\partial \theta} = 0, \quad \frac{\partial \ell}{\partial \sigma} = 0$$

we can obtain all possible stationary points of the likelihood surface. Now we investigate the properties of the likelihood surface at a stationary point.

Since a stationary point must satisfy $\partial \ell / \partial \sigma = 0$ we then have

$$\frac{n}{\sigma^2} = \sum_{j=1}^{n} \frac{2}{\sigma^2 + (y_j - \theta)^2} . \tag{5.109}$$

This relation implies that the second derivative of the log likelihood with respect to σ at any stationary point is always negative because

$$\frac{\partial^2 \ell}{\partial \sigma^2} = -\sum_{i=1}^{n} \frac{4(y_i - \theta)^2}{\left\{\sigma^2 + (y_i - \theta)^2\right\}^2} < 0 . \tag{5.110}$$

This shows that, at each θ, the likelihood surface is unimodal when it is considered as a function of σ alone by Proposition 4.11. This is because if two such stationary points occur, they must be separated by a third stationary point with $\partial^2 \ell / \partial \sigma^2 > 0$, which contradicts (5.110).

Similarly, we can also show the unimodality of the likelihood function when it is evaluated at the maximum likelihood estimate of σ (Copas, 1975). This is because (Copas, 1975)

$$\begin{aligned}
\frac{\partial^2 \ell}{\partial \theta^2} &= 2 \sum_{i=1}^{n} \frac{(y_i - \theta)^2 - \sigma^2}{\{\sigma^2 + (y_i - \theta)^2\}^2} \\
&= -\frac{1}{\sigma^2} \sum_{i=1}^{n} \frac{\{(y_i - \theta)^2 - \sigma^2\}^2}{\sigma^2 + (y_i - \theta)^2} \le 0
\end{aligned} \tag{5.111}$$

where the second equality is due to (5.109).

The likelihood surface will be unimodal if we can further show that the determinant of the Hessian matrix

$$\Delta = \left(\frac{\partial^2 \ell}{\partial \theta \partial \sigma}\right)^2 - \frac{\partial^2 \ell}{\partial \theta^2} \frac{\partial^2 \ell}{\partial \sigma^2}$$

is non-positive. Note that at any stationary point of the log-likelihood function we have

$$\begin{aligned}
\frac{\partial^2 \ell}{\partial \theta \partial \sigma} &= -4\sigma \sum_{i=1}^{n} \frac{y_i - \theta}{\{(y_i - \theta)^2 + \sigma^2\}^2} \\
&= \frac{2}{\sigma} \sum_{i=1}^{n} \frac{(y_i - \theta)\{(y_i - \theta)^2 - \sigma^2\}}{\{(y_i - \theta)^2 + \sigma^2\}^2}
\end{aligned}$$

and

$$\frac{\partial^2 \ell}{\partial \theta^2} = 2\sum_{i=1}^{n} \frac{(y_i-\theta)^2-\sigma^2}{\{(y_i-\theta)^2+\sigma^2\}^2}$$
$$= \frac{-1}{\sigma^2}\sum_{i=1}^{n} \frac{\{(y_i-\theta)^2-\sigma^2\}^2}{\{(y_i-\theta)^2+\sigma^2\}^2}\,.$$

Using these expressions we can then write $\sigma^2\Delta/4$ as follows:

$$\left[\sum_{i=1}^{n} \frac{(y_i-\theta)\{(y_i-\theta)^2-\sigma^2\}}{\{(y_i-\theta)^2+\sigma^2\}^2}\right]^2 - \sum_{i=1}^{n}\left\{\frac{y_i-\theta}{(y_i-\theta)^2+\sigma^2}\right\}^2 \sum_{i=1}^{n}\left\{\frac{(y_i-\theta)^2-\sigma^2}{(y_i-\theta)^2+\sigma^2}\right\}^2$$

which is non-positive by the Cauchy–Schwarz's inequality. It then follows that the determinant Δ is also non-positive.

Consequently by Proposition 4.11, the likelihood equation for the Cauchy location-scale model will have a unique solution. The technique of proving the uniqueness of a root to the likelihood equation demonstrated above may be applied to other situations when an estimating function is derived from some scalar objective function. To prove that the estimating equation has a unique root, it sometimes suffices to investigate the behaviours of the objective function at its possible stationary points.

Example 5.21 Estimating the correlation coefficient

Consider a set of independent observations (x_i, y_i), $i = 1, \ldots, n$, from a bivariate normal distribution, which is standardised to have means $\mu_x = \mu_y = 0$ and variances $\sigma_x^2 = \sigma_y^2 = 1$. We assume that there is an unknown correlation coefficient ρ between any x_i and y_i, and we wish to infer the value of ρ based on these observations. This problem was given a detailed analysis in Section 4.3. There we saw how catastrophic theories may shed light on the study of the multiple root issue. In Example 5.16 we considered the problem again, emphasising the merits and limitations of the approach by examining the asymptotic behaviours of available analytic formulae. Now we shall discuss this problem once again from the viewpoint of model enlargement.

The log likelihood function for the standardised bivariate normal distribution can be written as

$$\frac{\partial \ell}{\partial \rho} = \frac{n\rho}{1-\rho^2} - \frac{\rho}{(1-\rho^2)^2}\left(\sum_{j=1}^{n} x_j^2 - 2\rho\sum_{j=1}^{n} x_j y_j + \sum_{j=1}^{n} y_j^2\right) + \frac{1}{1-\rho^2}\sum_{j=1}^{n} x_j y_j\,.$$

The likelihood equation reduces to the cubic equation

$$P(\rho) = \rho\,(1-\rho^2) + (1+\rho^2)\sum_{j=1}^{n}\frac{1}{n}(x_j y_j) - \rho\,\sum_{j=1}^{n}\frac{1}{n}(x_j^2 + y_j^2) = 0$$

which can have as many as three real roots in the interval $(-1, 1)$.

A natural way of enlarging this model is to introduce an additional parameter

$$\sigma_x^2 = \sigma_y^2 = \sigma^2$$

based on the same normal model. The joint log likelihood for this model becomes

$$\ell = -\frac{n(\beta - 2\,\alpha\,\rho)}{2\sigma^2(1-\rho^2)} - \frac{n}{2}\log\left\{\sigma^4(1-\rho^2)\right\} - n\,\log(2\,\pi)$$

where α and β are given by

$$\alpha = \frac{1}{n}\sum xy \quad \text{and} \quad \beta = \frac{1}{n}\sum(x^2+y^2)$$

respectively. Setting $\partial\ell/\partial\rho$ and $\partial\ell/\partial\sigma^2$ to zero, we have

$$\frac{n}{\sigma^2(\rho^2-1)^2}\left[\alpha(1+\rho^2) - \rho\{\beta + \sigma^2(\rho^2-1)\}\right] = 0$$

$$-\frac{n}{2\sigma^4(\rho^2-1)}\{\beta - 2\alpha\rho + 2\sigma^2(\rho^2-1)\} = 0$$

which reduce to

$$\alpha(1+\rho^2) - \rho\{\beta + \sigma^2(\rho^2-1)\} = 0 \tag{5.112}$$

$$(\rho^2-1)[\beta - 2\alpha\rho + 2\sigma^2(\rho^2-1)] = 0\,. \tag{5.113}$$

Equations (5.112) and (5.113) jointly define a unique solution given by

$$\hat{\sigma}^2 = \frac{\beta}{2} = \frac{1}{2n}\sum(x^2+y^2), \quad \hat{\rho} = \frac{2\alpha}{\beta} = \frac{2\sum xy}{\sum(x^2+y^2)}. \tag{5.114}$$

In (5.114), $\hat{\sigma}^2$ is the pooled sample variance and $\hat{\rho}$ coincides with the usual moment estimator for the correlation coefficient.

5.13 Non-existence of roots

Up to now we have discussed varieties of methods for choosing an appropriate root from among roots to an estimating equation. Sometimes it may happen that there does not exist any (real) solution to a given estimating equation. By no solution we mean that there is no finite value in the parameter space, which solves the estimating equation. This occurs, for instance, when the concerned parameter space Θ is unbounded and an estimating equation cannot be satisfied by any point in the interior of Θ. The latter case often means that the observations do not contain enough information in order to reveal the structure of the mathematical model under consideration. A case in point is when there is some degeneracy occuring in the observed data.

It is difficult to obtain a general set of sufficient and necessary conditions for existence and uniqueness of solutions to a general estimating equation. For parametric models, however, such conditions can be obtained if one may assume the concavity of the log-likelihood function. Results along this line may be found in Haberman (1974), Wedderburn (1976), Silvapulle (1981) and Kaufmann (1988).

Consider, for instance, an experiment with binary outcomes. The data are obtained as frequencies y_j/m_j, expressing the fact that there are y_j 'patients' observed among m_j subjects in category j under some exposure X_j. Here X_j is a design or covariate matrix of dimension $p \times n$. Suppose that we wish to fit a logistic regression model

$$\operatorname{logit} \pi_j = \log \frac{\pi_j}{1 - \pi_j} = X_j^{\mathrm{t}} \beta$$

where π_j is the probability of 'contracting a disease' when exposed to X_j, and β is the vector parameter of interest of length p.

Wedderburn (1976) and Haberman (1977) show that if

$$0 < y_j < m_j \qquad \text{for} \qquad j = 1, \ldots, n$$

then the existence and uniqueness of the MLE are guaranteed. Now let

$$y = (y_1, \ldots, y_n)^{\mathrm{t}} \quad \text{and} \quad m = (m_1, \ldots, m_n)^{\mathrm{t}}.$$

Then the sufficient and necessary conditions for the existence and uniqueness of the MLE using the logistic model are that the following systems

$$\begin{aligned} \left(X^{\mathrm{t}} y\right) \beta &\geq 0 \\ \{X^{\mathrm{t}}(y - m)\}\beta &\leq 0 \end{aligned} \tag{5.115}$$

have only the trivial solution $\beta = 0$, where X is a covariate matrix. See Fahrmeir and Tutz (1994, p. 42) for more details on this point. In (5.115) the parameter is allowed to vary in the p-dimensional Euclidean space $\mathbb{R}^p$. When conditions (5.115) are violated, there does not exist any finite solution to the likelihood equation. Sample configurations violating (5.115) often occur in small observational studies.

In this section, we discuss briefly how one may overcome the problem of no solution caused by data degeneracy. We shall see that the Bayesian approach provides a useful remedy for this problem. In this approach, the problem of data degeneracy is compensated by the introduction of some prior knowledge, which acts as a penalty to the degenerate estimating equation. For this approach to work, it is important to ensure the appropriateness of the particularly chosen prior distribution.

Now we consider an example concerning the first urinary tract infections among young women. We consider a data set (Foxman *et al.*, 1997; available at www.cytel.com/examples/ch1.html) on first urinary tract infections (UTI), also known as cystitis or bladder infections, painful infections occurring frequently among women between late adolescence and menopause. In the US, 6.2 million yearly office visits are for such infections. The impact of these recurring infections on quality of life can be substantial, resulting in repeated physician visits and disability days. The major objective of UTI studies is to locate the potential risk sexual exposures, *vaginal intercourse* and *diaphragm use* being established ones, so that modification of them may effectively reduce the risk of UTI (Foxman *et al.*, 1997).

There are six binary covariates for this data set represented as 0 or 1 here: age($<$24), oral contraceptive use (OC), condom use (CM), lubricated condom use (LC), spermicide use (SE) and diaphragm use (DM). These covariates form 24 covariate classes, the observed frequencies being

$$\frac{y_j}{m_j}, \qquad j = 1, \ldots, 24 .$$

Among 239 subjects, 130 women had UTI. That is,

$$\sum_{j=1}^{24} m_j = 239, \quad \sum_{j=1}^{24} y_j = 130 .$$

One important characteristic of this data set is that

$$m_j = y_j \quad \text{for all} \quad x_{6j} = 1$$

where x_{6j}s are values corresponding to DM. In other words, all subjects with diaphragm use had UTI. We also note that for these covariate classes $m_j \leq 3$. Thus the sufficient and necessary conditions for the existence of the maximum likelihood estimate are violated by this data set: the likelihood function is maximised at the boundary of the parameter space when applying a usual logistic fit using all six covariates.

To analysis this data set we assume that the Y_j's are independent and

$$Y_j \sim \text{bin}\,(m_j, \pi_j) \qquad j = 1, \ldots, n = 24 . \tag{5.116}$$

A first attempt to analyse this degenerate data set may be to fit a usual logistic regression model

$$\log \frac{\pi_j}{1 - \pi_j} = X_j^{\text{t}} \beta = \beta_0 + x_{j1}\beta_1 + \cdots + x_{j5}\beta_5 \tag{5.117}$$

for $j = 1, \ldots, n$, by excluding the last covariate DM from the analysis. So doing, we find the MLEs for the odds ratios to be

$$(0.44,\ 2.83,\ 0.87,\ 10.69,\ 0.10,\ 0.50)\,.$$

Note that

$$\exp\{\hat{\beta}_2\} = 0.87 \approx 1$$

Suggesting that OC might have little effect in causing UTI. But since DM is a known risk factor, the estimates based on this artificial sub-model are not to be trusted unless evidence can be established otherwise.

An alternative approach is to replace the artificial sub-model (5.117) by the following model including all six covariates:

$$\text{logit}(\pi_j) = \beta_0 + x_{j1}\beta_1 + \cdots + x_{j5}\beta_5 + x_{j6}\beta_6 \tag{5.118}$$

for $j = 1, \ldots, n$, and assume that each π_j has a beta prior distribution

$$\pi_j \sim \frac{1}{B(a,\ a)}\pi_j^{a-1}(1-\pi_j)^{a-1} \tag{5.119}$$

where a is an unknown hyper-parameter. The three components (5.116), (5.118) and (5.119) define a *Bayesian logistic regression model*. Solving the posterior score estimating equation we obtain estimates for the odds ratios corresponding to all six covariates. These estimates of course depend on the value of the hyper-parameter a.

Figure 5.9 shows the estimate $\exp\{\hat{\beta}_2\}$ as a function of a. For the uniform prior ($a = 1$), the estimating equation, the posterior likelihood equation, is the same as the likelihood equation and is therefore ill conditioned. A slight increase in a

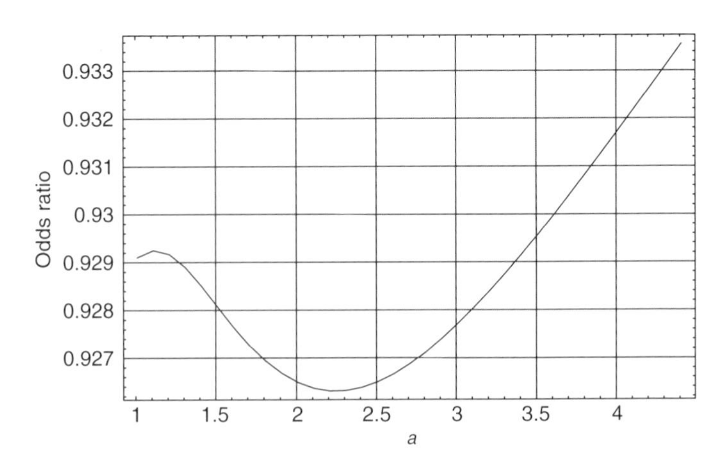

Fig. 5.9 Estimates of the odds ratio, $\exp\{\hat{\beta}_2\}$, as a function of a in the prior distribution $\beta(a, a)$.

pushes the estimates towards 1. A larger value of a implies a greater degree of prior belief about the parameter of interest. Figure 5.9 shows that the estimates are insensitive to the priors. The estimates are seen to be stable around a value much higher than the estimate obtained from the unjustified sub-model by excluding the covariate DM. The situations for the estimation of the other parameters are similar.

There still leaves unanswered the problem of choosing the *best* possible value of the hyper-parameter a for the Bayesian logistic regression model. There are many competing criteria for the purpose of *model selection*. The *cross-validatory approach* has the advantage over many other methods in that it requires little mathematical analysis so that relatively sophisticated models can be included into consideration. There are several versions of this approach, the *leave-one-out cross-validation* being the simplest and the most useful one.

To describe the technique of the leave-one-out cross-validation in our present context, suppose that we wish to consider a class of models containing (5.118) as the largest model. All other models in the class are obtained by excluding some covariates from the right-hand side of (5.118). For a particular model in the class, say the kth one, we fix the value of a, and compute the posterior mode $\hat{\beta}_{(j)}(a)$ without the use of data from the jth covariate class. So doing, we obtain n estimates

$$\hat{\beta}_{(1)}(a), \ldots, \hat{\beta}_{(n)}(a) .$$

Based on these values, we then evaluate the log-likelihood function

$$\mathrm{CV}(a) = \sum_{j=1}^{n} \log f\left(\hat{\beta}_{(j)}(a);\ y_j\right) \tag{5.120}$$

where $f(\beta;\ y)$ is the likelihood function, which in our problem is the binomial likelihood. Note that $\mathrm{CV}(a)$ is a function of the hyper-parameter a alone. Let $\hat{a}_k$ be the value that maximises (5.120). Repeating this procedure for each model in the class, we obtain the values

$$\mathrm{CV}(\hat{a}_1), \ldots, \mathrm{CV}(\hat{a}_M)$$

where M is the number of models in the class. The method of leave-one-out cross-validation proposes to choose the kth model in the class if $\mathrm{CV}(\hat{a}_k)$ is the largest among the M values.

Continuing our example of UTI study, we found that the optimal model is the one with OC removed and the corresponding optimal value of a is estimated to be

$$\hat{a} = 1.0049 .$$

This value of a corresponds to a prior distribution very close to the non-informative uniform distribution with $a = 1$ (see Figure 5.10).

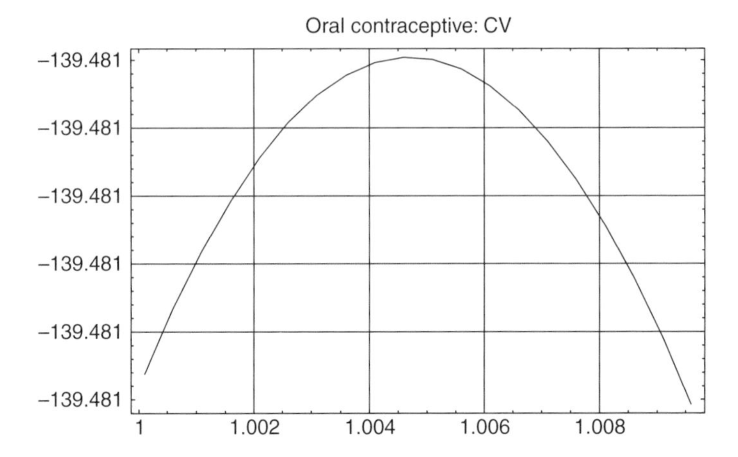

Fig. 5.10 Delete-one cross-validatory choice of the optimal value of the hyper-parameter a for the urinary tract infections example using the Bayesian logistic regression model. The horizontal axis: the value of the hyper-parameter; the vertical axis: the value of the delete-one cross-validation criterion.

5.14 Confidence intervals using the estimating function bootstrap

5.14.1 INTRODUCTION

Up to now we have concentrated on constructing a *point estimator* for a parameter of interest θ based on an unbiased estimating function $g(\theta, Y)$. In particular we discussed various techniques for choosing an appropriate estimator $\hat{\theta}$ when roots to estimating equation $g(\theta, Y) = 0$ have to be found numerically using some methods discussed in Chapter 3 and when the estimating equation gives rise to multiple solutions. More often, however, *confidence intervals* are used instead of point estimators for inferential purposes. In this section we discuss how confidence intervals may be constructed using an unbiased estimating function. In Section 5.14.2 we shall see that computational cost can be substantial for the purpose of setting an accurate confidence interval using the traditional *bootstrap methods* even though the original estimating equation admits a unique solution. The computational problem is seen to be avoided using the *estimating function bootstrap* introduced in Section 5.14.3. The latter confidence intervals have the same asymptotic accuracy as the traditional ones using the bootstrap-t, or other sophisticated bootstrap methods such as the $\mathrm{BC_a}$ method, and, in additional, have the nice *equivarience property*‡ under monotone transformations of the parameter.

‡ The alternative terminology *transformation-respecting* is also used frequently in the literature.

5.14.2 BOOTSTRAP-t INTERVALS

Throughout this section we shall assume that θ is a scalar. Most of the results however can be extended to the case of a vector parameter with little difficulty.

An interval I_α is said a *confidence interval* with *nominal coverage probability* α, or simply an α-level confidence interval, if for every $\theta \in \Theta$ the following holds

$$\lim_{n\to\infty} \mathrm{P}_\theta\{\theta \in I_\alpha\} = \alpha\,. \tag{5.121}$$

That is, when sample size n goes to infinity, the interval I_α will contain the parameter of interest θ with probability tending to the prescribed nominal level α. Given an unbiased estimating function $g(\theta, Y)$ and $\alpha \in (0, 1)$, how should we proceed to set such an interval I_α?

The rule of thumb is as follows. One first solves the estimating equation $g(\hat{\theta}, Y) = 0$ for an estimator $\hat{\theta}$, which is assumed to be the unique root, and then looks for an appropriate estimator, say $\hat{\sigma}^2/n$, for the variance of $\hat{\theta}$. Then an α-level confidence interval can be constructed by virtue of the asymptotic normality of the consistent root of g

$$\frac{\sqrt{n}(\hat{\theta}-\theta)}{\hat{\sigma}} \xrightarrow{\mathcal{L}} N(0, 1)\,. \tag{5.122}$$

For example, if a left-sided interval is appropriate for the problem at hand, then one may set I_α to be

$$I_\alpha = (-\infty,\ \hat{\theta} + n^{-1/2}\hat{\sigma} z_\alpha) \tag{5.123}$$

where z_α is the α-th percentile of the unit normal distribution, i.e. $\Phi(z_\alpha) = \alpha$, where $\Phi(\cdot)$ is the cumulative distribution function of $N(0, 1)$.

The normal limit in (5.122) can be shown to be accurate at $O(n^{-1/2})$ in approximating the distribution of the studentised quantity. That is,

$$\mathrm{P}_\theta\left\{\frac{\sqrt{n}(\hat{\theta}-\theta)}{\hat{\sigma}} \le t\right\} = \Phi(t) + O(n^{-1/2})\,. \tag{5.124}$$

It can be deduced then that the confidence interval I_α defined by (5.123) is *first-order accurate*, meaning that

$$\mathrm{P}_\theta\{\theta \in I_\alpha\} = \Pr\left\{\theta \le \hat{\theta} + n^{-1/2}\hat{\sigma} z_\alpha\right\} = \alpha + O(n^{-1/2})\,. \tag{5.125}$$

This formula says that the probability of I_α to miss θ is of order $O(n^{-1/2})$, a quantity tending to zero as fast as $n^{-1/2}$. This quantity is usually called the *error probability*, a criterion commonly used to judge the accuracy of a confidence interval. Equation (5.125) can be easily derived using (5.124). We leave this as an exercise for the reader. Formula (5.125) gives the order of accuracy of the asymptotically consistent confidence interval I_α in the sense of (5.121).

The first-order accurate confidence interval I_α can be improved using the bootstrap method. To apply the usual bootstrap technique we shall assume for the moment that the data $y = \{y_1, \ldots, y_n\}$ are realisations of independent and identically distributed random variables, $Y = \{Y_1, \ldots, Y_n\}$. Let $Y^* = \{Y_1^*, \ldots, Y_n^*\}$ be bootstrap samples drawn *with replacement* from $y_1, \ldots, y_n$, each datum having probability mass $1/n$. The bootstrap estimating function $g(\theta, Y^*)$ is defined exactly as the original one $g(\theta, y)$ except that the observations $y = \{y_1, \ldots, y_n\}$ are replaced by the bootstrap samples $Y^* = \{Y_1^*, \ldots, Y_n^*\}$. Solving the bootstrap estimating equation, $g(\hat{\theta}^*, Y^*) = 0$, which again is assumed to have a unique solution, we get a bootstrap version $\hat{\theta}^*$ of the estimate $\hat{\theta}$ which solves the equation $g(\hat{\theta}, y) = 0$. Let $\hat{\sigma}^*$ be defined exactly as $\hat{\sigma}$ except that the bootstrap sample Y^* is used instead of the original sample y. Now we define the *bootstrap-t variate* as follows:

$$T^* = \frac{\sqrt{n}(\hat{\theta}^* - \hat{\theta})}{\hat{\sigma}^*}.$$

and let w_α be the α-th quantile of the distribution of T^*. The *bootstrap-t method* proposes to replace the normal interval I_α by

$$I_\alpha^* = (-\infty, \hat{\theta} + n^{-1/2}\hat{\sigma}w_\alpha) \tag{5.126}$$

with the normal quantile z_α in I_α replaced by the bootstrap quantile w_α. The interval I_α^* is called a *bootstrap-t interval*.

The rationale behind the bootstrap-t method is that while the unit normal distribution approximates the distribution F of $\sqrt{n}(\hat{\theta} - \theta)/\hat{\sigma}$ at the rate $n^{-1/2}$, as stated in (5.124), the bootstrap distribution of T^* approximates F at the faster rate n^{-1}. That is,

$$\mathrm{P}_*\left\{\frac{\sqrt{n}(\hat{\theta}^* - \hat{\theta})}{\hat{\sigma}^*} \leq t\right\} - \mathrm{P}_\theta\left\{\frac{\sqrt{n}(\hat{\theta} - \theta)}{\hat{\sigma}} \leq t\right\} = O_p(n^{-1}) \tag{5.127}$$

where P_* denotes probability under the bootstrap sampling. By virtue of (5.127), one can then easily derive the result that

$$\mathrm{P}_\theta\{\theta \in I_\alpha^*\} = \mathrm{P}_\theta\{\theta \leq \hat{\theta} + n^{-1/2}\hat{\sigma}w_\alpha\} = \alpha + O(n^{-1}). \tag{5.128}$$

Again we omit the derivation of (5.128). An interval having coverage error of order $O(n^{-1})$ is called *second-order accurate*. Therefore, the bootstrap-t interval I_α^* is second-order accurate by (5.128), improving the normal interval I_α, which, by (5.125), is only first-order accurate.

The improved second-order accuracy of the bootstrap-t interval is achieved, nevertheless, at the expense of heavy computation. This is because the bootstrap estimates $\hat{\theta}^*$ have to be found numerically using some root-finding algorithm of Chapter 3 for *every* bootstrap replication. In a problem with small to medium sample size, the bootstrap estimating equations usually have to be solved for about 2000 times in order to approximate the distribution of the bootstrap-t statistic T^*. This process can be costly and time-consuming. More importantly, if an algorithm fails to converge in one or more cases, the bootstrap variance estimates $\hat{\sigma}^*$ can be affected badly. The success of the bootstrap-t method sensitively depends on the availability of a stable variance estimate; see Hall (1992), Efron and Tibshirani (1993) and Davison and Hinkley (1997), etc. Accordingly, when a root-finding algorithm fails to converge for even a small portion of times, the resulting bootstrap-t interval I_α can be quite misleading.

A second problem with the interval I_α^* concerns its theoretical property under reparameterisation. The interval I_α^*, as well as I_α, is not equivariant under reparametrisations, that is, $\xi(I_\alpha^*)$ is not the α-level bootstrap-t interval for a transformed parameter $\xi = \xi(\theta)$, where $\xi(\cdot)$ is a non-linear monotone function. Second-order accurate bootstrap intervals, such as a *bias-corrected accelerated* or a $\mathrm{BC_a}$ interval, do exist, which are equivariant under reparametrisations. To apply these methods, however, the computational problem for finding the roots of the bootstrap estimating equations is still essential. Therefore, the convergence issue of an algorithm is again crucial for the success of such methods. For example, to apply the $\mathrm{BC_a}$ method, one needs to estimate two quantities called the *bias-correction* and the *acceleration constant*. The usual practice is that the amount of bias-correction is estimated by the same bootstrap outputs and the acceleration constant is estimated using the jackknife method. These estimates can be badly affected by one or more cases when the algorithm fails to converge. There is a large literature on bootstrap confidence intervals, for details we refer the reader to Davison and Hinkley (1997) and references therein.

5.14.3 THE ESTIMATING FUNCTION BOOTSTRAP

The second method for obtaining a confidence interval is to use an asymptotically pivotal estimating function parallelling (5.122). The confidence intervals obtained using this method are equivariant under reparametrisations and avoid the computational problems mentioned in the previous subsection.

Now we shall assume that unbiased estimating function $g(\theta, Y)$ can be written in the form

$$g(\theta, Y) = \sum_{i=1}^{n} g_i(\theta, Y_i) \tag{5.129}$$

where $Y_1, \ldots, Y_n$ are independent but may have possibly different distributions as in a regression problem. In (5.129) each component $g_i(\theta, Y_i)$ is assumed to be unbiased as well.

Using the assumption of independence, the variance of $g(\theta, Y)$ can be written as

$$\mathrm{Var}\{g(\theta, Y)\} = \sum_{i=1}^{n} \mathrm{Var}\{g_i(\theta, Y_i)\}\,. \tag{5.130}$$

Several estimates are possible are possible for estimating this variance. One such an estimate is given by

$$V(\theta, y) = \sum_{i=1}^{n} \{g_i(\theta, y_i) - \bar{g}(\theta, y)\}^2 \tag{5.131}$$

where

$$\bar{g}(\theta, y) = \frac{1}{n} \sum g_i(\theta, y_i) .$$

Note that estimate $V(\theta, y)$ depends on θ. Similar to (5.122), under mild conditions (e.g. Billingsley, 1995, p. 357), we may apply the *central limit theorem* to $g(\theta, Y)$ to give

$$g_T(\theta, Y) = \frac{g(\theta, Y)}{\sqrt{V(\theta, Y)}} \xrightarrow{\mathcal{L}} N(0, 1) . \tag{5.132}$$

Now we make the further assumption that $g_T(\theta, Y)$ is monotone in θ. For concreteness, we shall assume for the rest of this subsection that it is monotonically decreasing in θ. By (5.132) we have

$$\Pr\{g_T(\theta, Y) \geq z_{1-\alpha}\} \approx \alpha$$

which yields the interval

$$J_\alpha = (-\infty, \hat{\theta}_\alpha) \tag{5.133}$$

with nominal coverage probability α, where $\hat{\theta}_\alpha$ solves the equation

$$g_T(\hat{\theta}_\alpha, y) = z_{1-\alpha} . \tag{5.134}$$

The interval J_α defined by (5.133) is a competitor of the interval I_α defined by (5.123). Under some conditions it can be shown that the normal approximation (5.132) is in general good to order $O(n^{-1/2})$. For some specific parametric models involving symmetric distributions this order, however, can be shown to be $O(n^{-1})$. One such an example is when g is the score from the Cauchy location model. Accordingly, the confidence interval J_α is first-order accurate in general, with occasional possibility to achieve second-order accuracy even though the corresponding I_α is first-order accurate.

The interval J_α has an important advantage over the interval I_α in that it is equivariant under reparametrisations. This can be easily appreciated by examining the studentised quantity $g_T(\theta, Y)$ defined by (5.132). Under a monotone transformation $\theta = \theta(\xi)$, the estimating function for ξ becomes

$$g(\xi, Y) = \left\{\frac{\partial}{\partial \xi}\theta(\xi)\right\} g(\theta(\xi), Y)$$

and the variance estimate for $g(\xi, Y)$ using formula (5.131) differs from that for $g(\theta, Y)$ by the square of the same factor, the constants being cancelled out to give

$$g_T(\xi, Y) = g_T(\theta, Y) .$$

So if the relation $\xi = \xi(\theta)$ is monotonely increasing, then the α-level normal confidence interval for ξ is given by $\xi(J_\alpha)$ using the normal approximation (5.132).

As in the previous subsection suppose that we wish to improve the interval J_α to construct a second-order accurate confidence interval. Mimicking the bootstrap-t method, this is possible if we can improve the normal limit in (5.132) using the distribution of a bootstrap version of $g_T(\theta, Y)$. To do so, let $\hat{\theta}$ be the unique solution to $g(\hat{\theta}, y) = 0$. The *estimating function bootstrap* method proposed by Hu and Kalbfleisch (2000) consists of the following steps.

1. Compute the observed summands $g_i(\hat{\theta}, y_i)$,
$$z_1 = g_1(\hat{\theta}, y_1), \ldots, z_n = g_n(\hat{\theta}, y_n)\,.$$
Note that $\sum z_i = 0$.
2. Let $\hat{F}$ be the empirical distribution putting probability mass $1/n$ at each z_i. Note that $\hat{F}$ has mean zero. Draw independent bootstrap samples from $\hat{F}$,
$$Z_1^*, \ldots, Z_n^* \sim \hat{F}\,. \tag{5.135}$$
3. Compute the bootstrap version of $g_T(\theta, y)$,
$$g_T^* = \frac{\sum_{i=1}^n Z_i^*}{\sqrt{\sum_{i=1}^n (Z_i^* - \bar{Z}^*)^2}} \tag{5.136}$$
where $\bar{Z}^* = n^{-1} \sum_{i=1}^n Z_i^*$.
4. Let $\hat{u}_{1-\alpha}$ be the $(1-\alpha)$th percentile of the distribution of g_T^*. Then a left-sided α-level confidence interval is given by
$$J_\alpha^* = (-\infty, \tilde{\theta}_\alpha) \tag{5.137}$$
where $\tilde{\theta}_\alpha$ solves the equation
$$g_T(\tilde{\theta}_\alpha, y) = \hat{u}_{1-\alpha}\,. \tag{5.138}$$

Note that in defining the bootstrap interval (5.137), one only need to solve equations twice for finding the estimate $\hat{\theta}$ and the confidence limit $\tilde{\theta}_\alpha$ defined by (5.138). Like the naive interval J_α, the estimating function bootstrap interval J_α^* is also equivariant under reparametrisations for the same reason as discussed previously. Moreover, Hu and Kalbfleisch (2000) shows that, under certain conditions, the following holds

$$\mathrm{P}_*\{g_T^* \le t\} - \mathrm{P}_\theta\{g_T(\theta, Y) \le t\} = O_p(n^{-1}) \tag{5.139}$$

a result that parallels (5.127), where P_* denotes probability under the bootstrap sampling (5.135). Accordingly, the interval J_α^* of (5.137) is second-order accurate. Although the estimating function bootstrap has the nice properties outlined in this section, it is not without its problem. One difficulty with this method is that it is not clear how to apply this method to situations when the involved estimating function is not a monotone function in the parameter of interest.

5.15 Bibliographical notes

A classical treatment of the problem of multiple roots may be found in Lehmann (1983, Section 6.4). A general discussion on estimation in parametric models, including non-uniqueness of maximum likelihood estimator, is given by Stuart and Ord (1991, Chap. 18), where the estimation of the correlation coefficient from binormal distribution is discussed in detail.

Barnett (1966) gave detailed discussions on how to numerically find all possible roots to a likelihood equation, and presented detailed Monte Carlo experiments on the case of Cauchy location model using various techniques including Daniels (1960) smoothed likelihood method (see Chapter 4). Solari (1969) and Copas (1972) discussed the problem of multiple solutions involving some structural equations.

Recent interests in the problem of multiple roots may be found in Stefanski and Carroll (1985, 1987), McCullagh (1991), Heyde (1997), Heyde and Morton (1998), Singh and Mantel (1998), and Small and Yang (1999), among others. A comprehensive review is given by Small, Wang and Yang (2000).

The roots to an estimating equation do not in general correspond to local extrema of any scalar objective function. This fact provides considerable difficulty to the problem of choosing an appropriate estimate from among roots to an estimating equation. The problem of multiple roots, along with other problems associated with an estimating function, has motivated a line of research for building an objective function for an estimating function; see, McLeish and Small (1992), Li (1993), Li and McCullagh (1994), Small and McLeish (1994), Barndorff-Nielsen (1995), Hanfelt and Liang (1995, 1997), and Wang (1999). For more information on this topic see Chapter 6, where various approaches for building an artificial likelihood function are described in detail.

The estimating function bootstrap for setting confidence intervals discussed in Section 5.14 is based on Hu and Kalbfleisch (2000). For more information on bootstrap confidence intervals, the interested reader may consult Hall (1992), Efron and Tibshirani (1993), Shao and Tu (1995), Davison and Hinkley (1997), among others. The idea of setting confidence intervals using unbiased estimating functions may be traced back to Wilks (1938), who showed that a two-sided version of J_α^* based on the score function has the shortest expected length for all unbiased estimating functions.

6 Artificial likelihoods and estimating functions

6.1 Introduction

In the previous chapters, we have seen that selecting an appropriate root as an estimate from roots to an estimating equation $g(\theta) = 0$ can provide a considerable challenge even for less complicated parametric models. The problem gets more entangled when there does not exist an objective scalar function $\lambda(\theta)$ such that $\dot{\lambda}(\theta) = g(\theta)$. When this is true, there does not exist a parametric model such that the estimating function $g(\theta)$ is the score function of the parametric model. This case can arise only when the dimension of the estimating function $g(\theta)$ is equal to or higher than two. This is because any scalar estimating function $g(\theta)$, under mild conditions, can always be integrated with respect to θ to give a scalar objective function. Geometrically, when there does not exist a scalar objective function for a vector-valued estimating function $g(\theta)$, we obtain a *non-conservative vector field* on the parameter space Θ defined by the estimating function $g(\theta)$. In this chapter we shall discuss various methods for building artificial objective functions for use in conjunction with unbiased estimating functions.

There are several advantages to constructing an objective function. For example, an objective function allows us to *weight* roots of an estimating function and thereby to provide an objective ordering on the roots to assist root selection. Additionally, an objective function may allow us to *pool information* from the data with prior information, be it subjective or empirical. For example, if expert opinions were elicited to obtain probability weights $\pi_1, \ldots, \pi_m$ for each of m roots $\hat{\theta}_1, \ldots, \hat{\theta}_m$, then an objective function λ, which behaves like a log-likelihood, could be combined with the expert weights to produce posterior weights of the form

$$\pi_1 \Lambda(\hat{\theta}_1), \ldots, \pi_m \Lambda(\hat{\theta}_m)$$

where $\Lambda(\theta) = \exp\{\lambda(\theta)\}$.

In this chapter, we shall consider how to build an objective function from several different perspectives and shall discuss the connection with root selection. In the absence of any knowledge about the likelihood function in a semi-parametric model, the first approach is to construct an objective function by projecting the true likelihood ratio into a carefully chosen subspace spanned by functions of the data and the parameter of interest θ. How such a subspace should be chosen depends on the degree of knowledge we possess about the data so that the projection can be practically computed. McLeish and Small (1992) show that when the first two moments can be specified there exists a natural subspace so that the projection can be computed. This approach is introduced in Section 6.2. The derived *artificial likelihood function* may be used in conjunction with the quasi-score.

In Section 6.3 we ask the more general question of how to build an objective function for an optimal estimating function based on a set of elementary estimating functions, which is not necessarily a quasi-score. The idea is to use a particular approximate form of the true log likelihood function so that the projection onto the space spanned by the elementary estimating functions can be calculated. Since an approximate form of the likelihood function is used in obtaining the projection, the generalised projected log likelihood function does not reduce to the projected objective function studied in Section 6.2 for quasi-scores. The two methods for quasi-scores are however locally equivalent.

In Section 6.4 we discuss the possible use of the method by performing a direct line integration of an optimal estimating function $g(\theta, y)$. When $g(\theta, y)$ defines a conservative vector field, there is no ambiguity in the definition. But in general line integration is path-dependent. The difficulty of the approach therefore lies in the choice of an appropriate path. There are two situations, however, when this method may be of some merit. The first is when $g(\theta, y)$ is nearly conservative. For example, if

$$E_\theta\left[\dot{g}(\theta, Y) - \{\dot{g}(\theta, Y)\}^{\mathrm{t}}\right] = 0$$

then $g(\theta, Y)$ may be considered as approximately conservative. The quasi-scores belong to this class of estimating functions. The second case is when local paths are of interest. The latter case is explored in this section.

In Section 6.5 we consider the problem for building an artificial objective function based on an unbiased estimating function $g(\theta)$, which may or may not have the forms assumed in the previous sections. We introduce a geometrical approach by regarding $g(\theta)$ as a vector field defined on the parameter space. The well known *Helmholtz decomposition* theorem is generalised for a vector-valued estimating function of any dimension. The estimating function $g(\theta, y)$ is decomposed into a conservative part and a residual part with zero divergence. This decomposition is not unique. Fortunately, there is a particularly interesting decomposition for a first-order local approximation of $g(\theta, y)$. The potential function associated with the conservative part of this decomposition is proposed as an artificial objective function for $g(\theta, y)$.

Finally in Section 6.6, we introduce a type of objective function associated with an unbiased estimating function using the idea of the generalised method of moments. This approach may be regarded as the generalised least squares method. The usefulness of this approach is demonstrated through an application to the longitudinal data analysis.

6.2 Projected likelihoods

6.2.1 INTRODUCTION

In this and the next section, we shall discuss two types of projected likelihood functions. These functions may be regarded as semiparametric counterparts of the true likelihood functions when only low order moments are available for the underlying random variables. One possible use of such an objective function is to interpret the values of the function evaluated at the roots of an estimating function as measures representing the plausibility of the respective roots. In particular, we shall be interested in projections of likelihood ratios

$$\lambda(\theta) = \frac{L(\theta)}{L(\eta)}$$

into an appropriate subspace, yielding a projected likelihood ratio $\hat{\lambda}(\theta, \eta)$. To be useful in this task, a projected likelihood ratio, say $\hat{\lambda}(\theta, \eta)$, must provide a consistent partial ordering of the parameter space. The following conditions are sufficient, and would appear to be natural:

1. The ordering should be *anti-symmetric*:
$$\hat{\lambda}(\theta, \eta) = -\hat{\lambda}(\eta, \theta).$$
2. The ordering should be *transitive*:
$$\hat{\lambda}(\theta, \xi) \geq 0 \quad \text{if } \hat{\lambda}(\theta, \eta) \geq 0 \text{ and } \hat{\lambda}(\eta, \xi) \geq 0.$$

The projected likelihood ratio discussed in the next subsection fails the anti-symmetry condition. By contrast, the second type of projected likelihood ratio to be introduced in the next section is constructed to be anti-symmetric. However, it is easy to find examples where the latter objective function is not transitive; see Small *et al.* (2000). So neither type of projected likelihoods can guarantee a consistent ordering of the parameter space.

Nevertheless, we do not need to compare all points in the parameter space in order to select roots. For example, if an estimating function has only two roots, then the issue of the transitivity of the projected likelihood is irrelevant to the problem of choosing one of these two values. In addition, the projected likelihood may well be transitive on the roots while not being transitive on the parameter space as a whole. Finally, even if it is not transitive or anti-symmetric on the full set of roots, a projected likelihood may still help us eliminate certain roots from consideration.

In the next section, we ask whether we can find a projection of the likelihood function or the likelihood ratio, instead of the score function, into a properly chosen space of functions. The elements of this space are functions of data and the parameter of interest. As in the case of quasi-score, for such a method to be useful, we should be able to define the projection in terms of the low order moments of the ys. This approach therefore allows us to directly construct an objective function from the start using the assumptions about lower moments.

6.2.2 PROJECTED LIKELIHOODS

Consider independent random variables, $Y_1, \ldots, Y_n$, each having means $\mu_i(\theta)$ and variances $\sigma_i^2(\theta)$, which are known functions of the parameter of interest θ. The quasi-score

$$q(\theta) = \sum_{i=1}^{n} \{\dot{\mu}_i(\theta)\}^{\mathrm{t}} \{\sigma_i^2(\theta)\}^{-1} \{Y_i - \mu_i(\theta)\}$$

may be regarded as the *projection* of the true score function onto the space of linear and unbiased estimating functions spanned by the functions

$$Y_1 - \mu_1(\theta), \ldots, Y_n - \mu_n(\theta).$$

The *inner product* is defined by

$$\langle g_1, g_2 \rangle_\theta = E_\theta(g_1 g_2)$$

between two unbiased estimating functions g_1 and g_2. See Small and McLeish (1994) and Section 2.5. As such, the quasi-score has a number of properties in common with a genuine score function. For instance, it inherits the first two Bartlett identities from the score function: it is unbiased and information-unbiased. This approach is available whenever we are able to specify the first two moments of y_i in order to compute the inner products.

Given such semiparametric information we now consider the projection of the true likelihood ratio

$$\lambda(\theta, \eta) = \frac{L(\eta)}{L(\theta)}$$

onto a proper linear space constituting all functions of the following form:

$$h(\theta; Y_1, \ldots, Y_n) = c(\theta) + \sum_{k, i_1 < \cdots < i_k} a_{i_1 \cdots i_k}(\theta)(Y_{i_1} - \mu_{i_1}) \cdots (Y_{i_k} - \mu_{i_k}) \quad (6.1)$$

where the sum is taken over all non-empty subsets of $\{1, 2, \ldots, n\}$ and $c(\theta)$ is functionally independent of $Y_1, \ldots, Y_n$. McLeish and Small (1992) show that the linear space consisting of functions (6.1) has certain maximality property. Any function $h(\theta; Y_1, \ldots, Y_n)$ of form (6.1) has a constant mean and variance over all distributions of independent random variables satisfying

$$E_\theta(Y_i) = \mu_i(\theta) \quad \text{and} \quad \mathrm{Var}(Y_i) = \sigma_i^2(\theta)$$

and *vice versa*. For more details see Small and McLeish (1994, Chapter 6). Projecting $\lambda(\theta, \eta)$ onto this linear space by minimising the squared distance

$$E_\theta \left[\lambda(\theta, \eta) - c(\theta) - \sum_{k, i_1 < \cdots < i_k} a_{i_1 \cdots i_k} \{Y_{i_1} - \mu_{i_1}(\theta)\} \cdots \{Y_{i_k} - \mu_{i_k}(\theta)\} \right]^2$$

gives the coefficients

$$c(\theta) = 1 \quad \text{and} \quad a_{i_1 \cdots i_k}(\theta) = \prod_{j=1}^{k} \frac{\mu_{i_j}(\eta) - \mu_{i_j}(\theta)}{\sigma_{i_j}^2(\theta)}.$$

Accordingly, the *projected artificial likelihood ratio* takes the form

$$\hat{\lambda}(\theta, \eta) = \prod_{i=1}^{n} \left[1 + \frac{\mu_i(\eta) - \mu_i(\theta)}{\sigma_i^2(\theta)} \{Y_i - \mu_i(\theta)\} \right] \tag{6.2}$$

which was first obtained by McLeish and Small (1992), and treated in more detail in Small and McLeish (1994). The artificial likelihood ratio is tangent to the quasi-score. That is,

$$\frac{\partial}{\partial \eta} \{\hat{\lambda}(\theta, \eta)\}\Big|_{\eta=\theta} = \sum_{i=1}^{n} \frac{\dot{\mu}_i(\theta)}{\sigma_i^2(\theta)} \{Y_i - \mu_i(\theta)\}.$$

Since we have that

$$E_\theta\{\hat{\lambda}(\theta, \eta)\} = 1$$

the function $\log\{\hat{\lambda}(\theta, \eta)\}$ is therefore an exact *local log-density* in the sense of Severini (1998). Moreover, the projected likelihood ratio has the property that

$$E_\eta\{\hat{\lambda}(\theta, \eta)\} \geq 1$$

with equality if and only if

$$\mu_i(\eta) = \mu_i(\theta) \quad \text{for all } i = 1, \ldots, n.$$

Note that in deriving $\hat{\lambda}(\theta, \eta)$ the value θ is assumed to be the true value of the parameter and η is assumed to lie in a neighbourhood of θ. Therefore, a further approximation to the likelihood ratio is by substituting the value of θ in (6.2) by an estimate $\hat{\theta}$ of θ. For instance, $\hat{\theta}$ may be obtained using the quasi-score. Finally, we arrive at the following semi-parametric approximation to the likelihood function:

$$\hat{L}(\theta) \propto \prod_{i=1}^{n} \left[1 + \frac{\mu_i(\theta) - \mu_i(\hat{\theta})}{\sigma_i^2(\hat{\theta})} \{Y_i - \mu_i(\hat{\theta})\} \right]. \tag{6.3}$$

For multivariate responses, the projected artificial likelihood function (6.3) takes the following generalised form:

$$\hat{L}(\theta) \propto \prod_{i=1}^{n} \left[1 + \{\mu_i(\theta) - \mu_i(\hat{\theta})\}\{\Sigma_i(\theta)\}^{-1}\{Y_i - \mu_i(\hat{\theta})\} \right] \tag{6.4}$$

where $\Sigma_i(\theta)$ is the covariance matrix of Y_i. The projected artificial likelihood function was applied to stable laws using the Toronto stock market data; for details see McLeish and Small (1992) and Small and McLeish (1994, Section 6.7).

6.3 Generalised projected artificial likelihoods

6.3.1 INTRODUCTION

The projection approach for constructing artificial likelihood functions of the last section is natural because the linear space onto which the likelihood ratio is projected is the maximal linear vector space allowing exact calculation of the projection under the first and the second moment specifications. In this section we extend this idea by projecting the likelihood ratio onto another subspace, which is associated with an optimal estimating function. For this approach to work, the central problem is again whether we can compute the desired projection using the semiparametric assumptions of moments. It turns out in general that such projection, using the usual inner product, cannot be calculated given the information that is used in specifying the optimal estimating function. To avoid this difficulty, we first approximate the likelihood ratio using an appropriate Taylor expansion.

Now consider the general situation where there are k conditions available involving a parameter θ of dimension $p \leq k$. Here the number of conditions k may or may not be the sample size. Suppose that the k conditions lead to an estimating function $g(\theta, y)$ satisfying the unbiasedness condition

$$E_\theta\{g(\theta, Y)\} = 0 \quad \text{for all } \theta \in \Theta. \tag{6.5}$$

An estimating function $g(\theta, y)$ satisfying (6.5), which is used as a building block for constructing other optimal estimating functions, is usually called an *elementary estimating function*.

An elementary estimating function, a column vector of length k, is usually constructed by putting together all available information about the underlying random variables. For instance, if we can specify the means $\mu_i(\theta)$ for the Y_i, then the ith component of $g(\theta, Y)$ can be simply taken as $Y_i - \mu_i(\theta)$. In this case the dimension of the elementary estimating function is equal to the sample size. In order to make inference about θ, we have to construct an p-dimensional estimating function $q(\theta, Y)$ using the elementary estimating function $g(\theta, Y)$. Ideally, we wish $q(\theta, Y)$ to be optimal in certain sense. In Chapter 2, we introduced several fixed sample optimality criteria of estimating functions. These criteria are stated in terms of the variance of the standardised estimating function. When the inferential focus is on estimators of the estimating equations, however, optimality criteria in terms of the asymptotic variances of the estimators are equally sensible. Under suitable conditions, the fixed sample optimality criterion based on the Godambe efficiency is equivalent to the asymptotic optimality criterion based on the asymptotic variance of the consistent root to the corresponding estimating equation; see Kale (1985) and Small and McLeish (1988). In the following discussion, by an optimal estimating function we shall mean that the consistent root to the estimating equation will have smallest asymptotic variance, in the sense of *Loewner ordering*, among all such estimators in the stated class.

Now we turn to the construction of an optimal estimating function for θ. If we can further specify the covariance matrix of the elementary estimating function g, then we can show (McCullagh and Nelder, 1989, Section 9.4.2) that the optimally weighted estimating function for θ takes the following form:

$$q(\theta, y) = D_\theta^{\mathrm{t}} V_\theta^{-1} g(\theta, y) \tag{6.6}$$

where

$$D_\theta = E_\theta\{-\dot{g}(\theta, Y)\}, \quad V_\theta = \mathrm{Cov}\{g(\theta, Y)\}.$$

Now the estimating function $q(\theta, y)$ of (6.6) has the same dimension p as that of θ. The asymptotic variance

$$\{D_\theta^{\mathrm{t}} V_\theta^{-1} D_\theta\}^{-1}$$

of a consistent root $\hat{\theta}$ to $q(\hat{\theta}, y) = 0$ can be shown to be minimal among all roots to estimating functions of the form

$$h(\theta, y) = A_\theta g(\theta, y)$$

where A_θ is a weight matrix of dimension $p \times k$, which is functionally independent of the data. Note that the optimal estimating function (6.6) reduces to the quasi-score if the ith component of g is $y_i - \mu_i$.

6.3.2 GENERALISED PROJECTED ARTIFICIAL LIKELIHOOD RATIOS

We have seen that the elementary estimating function $g(\theta, y)$ induces an optimal estimating function $q(\theta, y)$ for the parameter of interest θ given by (6.6). In general, we should not expect that there is a scalar objective function whose gradient is $q(\theta, y)$. However, similar to the case of the quasi-score, as studied in Section 6.2, we can proceed to build an artificial objective function for use in conjunction with the optimal estimating function $q(\theta, y)$. For the quasi-scores we focused on a vector space constructed using the elementary estimating functions

$$Y_i - \mu_j(\theta), \quad j = 1, \ldots, n.$$

Similarly, in the present case we shall consider a vector space spanned by the k elements of $g(\theta, y)$, which are assumed to be functionally independent. Since the inner products associated with the projection to be introduced shortly are defined locally in the parameter space Θ, accordingly, we shall only be interested in vector spaces defined locally in the parameter space Θ.

Let θ and η be two values in Θ. Locally at $\eta \in \Theta$, the elements of g span an k-dimensional vector space

$$\mathcal{L}_\eta = \{r = a^{\mathrm{t}} g(\eta, y) + b\ ;\ r\text{: square integrable}\}.$$

where a is an k-dimensional vector and b a scalar, both functionally independent of the observations but may depend on the parameter of interest. In the previous section, we considered the projection of the likelihood ratios into a proper linear space. For reasons which will become clear later, in this section, we shall consider projection of the logarithm of the likelihood ratios instead of the likelihood ratios. Let

$$R(\theta, \eta) = \log \frac{L(\theta)}{L(\eta)}$$

be the log-likelihood ratio between θ and η. Note that

$$R(\theta, \eta) = -\log\{\lambda(\theta, \eta)\}$$

where $\lambda(\theta, \eta)$ is the likelihood ratio defined in the previous section. Ideally, we would wish to project $R(\theta, \eta)$ into the linear space $\mathcal{L}_\eta$ using the inner product

$$\langle r, s\rangle_\eta = E_\eta(rs) \quad \text{for } r, s \in \mathcal{L}_\eta.$$

Such a projection is however not feasible in general using the semi-parametric assumptions about the random variables.

To avoid this difficulty, we consider instead an approximate form of the log likelihood in order to compute the projection for a general elementary estimating function g. For this purpose, we shall use the following convenient form. This form is derived using the following first-order Taylor expansion of $R(\theta, \eta)$ at η. In this form, we shall regard η as a reference point in the parameter space. Both the log-likelihood ratio and the approximate form are considered to be functions of θ. Let $\|\cdot\|$ denote the Euclidean norm. Then we have

$$\begin{aligned}
\log \frac{L(\theta)}{L(\eta)} &= \log L(\theta) - \log L(\eta) \\
&= \left\{\log L(\eta) + \frac{1}{L(\eta)}(\theta - \eta)^{\mathrm{t}}\{\dot{L}(\theta)\}_{\theta=\eta} + o(\|\theta - \eta\|)\right\} - \log L(\eta) \\
&= \frac{1}{L(\eta)}(\theta - \eta)^{\mathrm{t}}\{\dot{L}(\theta)\}_{\theta=\eta} + o(\|\theta - \eta\|) \\
&= \left\{\frac{L(\theta) - L(\eta)}{L(\eta)} + o(\|\theta - \eta\|)\right\} + o(\|\theta - \eta\|) \\
&= \frac{L(\theta) - L(\eta)}{L(\eta)} + o(\|\theta - \eta\|).
\end{aligned}$$

The last equality suggests that we may work on the *centred likelihood ratio*

$$R_\eta(\theta, \eta) = \frac{L(\theta)}{L(\eta)} - 1$$

when η and θ are close to one another. The subscript η in R_η indicates that the centred likelihood ratio is derived by an approximation of the true log-likelihood ratio $R(\theta, \eta)$ at η. The advantage of working with $R_\eta(\theta, \eta)$ is that we can find the exact projection of $R_\eta(\theta, \eta)$ onto $\mathcal{L}_\eta$ spanned by any elementary estimating function g.

Proposition 6.1 *Suppose that the first and the second moment of the unbiased elementary estimating function $g(\eta, y)$ can be calculated. Let*

$$C(\theta, \eta) = E_\theta\{g(\eta, Y)\} \quad \text{and} \quad V_\eta = \mathrm{Cov}_\eta\{g(\eta, Y)\}.$$

Then the projection of $R_\eta(\theta, \eta)$ onto $\mathcal{L}_\eta$ is given by

$$D_\eta(\theta, \eta) = \{C(\theta, \eta)\}^{\mathrm{t}} V_\eta^{-1} g(\eta, y). \tag{6.7}$$

Proof. It suffices to show that the following quantity:

$$R_\eta(\theta, \eta) - D_\eta(\theta, \eta)$$

is orthogonal to any element

$$r = a^{\mathrm{t}} g(\eta, Y) + b$$

in the linear space $\mathcal{L}_\eta$. Or equivalently, the following equality:

$$\langle R_\eta(\theta, \eta) - D_\eta(\theta, \eta),\ a^{\mathrm{t}} g(\eta, Y) + b\rangle_\eta = 0$$

holds for any constant vector a and scalar b. To see this, we let

$$A = \langle R_\eta(\theta, \eta),\ a^{\mathrm{t}} g(\eta, Y) + b\rangle_\eta$$

and note that

$$\begin{aligned} A &= \left\langle \frac{L(\theta)}{L(\eta)},\ a^{\mathrm{t}} g(\eta, Y) + b \right\rangle_\eta - \langle 1,\ a^{\mathrm{t}} g(\eta, Y) + b\rangle_\eta \\ &= E_\eta\left[\frac{L(\theta)}{L(\eta)}\{a^{\mathrm{t}} g(\eta, Y) + b\}\right] - E_\eta\{a^{\mathrm{t}} g(\eta, Y) + b\} \\ &= a^{\mathrm{t}} C(\theta, \eta). \end{aligned}$$

Similarly, if we let

$$B = \langle D_\eta(\theta, \eta),\ a^{\mathrm{t}} g(\eta, Y) + b\rangle_\eta$$

then we will have

$$\begin{aligned} B &= \langle \{C(\theta, \eta)\}^{\mathrm{t}} V_\eta^{-1} g(\eta, Y),\ a^{\mathrm{t}} g(\eta, Y) + b\rangle_\eta \\ &= \{C(\theta, \eta)\}^{\mathrm{t}} V_\eta^{-1} \langle g(\eta, Y),\ a^{\mathrm{t}} g(\eta, Y) + b\rangle_\eta \\ &= \{C(\theta, \eta)\}^{\mathrm{t}} a. \end{aligned}$$

□

In the last equality we used the fact that g is unbiased. The conclusion follows by combining the these calculations.

There are two noticeable problems, however, for the projected log likelihood ratio $D_\eta(\theta, \eta)$. First, the two parameters θ and η received asymmetrical treatments in the projection and in the process of approximating the log-likelihood function as well. Second, the centred likelihood ratio $R_\eta(\theta, \eta)$ is only first-order accurate to the genuine log likelihood ratio $R(\theta, \eta)$. Fortunately, the two issues can be dealt with simultaneously by considering an asymmetry correction to the centred likelihood ratio by using two first-order Taylor expansions of the log-likelihood ratio.

To see how we may correct the centred likelihood ratio, again we assume that θ and η are close to one another. First, we note that the second-order Taylor expansion of $\log\{L(\theta)/L(\eta)\}$ at η may be expressed as follows:

$$\log \frac{L(\theta)}{L(\eta)} = \frac{L(\theta) - L(\eta)}{L(\eta)} + \frac{1}{2}(\theta - \eta)^{\mathrm{t}} \left\{ \frac{\ddot{L}(\eta)}{L(\eta)} \right\} (\theta - \eta)$$
$$+ \frac{1}{2}(\theta - \eta)^{\mathrm{t}} \left\{ \frac{\partial^2}{\partial \theta^2} \log L(\theta) \right\}_{\theta=\eta} (\theta - \eta) + (\|\theta - \eta)^2\|). \tag{6.8}$$

On the other hand, by symmetry, we have a similar expression for $\log\{L(\eta)/L(\theta)\}$ at θ, namely

$$\log \frac{L(\eta)}{L(\theta)} = \frac{L(\eta) - L(\theta)}{L(\theta)} + \frac{1}{2}(\theta - \eta)^{\mathrm{t}} \left\{ \frac{\ddot{L}(\theta)}{L(\theta)} \right\} (\theta - \eta)$$
$$+ \frac{1}{2}(\theta - \eta)^{\mathrm{t}} \left\{ \frac{\partial^2}{\partial \eta^2} \log L(\eta) \right\}_{\eta=\theta} (\theta - \eta) + (\|\eta - \theta)^2\|). \tag{6.9}$$

Subtracting (6.9) from (6.8) and dividing by 2, we then obtain a second-order approximation to the log-likelihood ratio

$$\log \frac{L(\theta)}{L(\eta)} = \frac{1}{2} \frac{L(\theta) - L(\eta)}{L(\eta)} - \frac{1}{2} \frac{L(\eta) - L(\theta)}{L(\theta)} + o(\|\theta - \eta\|^2)$$
$$= \tfrac{1}{2} R_\eta(\theta, \eta) - \tfrac{1}{2} R_\theta(\eta, \theta) + o(\|\theta - \eta\|^2). \tag{6.10}$$

Now if we project the centred likelihood ratio $R_\theta(\eta, \theta)$ onto the linear space $\mathcal{L}_\theta$ with the inner product $\langle \cdot, \cdot \rangle_\theta$, then the projection by analogy will take the form

$$D_\theta(\eta, \theta) = \{C(\eta, \theta)\}^{\mathrm{t}} V_\theta^{-1} g(\theta, y). \tag{6.11}$$

Correcting either $D_\eta(\theta, \eta)$ or $D_\theta(\eta, \theta)$, the approximation (6.10) therefore suggests the following *generalised projected artificial log-likelihood ratio*:

$$D(\theta, \eta) \equiv \tfrac{1}{2} D_\eta(\theta, \eta) - \tfrac{1}{2} D_\theta(\eta, \theta)$$
$$= \tfrac{1}{2}\{C(\theta, \eta)\}^{\mathrm{t}} V_\eta^{-1} g(\eta, y) - \tfrac{1}{2}\{C(\eta, \theta)\}^{\mathrm{t}} V_\theta^{-1} g(\theta, y). \tag{6.12}$$

The artificial objective function $D(\theta, \eta)$ was proposed by Hanfelt and Liang (1995), who considered further refinements of $D(\theta, \eta)$ by introducing intermediate points between θ and η. Strictly speaking, the quantity $D(\theta, \eta)$ is not a projection but rather a sum of two local projections.

An important special case occurs when $y_i - \mu_i$ form the elementary estimating functions, so that we are concerned with constructing an objective function for the quasi-score

$$q(\theta, y) = \dot{\mu}_\theta^{\mathrm{t}} V_\theta^{-1}(y - \mu_\theta).$$

The formula (6.12) now takes the form

$$D(\theta, \eta) = \tfrac{1}{2}(\mu_\theta - \mu_\eta)^{\mathrm{t}}\{V_\theta^{-1}(y - \mu_\theta) + V_\eta^{-1}(y - \mu_\eta)\}. \tag{6.13}$$

The special case (6.13) was first derived by Li (1993).

Returning to general optimal estimating functions $q(\theta, y)$, we first note that $q(\theta, y)$ is also an information-unbiased estimating function. The Godambe efficiency, often known in this context as the *quasi-Fisher information*, of $q(\theta, y)$ has a particularly simple form

$$\begin{aligned} \mathrm{eff}_\theta(q) &= -E_\theta\{\dot{q}(\theta, Y)\} \\ &= D_\theta^{\mathrm{t}} V_\theta^{-1} D_\theta. \end{aligned} \tag{6.14}$$

Now we state some basic properties of the artificial objective function $D(\theta, \eta)$; see Hanfelt and Liang (1995) for a proof.

Proposition 6.2 *The following properties hold for the projected artificial log-likelihood ratios $D(\theta, \eta)$:*

1. *It is anti-symmetric:*
$$D(\theta, \eta) = -D(\eta, \theta).$$
2. *Suppose that $q(\theta, y)$, $C(\theta, \eta)$, $C(\eta, \theta)$ and their first partial derivatives with respect to θ are continuous in θ. Then $D(\theta, \eta)$ is locally tangent to the optimal estimating function. That is,*
$$\dot{D}(\theta, \eta) = q(\eta, y) + o(\|\theta - \eta\|). \tag{6.15}$$
In particular, if
$$g_i = y_i - \mu_i, \quad i = 1, \ldots, n$$
are the elementary estimating functions, then the gradient of $D(\theta, \eta)$ will be locally equal to the quasi-score.
3. *If the second partial derivatives of $q(\theta, y)$, $C(\theta, \eta)$, $C(\eta, \theta)$ with respect to θ are continuous in θ, then we have*
$$E_\eta\left\{-\frac{\partial^2}{\partial\theta^2} D(\theta, \eta)\right\}_{\theta=\eta} = \mathrm{eff}_\theta(q). \tag{6.16}$$

While the projected likelihood ratio of the last section is directly connected with the quasi-score, the above properties show that the generalised projected likelihood ratio $D(\theta, \eta)$ may be regarded as an artificial objective function associated with the optimal estimating function $q(\theta, y)$. Note that the anti-symmetry property of the projected likelihood ratio $D(\theta, \eta)$ does not hold for either $D_\eta(\theta, \eta)$ or $D_\theta(\eta, \theta)$. Note also that the second derivatives of the local projections $D_\eta(\theta, \eta)$ and $D_\theta(\eta, \theta)$ have zero expectations. The following properties further justify the fact that $D(\theta, \eta)$ may be regarded a semi-parametric analogue of a genuine log-likelihood ratio statistic; see Hanfelt and Liang (1995) for a proof.

Proposition 6.3 *Suppose that the optimal estimating equation $q(\theta, y) = 0$ has a consistent and asymptotically normal root $\hat{\theta}$. Let $\theta_0 \in \Theta$ be a fixed value. Then the following properties hold:*

1. *If θ_0 is the true value of the parameter, then we have*

$$2D(\hat{\theta}, \theta_0) \xrightarrow{\mathcal{L}} \chi_p^2.$$

2. *Let δ be a constant vector and suppose that*

$$\theta_1 = \theta_0 + \frac{1}{\sqrt{n}}\delta$$

is the true value of the parameter. Assume that the average quasi-Fisher information of $q(\theta, y)$ tends to a positive definite matrix asymptotically. That is

$$\lim_{n\to\infty} \mathrm{eff}_\theta(q) = \Omega$$

where Ω is a positive definite matrix. Then it holds that

$$2D(\hat{\theta}, \theta_0) \xrightarrow{\mathcal{L}} \chi_p^2(\delta^{\mathrm{t}}\Omega\delta)$$

where $\chi_p^2(a)$ denotes the chi-squared distribution with p degrees of freedom and non-central parameter a.

6.3.3 MULTIPLE ROOTS

Now we discuss the use of the projected artificial objective function when there exist more than one root to the optimal estimating equation $q(\theta, y) = 0$. Multiple roots can occur even for large samples. We have seen such an example involving the Cauchy location model; see Chapters 3 and 4. Another instance is when $q(\theta, y)$ satisfies the following property:

$$E_\eta\{q(\theta, Y)\} = 0 \quad \text{for some } \eta \neq \theta.$$

In Chapter 5, we defined an estimating function satisfying this property as an *irregular estimating function*. We considered some simple examples and demonstrated how the technique of *model embedding* may be used to overcome the multiple root problem involving irregular estimating functions. The projected artificial likelihood ratio statistic studied in this section provides an alternative method for choosing an estimate among the roots to such an irregular estimating function. The following result concerns a scalar parameter θ. This result was due to Li (1993) when $q(\theta, y)$ is the quasi-score. The general result when $q(\theta, y)$ is any optimal estimating function can be found in Hanfelt and Liang (1995).

Proposition 6.4 *Let θ_0 be the true value of θ. Let $\theta_1 \neq \theta_0$. Suppose that the optimal estimating function $q(\theta, y)$ satisfies the following conditions:*

1. *The estimating function satisfies the irregularity condition*

$$E_{\theta_0}\{q(\theta_1, Y)\} = 0.$$

2. *The limiting average* shifted information *is negative, namely*

$$\liminf_{n\to\infty} \frac{1}{n} E_{\theta_0}\{-\dot{q}(\theta_1, Y)\} < c < 0.$$

3. *The second derivative of $q(\theta, Y)$ in a neighbourhood of θ_1 is bounded in probability, i.e.,*

$$\frac{1}{n}|\ddot{q}(\theta, Y)| < M(Y)$$

where $E_{\theta_0}\{M(Y)\} < \infty$.

Then there exists a neighbourhood $\mathcal{N}$ of the false value θ_1 such that

$$\mathrm{P}_{\theta_0}\{D(\theta, \theta_1) > 0\} \to 1 \quad \textit{for } \theta \in \mathcal{N}. \tag{6.17}$$

By (6.17), a false value θ_1 behaves as if it is a local minimiser of the projected artificial likelihood function. On the other hand, at the true value of the parameter, the quasi-Fisher information is positive definite,

$$\liminf_{n\to\infty} \frac{1}{n} E_{\theta_0}\{-\dot{q}(\theta_0, Y)\} = \Omega > 0.$$

Consequently, in a punctured neighbourhood of θ_0, we have that

$$\mathrm{P}_{\theta_0}\{D(\theta_0, \theta) > 0\} \to 1. \tag{6.18}$$

In other words, the true value θ_0 behaves as if it is a local maximiser of the projected artificial likelihood function. For a large enough sample size, under the assumptions of Proposition 6.4, the sign of $D(\theta, \eta)$ may provide important information about the relative plausibility of the parameter values. That is, we may prefer a root $\hat{\theta}$ of $q(\hat{\theta}, y) = 0$ for which the inequality

$$D(\hat{\theta}, \theta) > 0$$

holds for any other values θ that is close to $\hat{\theta}$. Such a rule however does not exclude possibility of picking up an inconsistent root which behaves like a local maximiser. The method based on the quadratic artificial likelihood ratio test discussed in Section 6.5 avoids this difficulty.

We could also use the artificial likelihood ratio, $D(\theta, \eta)$, to directly compare two roots. Let θ_0 be the true value and $\theta_1 \neq \theta_0$. Then, as expected, the projected artificial likelihood ratio almost surely favours the true value over a false one in the limit. That is, when $n \to \infty$, we have

$$\mathrm{P}_{\theta_0}\{D(\theta_0, \theta_1) > 0\} \to 1. \tag{6.19}$$

For (6.19) to hold, we need the condition

$$\liminf_{n\to\infty} \frac{1}{n}\{C(\theta_0, \theta_1)\}^{\mathrm{t}} V_{\theta_1}^{-1} C(\theta_0, \theta_1) > c > 0.$$

The last condition has the demerit that it may not be easy to verify for a given problem.

6.3.4 EXAMPLE: ESTIMATING THE COEFFICIENT OF VARIATION

It is sometimes of interest to compare dispersions among different populations or strata using possibly different units. Two commonly used measures for this purpose are *Gini's coefficient of concentration* and Karl Pearson's *coefficient of variation*, both being unitless measures of dispersion. To illustrate the use of the projected artificial log-likelihood ratio, we now consider the problem for estimating the common coefficient of variation among K strata using independent observations,

$$y_{ij} \quad i = 1, \ldots, K; \quad j = 1, \ldots, n_i$$

where y_{ij} is the observation from the jth subject of the ith stratum. The population and the sample means of the i-stratum are denoted by

$$\mu_i = E(Y_{ij}) \quad \bar{y}_i = \frac{1}{n}\sum_{j=1}^{n_i} y_{ij}$$

respectively. We assume a constant *coefficient of variation* across the strata and shall let

$$\theta = \left(\frac{\sigma_i}{\mu_i}\right)^2$$

where σ_i^2 is the variance of the i-th stratum, the unbiased sample variance being given by

$$s_i^2 = \frac{1}{n_i - 1}\sum_{j=1}^{n_i}(y_{ij} - \bar{y}_i)^2.$$

For convenience, the parameter of interest θ is taken to be the square of the coefficient of variation.

A possible approach to this problem is to assume that the y_{ij} are from Gamma distributions and to eliminate the nuisance parameters μ_i from the likelihood function by conditioning on the $\bar{y}_i$, the complete sufficient statistics of the μ_i. Two problems may arise with this approach. First, the assumption of the Gamma laws may be incorrect. Second, even if the assumptions concerning the Gamma distributions are reasonable, conditioning arguments lead to a rather cumbersome conditional score function for θ.

A semi-parametric approach is not only robust against model misspecifications, but also rather transparent for the current problem. To see how we may construct an elementary estimating function, we first note that the traditional moment estimator, based on data from the i-th stratum, can be obtained by solving the following equation:

$$u(\theta, y_i) = s_i^2 - \theta \bar{y}_i^2 = 0, \quad i = 1, \dots, K. \tag{6.20}$$

The functions $u(\theta, y_i)$ are not unbiased because, by straightforward calculations, we have

$$E_\theta\{u(\theta, Y_i)\} = -\frac{1}{n}\theta\sigma_i^2, \quad i = 1, \dots, K. \tag{6.21}$$

Alternative estimators to the moment estimators defined by (6.20) can be obtained by correcting the biases in functions $u(\theta, y_i)$ using information (6.21). To do so we first look for a quantity, which is a function of the data and the parameter θ alone, so that the expectation of it will equal to the bias $E_\theta\{u(\theta, Y_i)\}$. Several such quantities are conceivable and a convenient one is given by

$$-\frac{1}{n}\theta s_i^2, \quad i = 1, \dots, K.$$

Using these quantities we then obtain an elementary estimating function for θ consisting of the following K components:

$$h(\theta, y_i) = s_i^2 - \theta\left(\bar{y}_i^2 - \frac{s_i^2}{n}\right), \quad i = 1, \dots, K. \tag{6.22}$$

To proceed further from the elementary estimating function

$$(h(\theta, y_i), \dots, h(\theta, y_K))^{\mathrm{t}}$$

to an optimal estimating function for θ using the theories of this section, we need to specify the variance of each $h(\theta, y_i)$, which, by (6.22), requires knowledge about the underlying distributions up to the fourth moments. Practically, it would usually be difficult, if not impossible, to specify the fourth moments as known functions of the parameter of interest.

One way to avoid the aforementioned difficulty is to use an alternative method to construct an alternative elementary estimating function as proposed by Yanagimoto and Yamamoto (1991). This method is based on the idea of conditioning. For our present example, it suggests to use

$$g(\theta, y_i) = s_i^2 - E_\theta(s_i^2|\bar{y}_i), \quad i = 1, \ldots, K \tag{6.23}$$

as the components of the elementary estimating function, provided that the conditional expectation $E_\theta(s_i^2|\bar{y}_i)$ can be specified as a function of the data and the parameter of interest θ alone. We note that the $g(\theta, y_i)$ are unbiased, by construction, conditional on the $\bar{y}_i$. That is

$$E_\theta\{g(\theta, y_i)|\bar{y}_i\} = 0. \tag{6.24}$$

It follows therefore that the $g(\theta, y_i)$ are unbiased in the usual sense as well, because we have

$$E_\theta\{g(\theta, y_i)\} = E\Big[E_\theta\{g(\theta, y_i)|\bar{y}_i\}\Big] = 0$$

where the outer expectation is taken with respect to $\bar{y}_i$. Accordingly, the elementary estimating function defined by (6.23) are unbiased both conditionally and unconditionally. See Section 8.3 for other concepts of unbiasedness of an estimating function in Bayesian contexts.

To complete the specification of the elementary estimating function of this approach we need to specify the conditional expectations

$$E_\theta(s_i^2|\bar{y}_i), \quad i = 1, \ldots, K$$

which require conditional knowledge up to the second moment about the underlying distributions. Here we shall consider families of distributions satisfying the following conditions:

$$E_\theta(s_i^2|\bar{y}_i) = \frac{n_i\theta}{n_i + \theta}\bar{y}_i^2, \quad i = 1, \ldots, K. \tag{6.25}$$

For the Gamma distribution, (6.25) may be derived using the fact that the random vector

$$\left(\frac{y_{i1}}{n\bar{y}_i}, \ldots, \frac{y_{in_i}}{n\bar{y}_i}\right)$$

has a Dirichlet distribution and is independent of $\bar{y}_i$. See Yanagimoto and Yamamoto (1991). Using the assumptions (6.25), we can then define the K components of an elementary estimating function, the ith component being given by

$$g_i(\theta, y_i) = s_i^2 - \frac{\theta n_i}{n_i + \theta}\bar{y}_i^2, \quad i = 1, \ldots, K. \tag{6.26}$$

A property special to the choice of $g_i(\theta, y_i)$ defined by (6.26) is that $g_i(\theta, y_i)$ are unbiased unconditionally for *any* distribution not necessarily satisfying the condition (6.25).

The advantage of the conditional approach by considering only distributions satisfying the property (6.25) is that it may be much easier to specify the conditional variance of $g_i(\theta, y_i)$ as a function of the data and the parameter of interest alone (without any *nuisance parameter*), rather than to specify the unconditional variance of an appropriate elementary estimating function. Hanfelt and Liang (1995) suggested the following choice of the conditional variance functions:

$$\begin{aligned} V_i &= \mathrm{Var}_\theta\{g_i(\theta, y_i) \mid \bar{y}_i\} \\ &= \mathrm{Var}_\theta(s_i^2 \mid \bar{y}_i) \\ &= \frac{2\theta^2(1+\theta)n_i^4\bar{y}_i^4}{(n_i-1)(n_i+3\theta)(n_i+2\theta)(n_i+\theta)^2}, \quad i = 1, \ldots, K. \end{aligned} \tag{6.27}$$

Using the variance functions (6.27), we finally arrive at the *conditionally optimal estimating function* for θ

$$q(\theta, y) = \sum_{i=1}^{K} D_i V_i^{-1} g_i(\theta, y_i) \tag{6.28}$$

where

$$D_i = \frac{n_i^2\bar{y}_i^2}{(n_i+\theta)^2}.$$

It is appropriate at this point to remark that the theories developed in this section remain valid if we replace the previous unconditional expectations with appropriate *conditional expectations*. The reader is referred to Hanfelt and Liang (1995) for details on the conditional version of the theories.

In the present case, the estimating function (6.28) is a scalar one, so we can integrate $q(\theta, y)$ with respect to θ to obtain the exact objective function of q. Ignoring the constant of integration, we get the following (unique) objective function

$$\begin{aligned} Q(\theta) = \sum_{i=1}^{K} \frac{(n_i-1)s_i^2}{2n_i^2\bar{y}_i^2} & \left\{ (n_i-2)(n_i-3)\log\frac{1+\theta}{\theta} + 6\log\theta - \frac{n_i^2}{\theta} \right\} \\ & + \sum_{i=1}^{K} \frac{1}{2n_i}\{(n_i-2)(n_i-3)\log(1+\theta) \\ & - n_i(n_i-1)\log\theta - 2n_i\log(n_i+\theta)\}. \end{aligned}$$

The projected artificial likelihood ratio $D(\theta, \lambda)$ now takes the form

$$D(\theta, \lambda) = \frac{1}{2}\sum \frac{n_i^2\bar{y}_i^2(\theta-\lambda)}{(n_i+\theta)(n_i+\lambda)}\left(V_{\theta_i}^{-1}A + V_{\lambda_i}^{-1}B\right) \tag{6.29}$$

where

$$A = s_i^2 - \frac{n_i\theta}{n_i+\theta}\bar{y}_i^2 \quad \text{and} \quad B = s_i^2 - \frac{n_i\lambda}{n_i+\lambda}\bar{y}_i^2.$$

Using the Gamma distributions, Hanfelt and Liang (1995) investigated through simulations the performance of $Q(\theta)$ and $D(\theta, \lambda)$ of (6.29) in constructing two-sided confidence intervals for θ. In terms of coverage errors, each method is observed to improve the Wald confidence intervals; see table 1 of Hanfelt and Liang (1995) for details.

6.4 Artificial likelihoods through integration

In the last section, we considered artificial objective functions through projection for a p-dimensional optimal estimating function

$$q(\theta, y) = D_\theta^{\mathrm{t}} V_\theta^{-1} g(\theta, y) \tag{6.30}$$

where $g(\theta, y)$ is a k-dimensional elementary estimating function. Perhaps, the most direct method for obtaining an artificial objective function is by performing a line integration for $q(\theta, y)$ in the parameter space Θ. A problem with this approach is that when the estimating function $q(\theta, y)$ does not define a conservative vector field in Θ, the line integrals will depend on the paths used in the integration. A sensitivity analysis on the dependence of paths is therefore highly recommended when applying this approach.

Before describing the general method, we first consider the quasi-score

$$s(\theta, y) = \{\dot{\mu}(\theta)\}^{\mathrm{t}} V^{-1}(\mu)\{y - \mu(\theta)\} \tag{6.31}$$

defined for a p-dimensional parameter θ using independent random variables $y = \{y_1, \ldots, y_n\}$ with mean vector $\mu(\theta) = (\mu_1, \ldots, \mu_n)^{\mathrm{t}}$. The covariance matrix $V(\mu)$ is an $n \times n$ diagonal matrix, the ii-th component being $V_i(\mu_i)$. That is,

$$V(\theta) = \mathrm{diag}\{V_1(\mu_1), \ldots, V_n(\mu_n)\}. \tag{6.32}$$

A particular feature of this covariance matrix is that each variance component $V_i(\mu_i)$ depends on the parameter of interest θ through its corresponding mean μ_i. Using the assumption (6.32), it can be verified that $\dot{s}(\theta, y)$ is a $p \times p$ symmetric matrix. Consequently, the line integration of $s(\theta, y)$ with respective to θ is independent of the paths. See McCullagh and Nelder (1989, Section 9.2) for details.

To obtain an artificial likelihood function for θ with the assumption (6.32), it suffices to work in the space of the means. In fact, Wedderburn (1974) defined a quantity $K(\mu_i; y_i)$ as the *quasi-likelihood* for θ arising from the ith observation y_i, where $K(\mu_i; y_i)$ is a solution to the following differential equation

$$\frac{\partial}{\partial \mu_i} K(\mu_i; y_i) = \frac{y_i - \mu_i}{V_i(\mu_i)}.$$

The solution to the above equation, ignoring the constant of integration, is given by

$$K(\mu_i; y_i) = \int^{\mu_i} \frac{y_i - \eta_i}{V_i(\eta_i)} \, \mathrm{d}\eta_i. \tag{6.33}$$

We note that the quasi-likelihood $K(\mu_i; y_i)$ of (6.33) depends on the parameter of interest θ through the mean μ_i. Wedderburn (1974) showed that $K(\mu_i; y_i)$ is the true log likelihood function when y_i comes from a one-parameter exponential family. In general, Wedderburn (1974) showed that the objective function $K(\mu_i; y_i)$ has a number of properties in common with a genuine log-likelihood function. Having defined the individual quasi-likelihood function for θ, the total quasi-likelihood $K(\mu; y)$ based on the whole data set is then the sum of each $K(\mu_i; y_i)$, namely

$$K(\mu; y) = \int^{\mu} \{V^{-1}(\xi)(y - \xi)\}^{\mathrm{t}}\, \mathrm{d}\xi + c \tag{6.34}$$

where $\xi = (\xi_1, \ldots, \xi_n)^{\mathrm{t}}$ and c is a constant of integration. Note that

$$\frac{\partial}{\partial\theta} K(\mu; y) = s(\theta, y)$$

where $s(\theta, y)$ is given by (6.31). Since the estimating function $s(\theta, y)$ defines a conservative vector field in Θ, an alternative expression for the quasi-likelihood $K(\mu; y)$ is through the following line integration in Θ-space:

$$K(\mu(\theta); y) = \int_{\xi(a)=\theta_0}^{\xi(b)=\theta} \{s(\xi(s), y)\}^{\mathrm{t}}\, \mathrm{d}\xi(s) + c \tag{6.35}$$

where $\xi(s)$ is a *regular curve* in Θ joining the two points θ_0 and θ. Here, a curve in the Θ-space means a set of points $\xi(s)$ for which

$$\xi(s) = (\xi_1(s), \ldots, \xi_p(s)) \quad a \leq s \leq b.$$

Such a curve is called *regular* if it has no double points and there exist a finite subintervals of $[a, b]$, such that in each subinterval we have $\xi_i(s) \in \mathrm{C}^1$ and

$$\sum_{i=1}^{p} \left\{ \frac{\mathrm{d}\xi_i(s)}{\mathrm{d}s} \right\}^2 \neq 0.$$

In (6.35), θ_0 is regarded as a reference point and $K(\mu(\theta); y)$ is interpreted as a function of θ.

Now we move from the quasi-score $s(\theta, y)$ to the optimal estimating function $q(\theta, y)$ of (6.30). A natural generalisation of the quasi-likelihood (6.35) is to consider the following line integration

$$Q(\theta, \eta) = \int_{\xi(a)=\theta}^{\xi(b)=\eta} \{q(\xi(s), y)\}^{\mathrm{t}}\, \mathrm{d}\xi(s) + c \tag{6.36}$$

where $\xi(t)$ is a regular curve in the Θ-space joining the two points θ and η. Unfortunately, ambiguities arise in the definition (6.36) when $q(\theta, y)$ does not define a conservative vector field in Θ. This is because the line integration $Q(\theta, \eta)$ for such an estimating function not only depends on the two end points θ and η but also on the particular path joining the two points. Non-conservative estimating functions abound. For example, the quasi-score $s(\theta, y)$ of (6.31) are usually not conservative when there are non-zero off-diagonal elements in the covariance matrix V. Or, even when V is diagonal, $s(\theta, y)$ may still be non-conservative if the assumption (6.32) fails to be satisfied.

Nevertheless, for an optimal estimating function $q(\theta, y)$, the effect of paths on the integration (6.36) may be small when θ and η are close to one another. Recall that an estimating function $h(\theta, y)$ will be conservative if and only if the quasi-Hessian $\dot{h}(\theta, y)$ is a symmetric matrix. The definition $Q(\theta, \eta)$ of (6.36) for a suitably chosen path may be used as an artificial objective function because $q(\theta, y)$ is approximately conservative in the sense that the expectation of the quasi-Hessian $\dot{q}(\theta, y)$ is symmetric. That is,

$$E_\theta\{\dot{q}(\theta, Y)\} = [E_\theta\{\dot{q}(\theta, Y)\}]^{\mathrm{t}}. \tag{6.37}$$

An estimating function satisfying the property (6.37) will be referred to as an *E-conservative estimating function*. Since, by (6.31), we have that

$$E_\theta\{\dot{q}(\theta, Y)\} = D_\theta^{\mathrm{t}} V_\theta^{-1} D_\theta$$

which is a symmetric matrix. Therefore, the optimal estimating function $q(\theta, y)$ is E-conservative. It is a consequence of the law of large numbers then that the quasi-Hessian of any E-conservative estimating function will be approximately symmetric when the sample size is large. The above discussion suggest that, when the sample size is large, one may use (6.36) to define an artificial likelihood function for a suitably chosen path.

Now let Γ be a regular curve in the parameter space such that

$$\Gamma = \{\xi(s) \colon a \le s \le b \text{ and } \theta = \xi(a), \eta = \xi(b)\}.$$

Then a *generalised quasi-likelihood ratio* (Hanfelt and Liang, 1995) between θ and η may be defined as follows

$$Q_\Gamma(\theta, \eta) = \int_a^b \{q(\xi(s), y)\}^{\mathrm{t}}\, \mathrm{d}\xi(s) + c. \tag{6.38}$$

The quantity $Q_\Gamma(\theta, \eta)$ is a scalar function depending on data y and the two points θ and η in Θ. Generally, $Q_\Gamma(\theta, \eta)$ also depends on the particular path Γ connecting θ and η. The scalar objective function $Q_\Gamma(\theta, \eta)$, as suggested by Hanfelt and Liang (1995), may be used as a semiparametric analogue of the log likelihood function. Similar to the projective artificial log-likelihood function $D(\theta, \eta)$ considered in the last section, the following finite-sample properties also hold for $Q_\Gamma(\theta, \eta)$; see Hanfelt and Liang (1995) for a proof.

Proposition 6.5 *The generalised quasi-likelihood ratio $Q_\Gamma(\theta, \eta)$, where Γ is an arbitrary regular curve in Θ-space, has the following properties.*

1. *It is* anti-symmetric:
$$Q_\Gamma(\theta, \eta) = -Q_\Gamma(\eta, \theta).$$

2. *It is* tangent *to the optimal estimating function $q(\theta, y)$. That is, if the following quantities*
$$q(\theta, y), \xi(s) \quad \text{and} \quad \mathrm{d}\xi(s)/\mathrm{d}s$$
are uniformly continuous in s and θ, then we have that
$$\frac{\partial}{\partial\theta}\{Q_\Gamma(\theta, \eta)\} = q(\eta, y) + o(\|\theta - \eta\|). \tag{6.39}$$

3. *If each component of the $p \times p \times p$ array $\ddot{q}(\theta, y)$ is uniformly continuous in θ, then locally at η*
$$-E_\eta\left\{\frac{\partial^2}{\partial\theta^2}Q_\Gamma(\theta, \eta)\right\}_{\theta=\eta} = i_\eta \tag{6.40}$$
where i_η is the quasi-Fisher information of $q(\eta, y)$ given by (6.14).

The above proposition shows that the generalised quasi-likelihood $Q_\Gamma(\theta, \eta)$ and the projected artificial log-likelihood $D(\theta, \eta)$ are locally equivalent to one another up to the order $o(\|\theta - \eta\|)$. Results on artificial likelihood ratio tests similar to those given in Proposition 6.3 can also be obtained using the artificial objective function $Q_\Gamma(\theta, \eta)$. Hanfelt and Liang (1995) show that results paralleling those given in Proposition 6.4 are also valid concerning the behaviour of the artificial likelihood ratio tests using $Q_\Gamma(\theta, \eta)$. Consequently, the generalised artificial likelihood function studied in this section provides an alternative way for choosing an appropriate root from the solutions to an estimate equation $q(\theta, y) = 0$.

Hanfelt and Liang (1995) also investigated the case of a logistic regression model when measurement errors are present. They considered the case for $p = 5$ using a modified *conditional score* estimating function proposed earlier by Stefanski and Carroll (1987). This is a case where one frequently finds multiple roots to the concerned estimating equation. A detailed description on the derivation of this estimating function was given in Section 4.7. Using an artificial data set with $n = 200$, Hanfelt and Liang (1995) performed a simulation study using $Q_\Gamma(\theta, \eta)$ to judge the plausibility of the three roots associated with the data set. They found that the root closest in Euclidean distance to the true value of the parameter is favoured by $Q_\Gamma(\theta, \eta)$ using two somewhat arbitrarily chosen paths. Similar findings were reported in Hanfelt and Liang (1997) based on more intensive simulation studies.

6.5 Quadratic artificial likelihoods

6.5.1 QUADRATIC ARTIFICIAL LIKELIHOODS

In this section, we study the problem of constructing an artificial objective function from a geometrical perspective using only the information necessary to define an optimal estimating function. Similar to the artificial objective functions discussed in previous sections, such as the projected artificial likelihood functions, the artificial log-likelihood function studied, in this section may also be regarded as a semiparametric analogue of a genuine log likelihood function.

Let $g(\theta, y)$ be an estimating function for θ, which is optimal in the sense of Godambe (Section 2.4) or asymptotically optimal in the sense of Section 6.3.1. It is therefore implicitly assumed that the dimension of $g(\theta, y)$ is equal to that of θ. In Section 6.3.1 we saw how such an optimal estimating function may be constructed using an elementary estimating function. In the next section, we shall again consider the case when the dimension of an estimating function is greater than that of the parameter of interest when we discuss the *generalised method of moments*. In this section we shall assume that the parameter of interest θ is vector-valued having dimension $p > 1$.

As we have seen in Chapter 4, a vector-valued estimating function $g(\theta, y)$ has a natural geometric interpretation as a *vector field* defined in the parameter space Θ. Imagine that the parameter space Θ is filled with some flowing liquid and a particle is placed in Θ. The locus of the particle, say Γ, then forms a smooth curve in Θ. The gradient of Γ is a p-tuple, which depends on the location θ. If this gradient vector, at θ, is given by $g(\theta, y)$, then the estimating function $g(\theta, y)$ is said to define the *vector field* of the flowing substance in Θ. This view of an estimating function helps us to find an artificial objective function for inference in conjunction with an optimal estimating function. There are other advantages obtained by examining the theory of estimating functions from a dynamic perspective. The reader is referred to Chapter 7 for a more systematic study of such a dynamic theory. Figure 6.1 plots the two-dimensional estimating function (5.99) as vector fields locally at the roots to the corresponding estimating equations for three artificially generated data sets.

There are two particularly important types of vector fields, namely, the conservative vector fields and the divergence-free vector fields. A vector field $g(\theta, y)$ is said to be *conservative* (or *irrotational*), if $\dot{g}(\theta, y)$ is a $p \times p$ symmetric matrix, or the following conditions are satisfied:

$$\frac{\partial g_i(\theta, y)}{\partial \theta_j} = \frac{\partial g_j(\theta, y)}{\partial \theta_i} \quad \text{for } i \neq j = 1, \ldots, p$$

where $g_j(\theta, y)$ is the jth component of $g(\theta, y)$. On the other hand, a vector field $g(\theta, y)$ is said *divergence-free* (or *solenoidal*), if the divergence of $g(\theta, y)$ vanishes

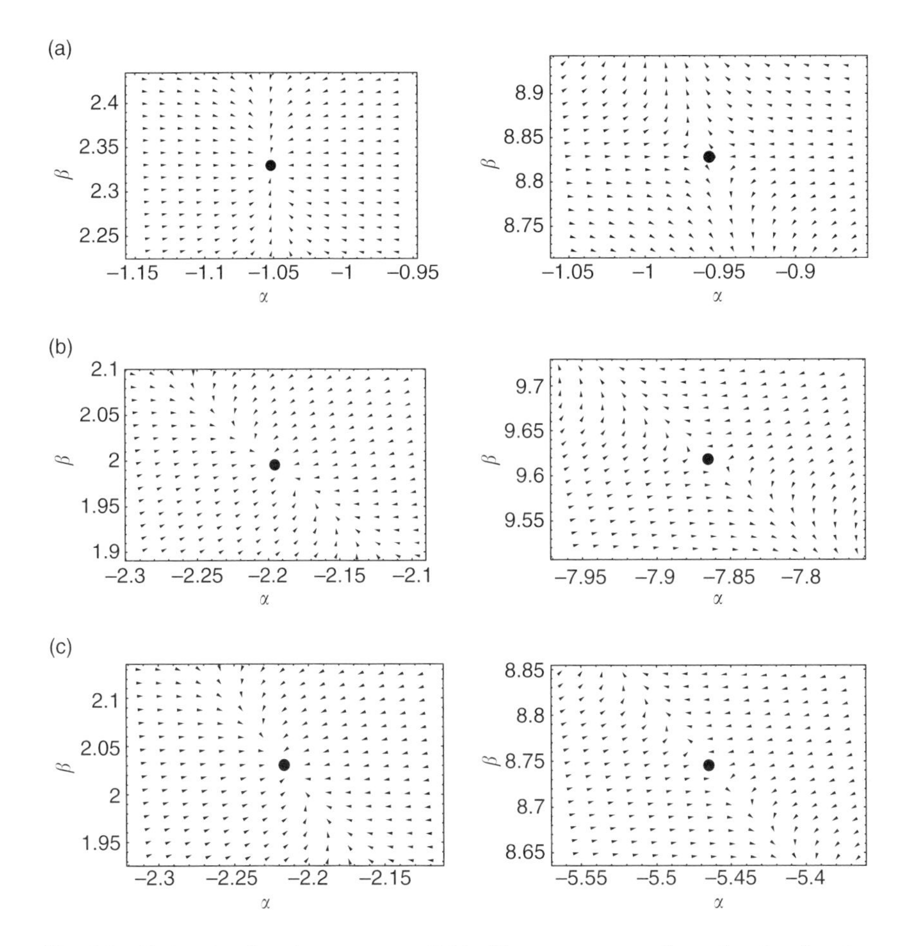

Fig. 6.1 Estimating functions as vector fields. The non-conservative estimating function (5.99) arising from a logistic regression model with errors in variables are plotted as vector fields in the (α, β)-plane. The vector fields are shown near two roots for each of the three simulated samples (marked a, b and c). The roots of the estimating equations, shown in black dots, are fixed points of the vector fields. The fixed points at the left panel are *sinks*, which resemble the MLEs. Which root for each sample should we choose as our estimate for the regression parameter? This problem was studied earlier in Section 5.10.

for any $\theta \in \Theta$. That is,

$$\operatorname{div}\{g(\theta, y)\} = \sum_{j=1}^{p} \frac{\partial g_j(\theta, y)}{\partial \theta_j} = 0.$$

Geometrically, therefore, the nonexistence of an objective function for an estimating function $g(\theta, y)$ implies that the vector field defined by $g(\theta, y)$ curls or rotates. Wang (1999) shows that for any $p > 1$, one can write

$$g(\theta, y) = g_c(\theta, y) + g_r(\theta, y) \tag{6.41}$$

where $g_c(\theta, y)$ is conservative and $g_r(\theta, y)$ is divergence-free. The decomposition (6.41) is a generalisation of the well-known *Helmholtz decomposition*, which only concerns vectors of dimension three.

Having found such a decomposition, one can construct an objective function by discarding the divergence-free component $g_r(\theta, y)$ and integrating the conservative part:

$$\ell(\theta, \xi) = \int_\theta^\xi \{g_c(\eta(s), y)\}^{\mathrm{t}} \, \mathrm{d}\eta(s) \tag{6.42}$$

along a suitable regular path $\eta(s)$ from θ to ξ. A problem with this definition of $\ell(\theta, \xi)$ is that the generalised Helmholtz decomposition (6.41) is not unique. For example, suppose that $g(\theta, y)$ is conservative to begin with. Then we would normally expect that the divergence-free part would be zero. However, with no additional guidance as to how to perform the decomposition, there is no reason for this to be true. Since we intend to discard the divergence-free part, it should contain as little statistical information as possible. This suggests two principles for the construction of the generalised Helmholtz decomposition:

1. If $\dot{g}(\theta, y)$ is symmetric, then we require
$$g_r(\theta, y) = 0. \tag{6.43}$$
2. If $g(\theta, y)$ is E-conservative, that is, if $E_\theta\{\dot{g}(\theta, Y)\}$ is symmetric, then we require
$$E_\theta\{g_r(\theta, Y)\} = 0. \tag{6.44}$$
The second principle is the expected version of the first one, so it is also a natural requirement.

The solution to this problem involves the use of the geometry of differential forms—a topic that we shall not discuss here. The interested reader is referred to Darling (1994). A form of the decomposition that would satisfy these two principles is

$$g(\theta, y) = d\{\ell(\theta)\} + * d[\alpha(\theta) * d\{g(\theta, y)\}] \tag{6.45}$$

where $\ell(\theta)$ is a scalar function that is intended to be used as our objective function, $\alpha(\theta)$ is an appropriate scalar function of θ only, $*$ is the *Hodges star operator*, and d here denotes the *exterior derivative* operator. Note that the vector field

$$g_r^o(\theta, y) = * d[\alpha(\theta) * d\{g(\theta, y)\}]$$

vanishes if $g(\theta, y)$ is conservative. Similarly, we have

$$E_\theta\{g_r^o(\theta, Y)\} = 0$$

if $g(\theta, y)$ is E-conservative.

In general, there is no guarantee that such a scalar function $\alpha(\theta)$ exists. We can show, however, that if $g(\theta, y)$ is linear in θ then there exists such a scalar function $\alpha(\theta)$ so that the following decomposition:

$$g(\theta, y) = d\{\ell(\theta)\} + g_r^o(\theta, y)$$

holds, with $\alpha(\theta)$ being explicitly given by

$$\alpha(\theta) = (-1)^k \tfrac{1}{4}\theta^{\mathrm{t}}\theta. \tag{6.46}$$

See Wang (1999) for further discussion on implications of the choice of (6.46).

The above arguments suggest that we may use a linear approximation to $g(\theta, y)$ locally at some parameter values of interest. Such points of interest include, but are not restricted to, the set of zeros of $g(\theta, y)$. Suppose that ξ is a known value in Θ. Applying (6.42), (6.45) and (6.46) to the linear approximation of $g(\theta, y)$ at ξ, we obtain the following *quadratic artificial log-likelihood function* for θ:

$$\ell(\theta, \xi) = -\tfrac{1}{2}(\theta - \xi)^{\mathrm{t}} J(\xi)(\theta - \xi) + \{g(\xi)\}^{\mathrm{t}}(\theta - \xi) \tag{6.47}$$

where

$$J(\xi) = -\tfrac{1}{2}\left[\dot{g}(\xi) + \{\dot{g}(\xi)\}^{\mathrm{t}}\right]$$

is the symmetrised quasi-Hessian. In defining (6.47) we also assumed that ξ is a *non-degenerate point* so that the determinant, det $\{g(\theta, y)\}$, evaluated at ξ will not vanish with probability one.

When $g(\theta, y)$ is the quasi-score, the quadratic artificial objective function (6.47) becomes

$$\ell(\theta, \xi) = -\frac{1}{2}\{\mu(\theta) - \mu(\xi)\}^{\mathrm{t}}\{\Sigma(\xi)\}^{-1}\Big[\{\mu(\theta) - \mu(\xi)\} - 2\{y - \mu(\xi)\}\Big]$$

a form closely resembling the projected artificial likelihood (6.2) and the generalised projected artificial likelihood (6.13).

6.5.2 QUADRATIC ARTIFICIAL LIKELIHOOD RATIO TESTS

The quadratic artificial log-likelihood function (6.47) depends on a value $\xi \in \Theta$, which is assumed to be given. We shall see how such a value may be chosen shortly. By analogy to the usual parametric likelihood ratio, we define the *quadratic artificial likelihood ratio* by

$$\gamma(\xi) = 2\{\ell(\tilde{\theta}, \xi) - \ell(\tilde{\theta}_0, \xi)\} \tag{6.48}$$

which depends on the data as well as the reference point ξ. In (6.48), $\tilde{\theta}_0$ maximises (6.47) under a null hypothesis

$$H_0 : \theta \in \Theta_0$$

and $\tilde{\theta}$ maximises (6.47) under the restriction

$$\theta \in \Theta = \Theta_A \cup \Theta_0$$

where Θ_A represents an appropriate alternative hypothesis.

We now describe some formal properties of tests using the criterion (6.48) in the context of testing a *general linear hypothesis*

$$H_0 : C(\theta) = R\theta - r = 0 \tag{6.49}$$

against the alternative $C(\theta) \neq 0$. In (6.49), R is an $m \times p$ matrix and r is an m-dimensional vector, m being an integer less or equal to the dimension p of the parameter of interest θ.

Let $\xi = \eta$ be a value satisfying $C(\eta) = 0$. Putting this value of ξ into (6.48), we then obtain

$$\gamma(\eta) = \{RJ^{-1}(\eta)g(\eta)\}^{\mathrm{t}}\{RJ^{-1}(\eta)R^{\mathrm{t}}\}^{-1}\{RJ^{-1}(\eta)g(\eta)\}.$$

Similarly, if we let $\xi = \hat{\theta}$ be a root to $g(\hat{\theta}) = 0$, then (6.48) specialises to

$$\gamma(\hat{\theta}) = C^{\mathrm{t}}(\hat{\theta})\{RJ^{-1}(\hat{\theta})R^{\mathrm{t}}\}^{-1}C(\hat{\theta}).$$

For a non-singular square matrix R, the general linear hypothesis (6.49) simplifies to

$$H_0 : \theta = \theta_0 = R^{-1}r$$

and $\gamma(\eta)$ and $\gamma(\hat{\theta})$ become

$$\gamma(\theta_0) = \{g(\theta_0, y)\}^{\mathrm{t}}J^{-1}(\theta_0)g(\theta_0, y), \tag{6.50}$$

$$\gamma(\hat{\theta}) = (\hat{\theta} - \theta_0)^{\mathrm{t}}J(\hat{\theta})(\hat{\theta} - \theta_0) \tag{6.51}$$

respectively.

Recall that when $g(\theta, y)$ is a score function and

$$I(\theta) = -E_\theta\{\dot{g}(\theta, Y)\}$$

is the total Fisher information, then

$$S_p = \{g(\theta_0, y)\}^{\mathrm{t}}\{I(\theta_0)\}^{-1}g(\theta_0, y) \tag{6.52}$$

defines the parametric *score statistic* (Cox and Hinkley, 1974, p. 315). An asymptotically equivalent form to S_p is the maximum likelihood test statistic, more popularly known as the *Wald test statistic* (Cox and Hinkley, 1974, p. 314)

$$W_p = (\hat{\theta} - \theta_0)^{\mathrm{t}}I(\hat{\theta})(\hat{\theta} - \theta_0). \tag{6.53}$$

Comparing (6.52) with (6.50) and (6.53) with (6.51), we see that the quantities $\gamma(\theta_0)$ and $\gamma(\hat{\theta})$ generalise the classical score and Wald test statistics, respectively.

Note that $\gamma(\theta_0)$ reduces to S_p and $\gamma(\hat{\theta})$ reduces to W_p when $g(\theta, y)$ is a genuine score and $J(\theta)$ is replaced by its expectation in the definitions for $\gamma(\theta_0)$ and $\gamma(\hat{\theta})$.

Returning to the problem of testing the general linear hypothesis (6.49), we may formally define the *generalised score test statistic* as

$$S(\xi) = \{\dot{\ell}(\tilde{\theta}_0, \xi)\}^{\mathrm{t}} J^{-1}(\xi) \dot{\ell}(\tilde{\theta}_0, \xi) \tag{6.54}$$

where $\ell(\theta, \xi)$ is the quadratic artificial log-likelihood function defined by (6.47), and $\tilde{\theta}_0$ is the same as in (6.48). Similarly, we can define the *generalised Wald test statistic* as follows:

$$W(\xi) = \{C(\tilde{\theta})\}^{\mathrm{t}} \{R J^{-1}(\xi) R^{\mathrm{t}}\}^{-1} C(\tilde{\theta}) \tag{6.55}$$

where the value of $\tilde{\theta}$ is also the same as in (6.48).

To investigate the properties of $S(\xi)$ and $W(\xi)$ and their relations with the quadratic artificial likelihood ratio test statistic (6.48), we first note that $\tilde{\theta}$ and $\tilde{\theta}_0$ can be expressed as

$$\begin{aligned} \tilde{\theta} &= \xi + J^{-1}(\xi) g(\xi, y), \\ \tilde{\theta}_0 &= \xi + J^{-1}(\xi)\{g(\xi, y) + R^{\mathrm{t}}\lambda\} \end{aligned}$$

where λ is the Lagrange multiplier

$$\lambda(\xi) = -\{R J^{-1}(\xi) R^{\mathrm{t}}\}^{-1}\{C(\xi) + R J^{-1}(\xi) g(\xi, y)\}.$$

Using these expressions for $\tilde{\theta}$ and $\tilde{\theta}_0$, we can first rewrite the quadratic artificial likelihood ratio (6.48) as

$$\gamma(\xi) = \{A(\xi)\}^{\mathrm{t}} \{R J^{-1}(\xi) R^{\mathrm{t}}\}^{-1} A(\xi) \tag{6.56}$$

where

$$A(\xi) = C(\xi) + R J^{-1}(\xi) g(\xi, y).$$

Similarly, using the relation

$$\dot{\ell}(\tilde{\theta}_0, \xi) = -R^{\mathrm{t}}\lambda$$

the generalised score statistic (6.54) can then be expressed as

$$\begin{aligned} S(\xi) &= \{\dot{\ell}(\tilde{\theta}_0, \xi)\}^{\mathrm{t}} J^{-1}(\xi) \dot{\ell}(\tilde{\theta}_0, \xi) \\ &= \lambda^{\mathrm{t}} R J^{-1}(\xi) R^{\mathrm{t}} \lambda. \end{aligned} \tag{6.57}$$

Simplifying (6.56) and (6.57), we conclude that

$$\gamma(\xi) = S(\xi) \quad \text{for any } \xi \in \Theta.$$

Similarly, we can show that

$$\gamma(\xi) = W(\xi) \quad \text{for any } \xi \in \Theta.$$

The likelihood ratio, the score and the Wald tests are known to be equivalent to one another when the log-likelihood function is exactly quadratic; see Engle (1981) and Buse (1982). The above results can therefore be regarded as a generalisation of this equivalence result in the classical parametric case. Now we state this result more formally.

Proposition 6.6 *For testing the general linear hypothesis (6.49), the quadratic artificial likelihood ratio statistic (6.48), the generalised score statistic (6.54) and the generalised Wald statistic (6.55), at any non-degenerate $\xi \in \Theta$, are equivalent to one another.*

6.5.3 PROFILED QUADRATIC ARTIFICIAL LIKELIHOODS

Suppose that θ is partitioned into

$$\theta^{\mathrm{t}} = (\psi^{\mathrm{t}}, \phi^{\mathrm{t}})$$

where both ψ and ϕ are vectors, ψ being the *parameter of interest* having dimension s, and ϕ being regarded as a *nuisance parameter* with dimension $k = p - s$. Suppose that for θ there is available an optimal estimating function $g(\theta) = g(\theta, y)$, which can be written as

$$g^{\mathrm{t}}(\theta) = (g_{\psi}^{\mathrm{t}}(\theta), g_{\phi}^{\mathrm{t}}(\theta)) \tag{6.58}$$

where $g_{\psi}(\theta)$ has dimension s and $g_{\phi}(\theta)$ has dimension k. For convenience, the dependence of estimating functions on data will be suppressed.

As in Godambe (1991), we shall assume that $g(\theta)$ satisfies the following two conditions:

1. Both $g_{\psi}(\theta)$ and $g_{\phi}(\theta)$ are unbiased.
2. The estimating function $g_{\psi}(\theta)$ is Godambe efficient for ψ given ϕ, and $g_{\phi}(\theta)$ is Godambe efficient for ϕ given ψ.

Suppose that $\hat{\phi}_{\psi}$ is the unique root solving the equation

$$g_{\phi}(\psi, \hat{\phi}_{\psi}) = 0$$

with the value of ψ fixed. Putting the value $\hat{\phi}_{\psi}$ into $g_{\psi}(\psi, \phi)$ gives an approximately unbiased estimating function for ψ

$$g_{\psi}(\psi) = g_{\psi}(\psi, \hat{\phi}_{\psi}). \tag{6.59}$$

Godambe (1991) argued that, given the above two conditions on g_{ψ} and g_{ϕ}, the estimating function $g_{\psi}(\psi)$ defined by (6.59) is an asymptotically efficient estimating function for ψ. For convenience, we shall refer to (6.59) as the *profiled estimating function*. We note, however, when $g(\theta)$ is the score for a parametric model,

$g_\psi(\psi)$ differs slightly from the score function derived from the usual *profile likelihood function*. Improvements upon $g_\psi(\psi)$ may be achieved by correcting its bias and forcing the bias-corrected function to be information-unbiased, a method introduced by McCullagh and Tibshirani (1990) in discussing adjustments of the profile likelihood. The same authors also pointed out that even in such parametric cases the resultant estimating function for ψ in vector cases may not be conservative.

Now suppose that we wish to test the hypothesis

$$H_0 : \psi = \psi_0 \tag{6.60}$$

against the alternative $H_A : \psi \neq \psi_0$. Note that this hypothesis is a special case of (6.49). Conversely, testing a general linear hypothesis (6.49) can be reduced to the problem of testing (6.60) by appropriately choosing the parameter of interest ψ and the nuisance parameter ϕ.

Now there are two apparently different ways to define the artificial likelihood ratios for testing (6.60). The first is to treat the problem of testing (6.60) as a special case of testing (6.49) with

$$R = (\mathrm{I}_{s\times s}, \mathrm{O}_{s\times k}) \quad \text{and} \quad r = \psi_0$$

where $\mathrm{I}_{s\times s}$ is the $s \times s$ identity matrix and $\mathrm{O}_{s\times k}$ is the $s \times k$ zero matrix. The second approach is to define the quadratic artificial likelihood function based on the profiled estimating function $g_\psi(\psi)$.

Fortunately, these two approaches are closely related to each other. To see this, we first establish some notation. Let

$$H = -\frac{1}{n} E_{\theta_0}\{\dot{g}(\theta_0, Y)\}, \quad \Sigma = \frac{1}{n} E_{\theta_0}\left[g(\theta_0, Y)\{g(\theta_0, Y)\}^{\mathrm{t}}\right]$$

be the average expected quasi-Hessian and the covariance matrix of $g(\theta, y)$. Let $\bar{H} = (H + H^{\mathrm{t}})/2$ be the symmetrised quasi-Hessian matrix. We shall use the following conventions for matrix partitions:

$$H = \begin{pmatrix} H_{\psi\psi} & H_{\phi\psi} \\ H_{\psi\phi} & H_{\phi\phi} \end{pmatrix} \quad \Sigma = \begin{pmatrix} \Sigma_{\psi\psi} & \Sigma_{\phi\psi} \\ \Sigma_{\psi\phi} & \Sigma_{\phi\phi} \end{pmatrix}$$

$$H^{-1} = \begin{pmatrix} H^{\psi\psi} & H^{\phi\psi} \\ H^{\psi\phi} & H^{\phi\phi} \end{pmatrix} \quad \Sigma^{-1} = \begin{pmatrix} \Sigma^{\psi\psi} & \Sigma^{\phi\psi} \\ \Sigma^{\psi\phi} & \Sigma^{\phi\phi} \end{pmatrix}$$

where, for instance, $H_{\psi\psi}$ is the $s \times s$ matrix corresponding to the expected quasi-Hessian of $g_\psi(\theta, y)$. Similarly, $H_{\phi\phi}$ corresponds to the expected quasi-Hessian of $g_\phi(\theta, y)$. The other sub-matrices can be interpreted in a similar fashion.

Proposition 6.7 *For testing (6.60), let $\hat{\phi}_0 = \hat{\phi}_{\psi_0}$ and $\hat{\theta} = (\hat{\psi}, \hat{\phi})$ be solutions to*

$$g_\phi(\psi_0, \hat{\phi}_{\psi_0}) = 0 \text{ and } g(\hat{\theta}) = g(\hat{\psi}, \hat{\phi}) = 0$$

respectively. Let $\theta_0^{\mathrm{t}} = (\psi_0^{\mathrm{t}}, \hat{\phi}_0^{\mathrm{t}})$.

Then, the quadratic artificial likelihood ratio $\gamma(\theta_0)$ based on $g(\theta)$ is equal to the that of $\gamma(\psi_0)$ based on $g_\psi(\psi)$ and is given by

$$\gamma(\psi_0) = \{g_\psi(\psi_0)\}^{\mathrm{t}} J^{\psi\psi}(\psi_0) g_\psi(\psi_0). \tag{6.61}$$

Similarly, the quadratic artificial likelihood ratio $\gamma(\hat{\theta})$ based on $g(\theta)$ is equal to that of $\gamma(\hat{\psi})$ based on $g_\psi(\psi)$ and is given by

$$\gamma(\hat{\theta}) = (\hat{\psi} - \psi_0)^{\mathrm{t}} \{J^{\psi\psi}(\hat{\theta})\}^{-1} (\hat{\psi} - \psi_0). \tag{6.62}$$

Proof. First, we consider the quadratic artificial likelihood ratio based on the p-dimensional estimating function $g(\theta, y)$ evaluated at the point $\xi = \theta_0$. If $\tilde{\theta}$ is the unrestricted maximiser of the quadratic artificial likelihood, then from the expression for $\tilde{\theta}$ obtained in Section 6.5.2, we have

$$\tilde{\theta} - \theta_0 = J^{-1}(\theta_0) g(\theta_0).$$

Similarly, for the maximiser $\tilde{\theta}_0$ under (6.60), we have

$$\tilde{\theta}_0 - \theta_0 = J^{-1}(\theta_0)\{g(\theta_0) - R^{\mathrm{t}} g_\psi(\theta_0)\} = 0$$

implying that $\tilde{\theta}_0 = \theta_0$. It therefore follows that

$$\begin{aligned} \gamma(\theta_0) &= \{g(\theta_0)\}^{\mathrm{t}} J^{-1}(\theta_0) g(\theta_0) \\ &= \{g_\psi(\theta_0)\}^{\mathrm{t}} J^{\psi\psi}(\theta_0) g_\psi(\theta_0) \end{aligned}$$

which implies (6.61).

Similarly, to study the quadratic artificial likelihood ratio based on $g(\theta, y)$ at the reference point $\xi = \hat{\theta}$, we note that $\tilde{\theta} = \hat{\theta}$ and

$$\tilde{\theta}_0 - \hat{\theta} = -J^{-1}(\hat{\theta}) R^{\mathrm{t}} \{J^{\psi\psi}(\hat{\theta})\}^{-1} (\hat{\psi} - \psi_0).$$

From these expressions we can then derive (6.62) in a straightforward manner.

On the other hand, to examine the quadratic artificial likelihood ratio based on the profiled estimating function $g_\psi(\psi)$, we first note that from the definition for $\hat{\phi}_\psi$

$$g_\phi(\psi, \hat{\phi}_\psi) = 0$$

we can show that

$$\frac{\partial}{\partial \psi} \hat{\phi}_\psi = -g_{\phi\phi}^{-1} g_{\phi\psi} \tag{6.63}$$

where the matrix $\dot{g}(\theta)$ is partitioned using the following notation

$$\dot{g}(\theta) = \{\{g_{\psi\psi}, g_{\psi\phi}\}, \{g_{\phi\psi}, g_{\phi\phi}\}\}.$$

The inverse of $\dot{g}(\theta)$ is partitioned similarly except that superscripts are used instead of subscripts.

By definition, the quadratic artificial likelihood for ψ at ξ_ψ is given by

$$\ell(\psi, \xi_\psi) = -\tfrac{1}{2}(\psi - \xi_\psi)^{\mathrm{t}} H_{\psi\psi}(\xi_\psi)(\psi - \xi_\psi) + \{g_\psi(\xi_\psi)\}^{\mathrm{t}}(\psi - \xi_\psi) \tag{6.64}$$

where $H_{\psi\psi}$ can be written as

$$H_{\psi\psi} = J_{\psi\psi} - J_{\psi\phi} J_{\phi\phi}^{-1} J_{\phi\psi}$$

using the expression (6.63). Evaluated at ψ_0 and $\hat{\psi}$, the quadratic likelihood ratio

$$\gamma(\xi_\psi) = 2\{\ell(\tilde{\psi}, \xi_\psi) - \ell(\tilde{\psi}_0, \xi_\psi)\} \tag{6.65}$$

by (6.50), (6.51) and Proposition 6.6, can be written as

$$\gamma(\psi_0) = \{g_\psi(\psi_0)\}^{\mathrm{t}}\{H_{\psi\psi}(\psi_0)\}^{-1} g_\psi(\psi_0),$$
$$\gamma(\hat{\psi}) = (\hat{\psi} - \psi_0)^{\mathrm{t}} H_{\psi\psi}(\hat{\theta})(\hat{\psi} - \psi_0)$$

respectively. Finally, by the matrix identity

$$H_{\psi\psi}^{-1} = J^{\psi\psi} = J_{\psi\psi}^{-1} + J_{\psi\psi}^{-1} J_{\psi\phi} \left\{ J_{\phi\phi} - J_{\phi\psi} J_{\psi\psi}^{-1} J_{\psi\phi} \right\}^{-1} J_{\phi\psi} J_{\psi\psi}^{-1}$$

we can show that $\gamma(\psi_0)$ and $\gamma(\hat{\psi})$ equals (6.61) and (6.62), respectively. This completes the proof. □

6.5.4 ASYMPTOTIC DISTRIBUTIONS

Now we study the distributional properties of the quadratic artificial likelihood ratios. Let θ_0 be a true or hypothetical value of θ. Let $\hat{\theta}$ be a zero of $g(\theta)$. The following conditions will be frequently used later on.

[C1] There exists a neighbourhood of θ_0 so that $g(\theta)$ is twice differentiable with respect to θ.

[C2] The root $\hat{\theta}$ is $\sqrt{n}$-consistent, i.e.,

$$\hat{\theta} = \theta_0 + O_p(n^{-1/2}).$$

[C3] The average of the second derivative of $g(\theta)$ is bounded in probability. That is,

$$\frac{\partial^2}{\partial\theta_j \partial\theta_k} \left\{ \frac{1}{n} g_i(\theta) \right\} = O_p(1) \quad \text{for } i, j, k = 1, \ldots, p$$

where $g_i(\theta)$ is the ith component of $g(\theta)$.

[C4] The weak law of large numbers applies to $\dot{g}(\theta)$ for any $\theta \in \Theta$. In particular,

$$-\frac{1}{n}\dot{g}(\theta_0) = H + o_p(1)$$

where

$$H = -\frac{1}{n}E_{\theta_0}\{\dot{g}(\theta_0)\} = O(1).$$

[C5] The weak law of large numbers applies to $g(\theta)\{g(\theta)\}^{\mathrm{t}}$ for any $\theta \in \Theta$. In particular,

$$\frac{1}{n}g(\theta_0)\{g(\theta_0)\}^{\mathrm{t}} = \Sigma + o_p(1)$$

where

$$\Sigma = E_{\theta_0}\left[\frac{1}{n}g(\theta_0)\{g(\theta_0)\}^{\mathrm{t}}\right] = O(1).$$

[C6] The central limit theorem applies to $g(\theta)$ for any $\theta \in \Theta$. In particular,

$$\frac{1}{\sqrt{n}}g(\theta_0) \xrightarrow{\mathcal{L}} N(0, \Sigma).$$

[C7] An estimating function $g(\theta)$ is said to be *first-order conservative* if the following relation:

$$\dot{g}(\theta) = \ddot{\lambda}(\theta) + O_p(n^{1/2})$$

holds for a scalar function $\lambda(\theta)$. Note that the concepts of E-conservativeness and the first-order conservativeness are closely related to one another. An E-conservative additive estimating function is usually first-order conservative. Conversely, an unbiased first-order conservative estimating function is often E-conservative as well.

Note that conservative estimating functions are automatically first-order conservative. So scores are first-order conservative. That quasi-scores are also first-order conservative can be derived from the weak law of large numbers

$$\frac{1}{n}\dot{g}(\theta) = \frac{1}{n}E_\theta\{\dot{g}(\theta)\} + o_p(1)$$

combined with the fact that the expected quasi-Hessian

$$E_\theta\{\dot{g}(\theta)\} = -\dot{\mu}^{\mathrm{t}}(\theta)V^{-1}(\theta)\dot{\mu}(\theta) = O(n).$$

is a symmetrical matrix.

Now we consider the distributions of the quadratic likelihood ratios (6.61) and (6.62). The special cases are (6.50) and (6.51) for testing $H_0 : \theta = \theta_0$ against $H_A : \theta \neq \theta_0$. We consider the latter cases first.

Proposition 6.8 *Assume that conditions* [C1] *through* [C4] *hold. Then we have the following results.*

(i) The test statistics (6.50) and (6.51) can be rewritten respectively as

$$\gamma(\theta_0) = \left\{\frac{1}{\sqrt{n}} g^{\mathrm{t}}(\theta_0)\right\} \bar{H}^{-1} \left\{\frac{1}{\sqrt{n}} g(\theta_0)\right\} + o_p(1), \tag{6.66}$$

$$\gamma(\hat{\theta}) = \left\{\frac{1}{\sqrt{n}} g^{\mathrm{t}}(\theta_0)\right\} (H^{-1})^{\mathrm{t}} \bar{H} H^{-1} \left\{\frac{1}{\sqrt{n}} g(\theta_0)\right\} + o_p(1) \tag{6.67}$$

where $\bar{H} = (H + H^{\mathrm{t}})/2$.

(ii) If [C7] *holds, then* $\gamma(\theta_0)$ *and* $\gamma(\hat{\theta})$ *are asymptotically equivalent to*

$$\left\{\frac{1}{\sqrt{n}} g^{\mathrm{t}}(\theta_0)\right\} H^{-1} \left\{\frac{1}{\sqrt{n}} g(\theta_0)\right\}. \tag{6.68}$$

(iii) Suppose in addition that [C5] *and* [C6] *hold. Then both* $\gamma(\theta_0)$ *and* $\gamma(\hat{\theta})$ *have the same limiting null distribution*

$$\gamma(\theta_0) \xrightarrow{\mathcal{L}} \sum_{j=1}^{p} \lambda_j Z_j^2 \tag{6.69}$$

where Z_j *are independent unit normal variates and* λ_j *are the eigenvalues of* ΣH^{-1}. *When* $g(\theta)$ *is information-unbiased so that* $\Sigma = H$, *then (6.69) reduces to the chi-squared distribution* $\chi^2(p)$ *with* p *degrees of freedom.*

Proof. Only an outline of the proof will be given. By [C1] and [C3], a second-order Taylor expansion of $g(\hat{\theta})$ at θ_0 gives the expression

$$-\dot{g}(\theta_0)(\hat{\theta} - \theta_0) = g(\theta_0) + O_p(1)$$

which, by [C4], implies the following relation:

$$\sqrt{n}(\hat{\theta} - \theta_0) = H^{-1} \left\{\frac{1}{\sqrt{n}} g(\theta_0)\right\} + o_p(1).$$

On the other hand, [C1] to [C4] imply that

$$-\frac{1}{n}\dot{g}(\hat{\theta}) = H + o_p(1) \quad \text{and} \quad \frac{1}{n} J(\hat{\theta}) = \bar{H} + o_p(1).$$

Using these relations, we can write $\gamma(\hat{\theta})$ as follows:

$$\begin{aligned}
\gamma(\hat{\theta}) &= (\hat{\theta} - \theta_0)^{\mathrm{t}} J(\hat{\theta})(\hat{\theta} - \theta_0) \\
&= \left\{\frac{1}{\sqrt{n}} H^{-1} g(\theta_0) + o_p(1)\right\}^{\mathrm{t}} \{\bar{H} + o_p(1)\} \left\{\frac{1}{\sqrt{n}} H^{-1} g(\theta_0) + o_p(1)\right\} \\
&= \left\{\frac{1}{\sqrt{n}} g^{\mathrm{t}}(\theta_0)\right\} (H^{-1})^{\mathrm{t}} \bar{H} H^{-1} \left\{\frac{1}{\sqrt{n}} g(\theta_0)\right\} + o_p(1).
\end{aligned}$$

The last expression gives (6.67). The formula (6.66) can be derived along similar lines. This proves part (i) of the proposition.

The simplification (6.68) occurs because the asymmetric part of $\dot{g}(\theta)$, by [C7], is negligible in the above arguments. So we have part (ii). Finally, with the assumptions [C5] and [C6], the asymptotic distribution (6.69) can be derived using standard arguments such as those given by Johnson and Kotz (1970, p. 149). This completes the proof. □

Now we state the results when nuisance parameters are present. Let $\hat{\theta}$, θ_0 be defined as in Proposition 6.7.

Proposition 6.9 *Assume that conditions* [C1] *through* [C4] *hold. Then we have the following results:*

(i) The test statistics (6.61) and (6.62) can be expressed as

$$\gamma(\theta_0) = \left\{\frac{1}{\sqrt{n}} g_\psi^{\mathrm{t}}(\theta_0)\right\} \bar{H}^{\psi\psi} \left\{\frac{1}{\sqrt{n}} g_\psi(\theta_0)\right\} + o_p(1),$$

$$\gamma(\hat{\theta}) = \left\{\frac{1}{\sqrt{n}} g_\psi^{\mathrm{t}}(\theta_0)\right\} \{H^{\psi\psi}(\theta_0)\}^{\mathrm{t}} \{\bar{H}^{\psi\psi}(\theta_0)\}^{-1} H^{\psi\psi}(\theta_0) \times \left\{\frac{1}{\sqrt{n}} g_\psi(\theta_0)\right\} + o_p(1)$$

respectively.

(ii) If [C7] *holds, then $\gamma(\theta_0)$ and $\gamma(\hat{\theta})$ are asymptotically equivalent to*

$$\left\{\frac{1}{\sqrt{n}} g_\psi^{\mathrm{t}}(\theta_0)\right\} H^{\psi\psi}(\theta_0) \left\{\frac{1}{\sqrt{n}} g_\psi(\theta_0)\right\}.$$

(iii) Assume further that [C5] *and* [C6] *hold. Then both $\gamma(\theta_0)$ and $\gamma(\hat{\theta})$ have the same limiting null distribution*

$$\gamma(\theta_0) \xrightarrow{\mathcal{L}} \sum_{j=1}^{s} \lambda_j Z_j^2 \tag{6.70}$$

where Z_j are independent unit normal variates and λ_j the eigenvalues of

$$\Sigma_{\psi\psi}(\psi_0) H^{\psi\psi}(\psi_0).$$

The simplification of the chi-squared distribution $\chi^2(s)$ with s degrees of freedom occurs if $g(\theta)$ is information-unbiased.

We shall omit the proof of Proposition 6.9, which is similar to that of Proposition 6.8.

The quadratic likelihood ratio tests are also *consistent*. To appreciate this point, let us consider, for instance, the case (6.50). Let $\theta_A \neq \theta_0$ be a point belonging to the alternative hypothesis. It is easy to check that the following relations

$$g(\theta_0) = g(\theta_A) + O_p(n)$$

and

$$J^{-1}(\theta_0) = J^{-1}(\theta_A) + o_p(1)$$

hold. The quadratic artificial likelihood ratio at $\xi = \theta_0$ therefore can be written as

$$\gamma(\theta_0) = \{g(\theta_A)\}^{\mathrm{t}} J^{-1}(\theta_A) g(\theta_A) + O_p(n). \tag{6.71}$$

If the alternative hypothesis with $\theta = \theta_A$ is true, then the first term of (6.71), by Proposition 6.8, has chi-squared distribution. Therefore (6.71) must go to infinity under H_A when n goes to infinity, implying that the test is consistent.

Note also that in Proposition 6.7, $\theta_0 = (\psi_0, \hat{\phi}_0)$, where $\phi_0 = \hat{\phi}_{\psi_0}$ is determined from $g_\phi(\psi_0, \hat{\phi}_0) = 0$. So the eigenvalues appearing in (6.70) depend only on ψ_0, implying that the quadratic likelihood ratio tests are also *similar*.

6.5.5 NUMERICAL STUDIES

To see how the quadratic artificial likelihood ratios will behave, we now consider a small simulation study involving a Bayesian logistic regression model similar to that considered in Section 5.13. We imagine that a health survey is performed, which involves six risk factors. Assume that these factors form $n = 24$ covariate classes. In each covariate class, there are y_j *cases* among m_j subjects, $j = 1, \ldots, n$.

The y_j are assumed to have binomial distributions

$$y_j \sim \binom{m_j}{y_j} \pi_j^{y_j} (1 - \pi_j)^{m_j - y_j}, \quad y_j = 0, 1, \ldots, m_j$$

with index parameters m_j and means $m_j \pi_j$. The response probabilities π_j are assumed to be 'canonically' related to the regression parameter θ in the following way:

$$\log \frac{\pi_j}{1 - \pi_j} = \sum_{i=0}^{6} x_{ij} \theta_i .$$

We assume that the covariates x_{ij} are also binary, assuming the values 1 or 0 according to the presence or absence of the ith risk factor for the jth class. We further make the assumption that the π_j have a common beta distribution,

$$\pi_j \sim \frac{1}{B(\lambda, \lambda)} \pi_j^{\lambda - 1} (1 - \pi_j)^{\lambda - 1}$$

where $\lambda > 0$ is an unknown hyper-parameter. This Bayesian approach can be useful when, for instance, some of the counts m_j are small.

Let X be the 24×7 binary design matrix, the first column being all 1's. We shall study the behaviour of the quadratic likelihood ratios based on the posterior score,

$$g(\theta, y) = X^{\mathrm{t}} \{y - \mu + (\lambda - 1) q\} \tag{6.72}$$

where

$$\mu = (m_1\pi_1, \ldots, m_n\pi_n)^{\mathrm{t}} \quad \text{and} \quad q = (1 - 2\pi_1, \ldots, 1 - 2\pi_n)^{\mathrm{t}}.$$

Note that (6.72) reduces to the score function for $\lambda = 1$ corresponding to the uniform prior.

For a particular design matrix X and an index vector, $(m_1, \ldots, m_n)$, we considered two cases. First, we set $\psi = \theta_2$ and wish to test the hypothesis that $\psi = \psi_0 = 0$. So $s = 1$ in (6.70). The 'true' value of the parameter is set to

$$\theta = (-0.80, 0.93, 0.00, 1.83, -1.70, -0.69, 1.48).$$

In the second case, we let $\psi = (\theta_2, \theta_6)$ and wish to test the hypothesis that $\psi = \psi_0 = 0$. So $s = 2$ in this case. The 'true' value is set to

$$\theta = (-0.80, 0.93, 0.00, 1.83, -1.70, -0.69, 0.00).$$

In our numerical studies, the design matrix, the index vector and the above parameter values were chosen similar to those in the real data analysis concerning the urinary tract infections among young women studies in Section 5.13. In this study, some of the m_j were small and the Bayesian logistic model was found appropriate. In our simulations, the hyper-parameter was set to $\lambda = 1.05$ for the first case and $\lambda = 1.5$ for the second case.

Figure 6.2 compares the simulated null distributions of the quadratic likelihood ratios with the asymptotic approximations (6.70) for the two cases described. The weights appearing in (6.70) were $\lambda_1 = 0.96$ for the first case (top) and $\lambda_1 = 0.92$, $\lambda_2 = 0.62$ for the second case (bottom). The asymptotic distributions (broken curves) are in good agreement with the simulated true distributions (solid curves). The true distributions were simulated using 1000 pseudo random samples in each case. Other results (not shown here) confirm the results found here.

6.6 Quadratic inference functions

The last approach for constructing an objective function associated with an estimating function is intrinsically linked to the *generalised method of moments*, a method popular among econometricians. We shall therefore describe the generalised method of moments first. The last section contains an application to longitudinal data analyses.

6.6.1 THE GENERALISED METHOD OF MOMENTS

Let Y be an r-dimensional random variable with sample space $\mathcal{Y} \in \mathbb{R}^r$. Let θ be a vector parameter belonging to a parameter space $\Theta \in \mathbb{R}^p$ for $p \geq 1$. Let

$$g(\theta, Y) : \Theta \times \mathcal{Y} \to \mathbb{R}^k$$

be a k-dimensional estimating function which is unbiased is the usual sense,

$$E_\theta\{g(Y, \theta)\} = 0, \quad \theta \in \Theta.$$

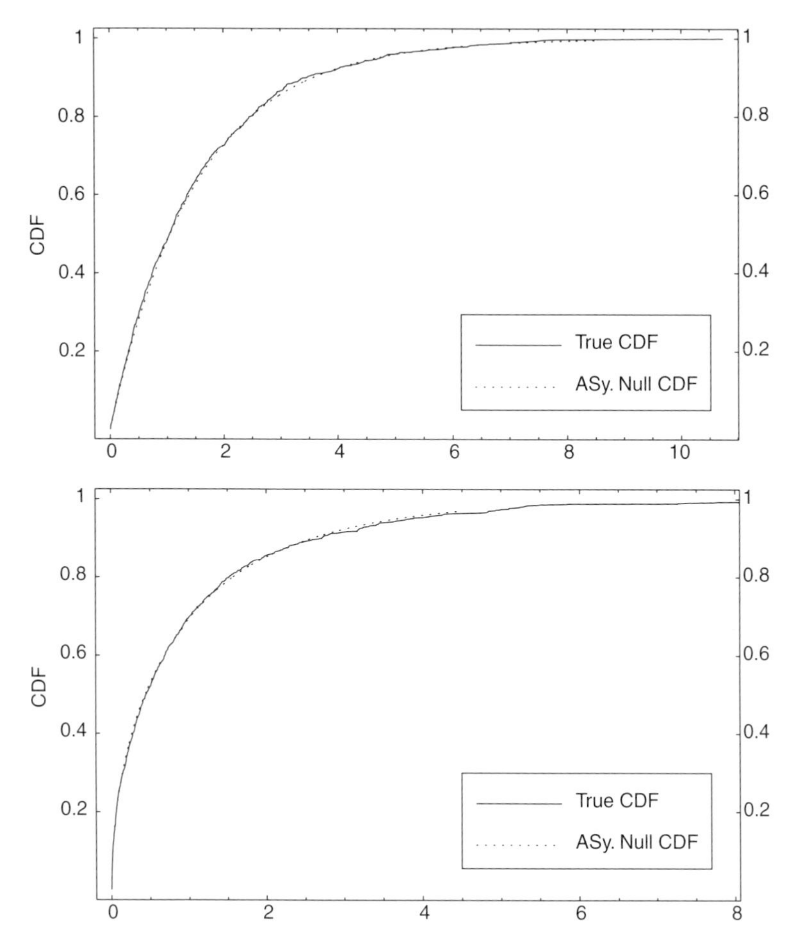

Fig. 6.2 Quadratic likelihood ratio tests. The behaviour of the quadratic likelihood ratios is investigated in a Bayesian logistic regression model having seven parameters. Two cases are considered. In the first case (top), the parameter of interest is a scalar, and in the second case (bottom), the parameter of interest is a two-dimensional vector. The solid curves show the distributions of the quadratic likelihood ratios based on 1000 simulated samples. The broken curves stand for the asymptotic null distributions of the weighted chi-squared distributions with weights $\lambda_1 = 0.96$ (top), and $\lambda_1 = 0.92$, $\lambda_2 = 0.62$ (bottom), respectively.

In this section we shall consider the general situation where the dimension of the estimating function g may be greater than the dimension of the parameter θ. That is, $k \geq p$.

In the following, we shall suppose that g is twice continuously differentiable in θ and that the following matrices

$$\Sigma(\theta) = E_\theta\big[g(\theta, Y)\{g(\theta, Y)\}^{\mathrm{t}}\big], \quad \Gamma(\theta) = E_\theta\{\dot{g}(\theta, Y)\}$$

are of full rank. Note that Σ and Γ have dimensions $k \times k$ and $k \times p$, respectively. Note also that even when $k = p$, $\Gamma(\theta)$ need not be symmetric.

Let $h(\theta, Y)$ be a k-dimensional elementary estimating function. To obtain a consistent estimator, we consider the following extension of the weighted least squares criterion:

$$Q_C(Y, \theta) = \{h(\theta, Y)\}^{\mathrm{t}} C h(\theta, Y) \tag{6.73}$$

where C is an appropriate $k \times k$ symmetric positive definite matrix. For independent random variables, $Y_1, \ldots, Y_n$, the formula (6.73) corresponds to the familiar weighted least squares criterion if we choose the jth component of h to be

$$h_j = Y_j - E(Y_j) \quad j = 1, \ldots, k$$

and the weighting matrix to be

$$C^{-1} = \mathrm{Cov}\{h(\theta, Y)\}.$$

For a general elementary estimating function h and a positive definite matrix C, we formally define the minimiser $\hat{\theta}$ of (6.73) as our estimator for θ. This method is considered by Hansen (1982), which is known as the *generalised method of moments* (GMM).

Under the regularity conditions outlined above, we can show that GMM solution $\hat{\theta}$ is a consistent estimator of θ. It is remarkable that the consistency of $\hat{\theta}$ is independent of the choice of C, provided that C is positive definite. For example, we may let $C = I_{k \times k}$, the identity matrix. Then (6.73) simply becomes the sum of squares

$$Q_I(Y, \theta) = \sum_{i=1}^{k} h_i^2(\theta, Y).$$

With arbitrary chosen C, however, loss of efficiency for the GMM estimator $\hat{\theta}$ is to be anticipated.

Under the usual regularity conditions, the asymptotic normality of $\hat{\theta}$ may also be established:

$$\sqrt{n}(\hat{\theta} - \theta_0) \xrightarrow{\mathcal{L}} \mathcal{N}(0, \Sigma_A) \tag{6.74}$$

where

$$\Sigma_A = (\Gamma^{\mathrm{t}} C \Gamma)^{-1} \Gamma^{\mathrm{t}} C \Sigma C \Gamma (\Gamma^{\mathrm{t}} C \Gamma)^{-1}.$$

Note that the asymptotic variance Σ_A in general depends on C. In the *just-identified case* when $k = p$, however, this dependency is redundant. This is because Γ and C are both invertible square matrices in the just-identified case, so it turns out that

$$\Sigma_A = (\Gamma^{\mathrm{t}} \Sigma^{-1} \Gamma)^{-1}$$

is functionally independent of C.

6.6.2 QUADRATIC INFERENCE FUNCTIONS

We have seen that the first-order asymptotic properties of the GMM estimator in the just-identified case are independent of the weight matrix. An equivalent approach in this case, to the first order, is to consider the following estimating equation:

$$h(\theta, Y) = 0. \tag{6.75}$$

The root solving (6.75), assumed unique, has the same asymptotic distribution as that of the GMM estimator, the inverse of the asymptotic covariance matrix being equal to

$$\mathrm{eff}_\theta(h) = \Sigma_A^{-1} = \Gamma^{\mathrm{t}}\Sigma^{-1}\Gamma. \tag{6.76}$$

The quantity defined by (6.76) is simply the Godambe efficiency of the estimating function $h(\theta, Y)$. Note that the Godambe efficiency of (6.76) is well defined in both the *just-identified* case when $k = p$ and in the *over-identified* case when $k > p$.

For the over-identified case, a parallel estimating function theory to the GMM methodology starts from considering a linear transformation of h by Hh, where H is a $p \times k$ full rank matrix. We assume that H does not depend on data. For unbiased estimating function h we then have that

$$\mathrm{eff}_\theta(Hh) = (H\Gamma)^{\mathrm{t}}(H\Sigma H^{\mathrm{t}})^{-1}(H\Gamma) \tag{6.77}$$

which reduced to $\mathrm{eff}_\theta(h)$ of (6.76) when H is an invertible square matrix. The Godambe efficiency is therefore invariant under a linear transformation in the just-identified case but depends on the transformation in the over-identified case.

If we let

$$H = \Gamma^{\mathrm{t}}C$$

then we will have

$$\mathrm{eff}_\theta(Hh) = \Sigma_A^{-1}.$$

So again, to the first order, the GMM is equivalent to the estimating function theory by considering the following class of estimating functions:

$$H(\theta)h(\theta, Y) = 0 \tag{6.78}$$

where H is the same as that appearing in (6.77).

Now let

$$H = \Gamma^{\mathrm{t}}\Sigma^{-1}$$

and consider the following estimating equation:

$$h^*(Y, \theta) = \Gamma^{\mathrm{t}}\Sigma^{-1}h(\theta, Y) = 0. \tag{6.79}$$

Using (6.77), we see that

$$\mathrm{eff}_\theta(h^*) = \Gamma^{\mathrm{t}}\Sigma^{-1}\Gamma$$

which is the Godambe efficiency in the just-identified case given by (6.76). The weight matrix of GMM corresponding to h^* is $C = \Sigma^{-1}$. This choice of weight is efficient in the sense that the GMM estimator, the minimiser of (6.73), has the smallest asymptotic variance at

$$C = \Sigma^{-1}.$$

That is,

$$(\Gamma^t C \Gamma)^{-1} \Gamma^t C \Sigma C \Gamma (\Gamma^t C \Gamma)^{-1} - (\Gamma^t \Sigma^{-1} \Gamma)^{-1}$$

is non-negative definite. Or, equivalently, the Godambe efficiency (6.77) is maximised at

$$H = \Gamma^t \Sigma^{-1}$$

achieving the value given by (6.76) in the class of estimating functions defined by (6.78). That is

$$\Gamma^t \Sigma^{-1} \Gamma - (H\Gamma)^t (H \Sigma H^t)^{-1} (H\Gamma)$$

is non-negative definite. The techniques of McCullagh and Nelder (1989, Section 9.5) can be used to prove the above statements.

The above argument suggests the use of the following *quadratic inference function* (Qu *et al.*, 2000):

$$Q(\theta; Y) = \{h(\theta, Y)\}^t \Sigma^{-1} h(\theta, Y) \tag{6.80}$$

for inferential purposes based on the unbiased estimating function h. The inference function (6.80) resembles Rao' (1947) score test statistic except that h is not a genuine score and the dimension of h is in general larger than that of θ. Hansen (1982) shows that, if $h(\theta, y)$ is unbiased and $\hat{\theta}$ minimises (6.80), then the quadratic inference function $Q(\hat{\theta}, Y)$ evaluated at $\hat{\theta}$ will asymptotically follow a χ^2 distribution with $k - p$ degrees of freedom, where p is the dimension of the parameter θ and k is the dimension of h.

6.6.3 GENERALISED ESTIMATING EQUATIONS

Now we apply the method of quadratic inference functions to longitudinal data analyses using the generalised estimating equations. In a typical experiment involving repeated measurements we have M subjects, each producing n_j responses

$$y_{jt}, \quad t = 1, \ldots, n_j; \quad j = 1, \ldots, M.$$

The y_{jt}'s are correlated for fixed subjects but independent across the subjects. As in the generalised linear models we assume that the mean response of y_{jt} is a nonlinear function of the linear predictor, i.e.,

$$\mu_{jt} = E(y_{jt}) = h(x_{jt}^t \beta).$$

The parameter of interest is β.

Denote the responses of the subjects by

$$y_j = (y_{j1}, \ldots, y_{jn_j})^{\mathrm{t}}$$

and their means by

$$\mu_j = (\mu_{j1}, \ldots, \mu_{jn_j})^{\mathrm{t}}$$

respectively, where $j = 1, \ldots, M$. The following estimating function:

$$g(\beta, y) = ((y_1 - \mu_1)^{\mathrm{t}}, \ldots, (y_M - \mu_M)^{\mathrm{t}})^{\mathrm{t}}.$$

is unbiased and has dimension $N = \sum_{j=1}^{M} n_j$. The estimating function $g(\beta, y)$ will play the role as the elementary estimating function in analysing the repeated measurements. Let

$$V_j = E\{(Y_j - \mu_j)^{\mathrm{t}}(Y_j - \mu_j)\}$$

to be the covariance matrix of Y_j. Because Y_i and Y_j are independent for $i \neq j$, the covariance matrix of $g(\beta, Y)$ will be block diagonal

$$\begin{aligned}\Sigma_g &= E\big[g(y, \beta)\{g(\beta, y)\}^{\mathrm{t}}\big] \\ &= \mathrm{diag}\{V_1, \ldots, V_M\}.\end{aligned}$$

Note also that

$$\begin{aligned}\Gamma_g^{\mathrm{t}} &= -E\{\dot{g}^{\mathrm{t}}(\beta, Y)\} \\ &= (\dot{\mu}_1^{\mathrm{t}}(\beta), \ldots, \dot{\mu}_M^{\mathrm{t}}(\beta)).\end{aligned}$$

So the Godambe efficient estimating function for β using (6.79) is given by

$$\begin{aligned}q(\beta, y) &= \Gamma_g^{\mathrm{t}} \Sigma_g^{-1} g \\ &= \sum_{j=1}^{M} \dot{\mu}_j^{\mathrm{t}} V_j^{-1}(y_j - \mu_j)\end{aligned} \tag{6.81}$$

which is the well-known *generalised estimating equation* proposed by Liang and Zeger (1986) for analysing longitudinal data. The generalised estimating equation of Liang and Zeger (1986) can be recognised as a special case of the quasi-score functions discussed throughout this book.

The quadratic inference function corresponding to (6.81) now takes the form

$$Q(\beta, y) = \sum_{j=1}^{M} (y_j - \mu_j)^{\mathrm{t}} V_j^{-1} (y_j - \mu_j). \tag{6.82}$$

One practical barrier in using the fully Godambe efficient estimating function (6.81), or the corresponding inference function (6.82), is that the V_j's are often of complicated nature due to data correlation within each subject. To reduce the large number of nuisance parameters, Liang and Zeger (1986) proposed to model V_j across the subjects using some 'working' correlation matrices $R_j(\alpha)$, which depend on a low-dimensional nuisance parameter vector α:

$$W_j = A_j^{1/2} R_j(\alpha) A_j^{1/2}, \quad j = 1, \ldots, M, \tag{6.83}$$

where A_j is the diagonal marginal covariance for the j-th subject. Replacing V_j by W_j in (6.82), we then arrive at a 'working' version of the quadratic inference function

$$Q(\beta, y) = \sum_{j=1}^{M} (y_j - \mu_j)^{\mathrm{t}} A_j^{-1/2} R_j^{-1}(\alpha) A_j^{-1/2} (y_j - \mu_j). \tag{6.84}$$

The inference function (6.84) is no longer efficient since the covariance matrices are in general misspecified. To improve efficiency, Qu *et al.* (2000) recently proposed an alternative idea for modelling the correlation structure. Consider the balanced case with each subject measured $n_j = n$ times, $j = 1, \ldots, M$. Assume that each subject has the same correlation structure. Now we model the common correlation matrix R using B given basis matrices as follows:

$$R^{-1} = a_1 J_1 + \cdots + a_B J_B \tag{6.85}$$

where the basis matrices J_i are known and the coefficients a are unknown and to be determined. As in the specification (6.83), the representation R given by (6.85) may not contain the true correlation matrix. The class of matrices (6.85) however is sufficiently rich to represent many practical correlation structures of interest provided that we choose the basis matrices appropriately. Using the correlation structure (6.85), we can then form the following unbiased estimating function:

$$\sum_{j=1}^{M} \dot{\mu}_j^{\mathrm{t}} A_j^{-1/2} \left(\sum_{\alpha=1}^{B} a_\alpha J_\alpha \right) A_j^{-1/2} (y_j - \mu_j). \tag{6.86}$$

In (6.86) the unknown coefficients a play the same role as the nuisance parameters α in (6.84).

To construct the optimal quadratic inference function based on (6.86), we need to eliminate the nuisance parameters a. To do so, we construct an unbiased estimating functions so that (6.86) is a linear transformation of that function. An obvious choice of such an unbiased estimating function is as follows:

$$q(\beta) = 1/M \sum_{j=1}^{M} q_j(\beta) = \begin{pmatrix} 1/M \sum_{j=1}^{M} \dot{\mu}_j^{\mathrm{t}} A_j^{-1/2} J_1 A_j^{-1/2} (y_j - \mu_j) \\ \vdots \\ 1/M \sum_{j=1}^{M} \dot{\mu}_j^{\mathrm{t}} A_j^{-1/2} J_B A_j^{-1/2} (y_j - \mu_j) \end{pmatrix}. \tag{6.87}$$

The elementary estimating function (6.87) has dimension $B \times p$.

To complete the specification of the the quadratic inference function (6.80), we need an estimate for the covariance matrix of $q(\beta)$. A natural such an estimator is given by

$$\tilde{\Sigma} = \frac{1}{M^2}\sum_{j=1}^{M} q_j^{\mathrm{t}}(\beta) q_j(\beta).$$

Finally, we arrive at the following quadratic inference function for analysing longitudinal data (Qu *et al.*, 2000):

$$Q(\beta) = \{q(\beta)\}^{\mathrm{t}}\tilde{\Sigma}^{-1}q(\beta). \tag{6.88}$$

Now we can estimate the regression parameter β by minimising (6.88). The minimiser of (6.88), to the first order, is asymptotically equivalent to the consistent root of the Godambe efficient estimating function (6.86). The method based on the objective function (6.88) however has the merit over the method using the estimating function (6.86) when the latter estimating function has multiple roots. Using the mean squared error criterion, Qu *et al.* (2000) also show in their simulation studies that the estimators by minimising the inference function of (6.88) have nearly the same efficiency as the solutions to the generalised estimating equations with correctly specified working correlations. However, when misspecifications occur for the working correlations in the generalised estimating equations, the quadratic inference function approach demonstrates some advantage in terms of efficiency.

Qu *et al.* (2000) also established the limiting χ^2 properties of tests based on the quadratic inference function (6.88). Consider the case when nuisance parameters are present. So the parameter β is partitioned into (ψ, ϕ), where ψ is the parameter of interest and ϕ is regarded as a nuisance parameter. Let $\hat{\beta}$ be the minimiser of (6.88). For testing

$$H_0 : \psi = \psi_0$$

let $\hat{\phi}_{\psi_0}$ be the minimiser of (6.88) with ψ fixed at $\psi = \psi_0$. Then, under the null hypothesis, the quantity

$$Q(\psi_0, \hat{\phi}_{\psi_0}) - Q(\hat{\beta})$$

will have the asymptotic χ^2 distribution with s degrees of freedom, where s is the dimension of ψ. The quadratic inference function under a local alternative has a noncentral χ^2 distribution with an appropriate noncentrality parameter.

6.7 Bibliographical notes

The projected artificial likelihood ratio (6.2) is due to McLeish and Small (1992) and treated more systematically in Small and McLeish (1994). The projected log likelihood ratio for the quasi-score (6.13) was studied in Li (1993). Hanfelt and Liang (1995) generalised this result and obtained the artificial objective function (6.12). Barndorff-Nielsen (1995) discussed the directed likelihoods when the parameter of interest is a scalar.

A different projection approach was taken in Li and McCullagh (1994), who proposed to project the score function onto a subspace of conservative estimating functions. A practical problem with this approach is that, given knowledge of the first few moments of a random variable, it is usually difficult to construct the projected conservative estimating functions. Such estimating functions usually have to be found using some approximate methods.

The quadratic artificial likelihood discussed in Section 6.5 was considered in Wang (1999). The equivalence between the quadratic artificial likelihood ratio and the generalised Wald statistic may be interpreted as a partial justification for the objective function (6.47) to be regarded as an artificial log-likelihood function. This is because Wald statistic for the score function is approximately equal to the Kullback–Leibler divergence (Kullback, 1978, p. 28). Le Cam (1990*a*) also discussed the point that the Wald statistic may be regarded as a contrast measure between two distributions. The root selection test introduced in Section 5.8 is based on a bootstrap version of the quadratic artificial likelihood ratio tests studied in Section 6.5.

The quadratic inference function introduced in Section 6.6 is based on Qu *et al.* (2000). This approach suggests removing the difficulty in estimating the working parameters α in the generalised estimating equations of Liang and Zeger (1986) by representing the correlation matrix as a linear combination of certain base matrices. These coefficients may then be eliminated using the generalised method of moments technique proposed by Hansen (1982). The generalised method of moments approach may be regarded as an extension of the generalised least squares method. It is also closely related to Neyman's minimum χ^2 method, a detailed account of which can be found in Ferguson (1958). Cramér (1946) gave an argument showing that the maximum likelihood estimator can be regarded as an approximation to the minimum chi-squared estimator.

7
Root selection and dynamical systems

In Chapter 3, we considered several iterative methods for finding roots of an estimating equation. Typically, such a method starts from a point $\theta^{(0)}$ in the parameter space Θ and iterates as $\theta^{(j)} \rightarrow \theta^{(j+1)}$ according to some rule. In this chapter, we shall regard such an algorithm as a dynamical system on the parameter space, which we can regard as a phase space for the dynamical system. We can also consider related algorithms moving in continuous time as the natural limiting form of the discrete time algorithms. This shift in perspective will lead us to new methods for root selection by regarding an estimating function as a dynamical system. In fact, the modified Newton's methods studied in Sections 5.5 and 5.6 have been motivated by such considerations. The purpose of this chapter is to give a systematic overview of dynamical aspects of the theory of estimating functions.

7.1 Dynamical estimating systems

Suppose that we are interested in estimating a p-dimensional parameter $\theta \in \Theta \subset I\!R^p$ based on a random sample $y = \{y_1, \ldots, y_n\}$. We assume that each y_i is generated from a probability distribution depending on θ. For instance, we may assume that the mean of Y_i is a known function of θ, such as the case in generalised linear models. Inference for θ is to be based on an estimating function $g(\theta, y)$. We shall assume that, unless stated otherwise, the dimension of the estimating function g is the same as that of θ.

From the dynamical viewpoint, we shall regard the parameter space Θ as a *phase space*, and the parameter $\theta(t)$ as the *state* of the random mechanism at *time* t, where the variable t may not have any physical interpretation. The crucial assumption concerning the random mechanism is that the state $\theta(t)$ develops according to the following *time-evolution law*:

$$\frac{\mathrm{d}}{\mathrm{d}t}\theta(t) = g(\theta(t), y) \tag{7.1}$$

where t appears on the right-hand side of (7.1) only through θ. Given an observation vector y, equation (7.1) defines a *continuous time dynamical system*. The dynamical system (7.1) is in fact *autonomous* (Gilmore, 1981, Ch. 19), because the estimating function $g(\theta(t), y)$ depends on t only through the state θ. Since $g(\theta, y)$ is an estimating function, for convenience we shall call (7.1) as a *dynamical estimating system* induced by g.

The autonomous system (7.1) induces a continuous map

$$\pi : R \times \Theta \to \Theta$$

satisfying the following properties:

$$\pi(t + s, \theta) = \pi(s, \pi(t, \theta)), \tag{7.2}$$

$$\pi(0, \theta) = \theta. \tag{7.3}$$

We shall call a map π satisfying (7.2) and (7.3) the *flow* of the dynamical system (7.1). The relationship between the flow π and the state $\theta(t)$ can be expressed by following equations:

$$\pi(t, \theta_0) = \theta(t), \quad \text{where } \theta_0 = \theta(0).$$

Mathematically, it is sometimes convenient to take properties (7.2) and (7.3) as axioms and define the continuous map π as a dynamical system defined on Θ. Note that in defining a mathematical dynamical system, the phase space Θ may be replaced by any topological space and R by any topological group. The most important benefit by regarding the flow as a dynamical system is that both continuous and discrete systems can be studied in the same framework. In the discrete case, one simply replaces R by Z, the integer group. There are, however, good reasons not to explore this advantage in the present book. The primary reason for not doing so is that the only discrete type dynamical systems we shall discuss will be in the form of iterative algorithms, and it is more illuminating for us to study such systems in their own right.

The flow of a dynamical system may be visualised as the path followed by a particle in a *fluid field*. Under certain conditions, a theorem of Picard guarantees local existence of the flow or the *integral curve* of a given smooth fluid field. Given a flow, we may obtain the notion of *velocity* at a given point of the flow on the phase (parameter) space. The velocity of a flow π is said to form a *vector field* on Θ, which is formally defined as a map

$$v : \Theta \to T\Theta$$

associating each point $\theta \in \Theta$ with a vector $v(\theta) \in T_\theta\Theta$, where $T_\theta\Theta$ is the *tangent space* to Θ at θ. In our present case, $v(\theta) = g(\theta, y)$ given y. Now it is almost a tauto logy to state that an estimating function $g(\theta, y)$ defines a dynamical system (7.1) with the associated vector field defined by the same estimating function $g(\theta, y)$.

For a dynamical system such as (7.1), we are particularly interested in its *time-asymptotic* recurrent behaviour, namely, the behaviour of the system when $t \to \infty$. Among the most important such properties are the properties of the fixed points of a dynamical system. The *fixed points* of the dynamical system (7.1) are defined as the solutions to the following equation:

$$\frac{\mathrm{d}}{\mathrm{d}t}\theta(t) = 0. \tag{7.4}$$

The fixed points of (7.1) defined by (7.4) are exactly the points θ satisfying the relation

$$g(\theta, y) = 0. \tag{7.5}$$

Consequently, the *fixed points* of the dynamical system (7.1) coincide with the *roots* to the estimating equation (7.5). If g is a score function, then the maximum likelihood estimator, under the smoothness condition, is one of the fixed points of the dynamical system induced by the score function. The above discussion suggests that certain properties of otherwise static estimators, such as the maximum likelihood estimators, may be investigated by studying the behaviour of the corresponding fixed points of the dynamical estimating system induced by the relevant estimating function.

7.2 Linear dynamical systems

7.2.1 LIAPUNOV STABILITY, LOCAL STABILITY

The stability property is the most important feature of a fixed point. Convergence properties of iterative algorithms depend critically on this property. By regarding an estimating function as a dynamical system, the stability of estimators obtained as roots to the corresponding estimating equation can be studied through the associated dynamical estimating system. To motivate the concept of stability, first let us consider a simple example.

Suppose that a jar is filled with a nutritive solution and contaminated by some bacteria. Let the state of the jar, $\theta(t)$, be the number of the bacteria at time t, which progresses according to the following rule:

$$\frac{\mathrm{d}}{\mathrm{d}t}\theta(t) = b\theta - p\theta^2 \tag{7.6}$$

where b and p denote birth and death rate respectively, both assumed to be a positive constant functionally independent of t. The dynamical system (7.6) has two fixed points, $\theta_1 = 0$ and $\theta_2 = b/p$.

Which (if either) of the two fixed points is *stable* or *unstable*? A state of a dynamical system is stable if the system developing from any state close to it, as time evolves, ends up at the same state. In other words, if a system, when perturbed slightly from a particular state, will be 'pulled back' to the same state, then the state is stable. A fixed point, which is not stable, is said unstable. In the case of the contaminated jar, a slight change in state from $\theta_1 = 0$, the jar being slightly polluted, then the bacteria will begin to reproduce. So the fixed point θ_1 is unstable. However, when the bacteria over-reproduce, some begin to die. Eventually an equilibrium state will be reached so that birth and death will balance each other, implying that θ_2 is a stable fixed point.

The concept of stability is an abstraction of the above discussion on the microbe system.

Definition 7.1 *A fixed point $\hat{\theta}$ of a dynamical estimating system*

$$\frac{\mathrm{d}}{\mathrm{d}t}\theta(t) = g(\theta(t), y)$$

is said to be Liapunov stable, *if for every neighbourhood U of $\hat{\theta}$ there exists a neighbourhood V of $\hat{\theta}$ such that $\pi(t, \theta_0) \in U$ for all $t \geq 0$ whenever $\theta_0 \in V$, where $\pi(t, \theta)$ is the flow function defined by (7.2) and (7.3).*

The interested reader may consult Irwin (1980, p.128) for a definition of the Liapunov* stability in a more general setting. A system therefore will stay as close as desired to a Liapunov stable fixed point if the system evolves from any state not too far away from that fixed point. A concept stronger than Liapunov stability is the local stability.

Definition 7.2 *A fixed point $\hat{\theta}$ of a dynamical estimating system*

$$\frac{\mathrm{d}}{\mathrm{d}t}\theta(t) = g(\theta(t), y)$$

is said to be locally stable, *or simply* stable, *if it is Liapunov stable and if for some neighbourhood W of $\hat{\theta}$, $\pi(t, \theta_0)$ converges to $\hat{\theta}$ for all $\theta_0 \in W$ as $t \to \infty$. That is, for any $\theta_0 \in W$, we have*

$$\pi(t, \theta_0) \quad \to \quad \hat{\theta} \quad \text{as } t \to \infty.$$

While Liapunov stability only ensures that a system near a Liapunov stable fixed point will not escape, a locally stable fixed point has the property that the system developing from any state close enough to a locally stable fixed point will eventually end up at the fixed point.

Definition 7.3 *A fixed point is* unstable *if it is neither stable nor Liapunov stable.*

7.2.2 LINEAR DYNAMICAL SYSTEMS

When is a fixed point stable? There is in general no simple criterion to test the stability of a fixed point. The issue of stability, however, can be relatively easily settled for a *linear* (or *affine*) *dynamical system*

$$\frac{\mathrm{d}}{\mathrm{d}t}\theta(t) = A\theta + b \tag{7.7}$$

* The alternative spelling *Lyapunov* also frequently appears in the literature.

where A is a $p \times p$ matrix and b a p-vector, both functionally independent of θ and t. When the matrix A is not symmetric, the linear dynamical system (7.7) induces a non-conservative vector field for $p \geq 2$. The linear dynamical system (7.7) is simple enough to allow us write down the explicit formulae (Scheinerman, 1996, Ch. 2) for the trajectory of (7.7), so we can proceed to discuss the time-asymptotic behaviour of the system. We shall, however, not go into the details but only recall the basic conclusions here.

Let λ_j, $j = 1, \ldots, p$, be the complex eigenvalues of the matrix A appearing on the right-hand side of (7.7). Let $\Re(\lambda_j)$ be the real part of λ_j. Assume that A is invertible. So

$$\hat{\theta} = -A^{-1}b$$

is the unique fixed point of (7.7).

Proposition 7.4 *Consider the linear dynamical system (7.7).*

(i) If

$$\Re(\lambda_j) < 0, \quad \textit{for all } j = 1, \ldots, p$$

then $\hat{\theta}$ is (locally) stable.

(ii) If

$$\Re(\lambda_j) > 0, \quad \textit{for some } j$$

then $\hat{\theta}$ is unstable.

We note that even when $\Re(\lambda_j) > 0$ for some j, there are still possibilities for the system to converge to an unstable fixed point $\hat{\theta}$, provided that the initial states θ_0 fall into the special subset of the phase space. Let ξ_j be the eigenvector of λ_j, $j = 1, \ldots, p$. Assume that ξ_j are independent. If we let $(a_1, \ldots, a_p)$ be the coordinate of an initial state θ_0 with respect to the vectors ξ_j, $j = 1, \ldots, p$, then the system will converge to the unstable fixed point $\hat{\theta}$ starting from θ_0 if $a_j = 0$ for $\Re(\lambda_j) > 0$. But the system will *typically* blow up for a *randomly* chosen initial point in the phase space.

In Section 4.9 we informally introduced the concept of a vector field associated with an estimating function and were able to distinguish the roots of an estimating equation according to the properties of the associated vector field locally at the respective fixed points. Now we give a somewhat more formal description of the classification of the fixed points for a linear dynamical system. The classification extends to non-linear dynamical systems, as was done in Section 4.9, if we locally linearise the dynamical system at the fixed points of interest.

Definition 7.5 *Consider the linear dynamical system (7.7) with an invertible matrix A. Let $\hat{\theta}$ be the unique fixed point of (7.7). Let λ_j, $j = 1, \ldots, p$, be the complex eigenvalues of A. Then the fixed points may be classified as follows.*

(i) If $\Re(\lambda_j) < 0$ for all j, then $\hat{\theta}$ is a stable fixed point and is called a sink *of the vector field associated with (7.7).*

(ii) *If* $\Re(\lambda_j) > 0$ *for* $j = 1, \ldots, p$, *then* $\hat{\theta}$ *is an unstable fixed point and is called a* source *of the vector field associated with (7.7).*

(iii) *If* $\Re(\lambda_j) > 0$ *for some* j *and* $\Re(\lambda_k) < 0$ *for other* k, *then* $\hat{\theta}$ *is an unstable fixed point and is called a* saddle point *of the vector field associated with (7.7).*

(iv) *If* $\Re(\lambda_j) \leq 0$ *for all* j *and* $\Re(\lambda_j) = 0$ *for some* j, *then* $\hat{\theta}$ *is Liapunov stable but not locally stable, so the system typically neither explodes nor approaches the fixed point.*

Now we give some simple examples to illustrate the various concepts introduced so far. The systems to be discussed in the following examples all have the form

$$\frac{\mathrm{d}}{\mathrm{d}t}\theta(t) = A\theta \tag{7.8}$$

where θ is a two-dimensional state vector and A a 2×2 invertible matrix. There is a unique fixed point, $\hat{\theta} = (0, 0)$, at the origin for any invertible A. In the following examples we shall also plot some 'typical' flows passing through some given states.

Example 7.6 A linear dynamical system with a sink

The linear dynamical system (7.8) with

$$A = \begin{pmatrix} -0.2 & 0 \\ 1 & -0.3 \end{pmatrix}$$

has a stable fixed point at the origin because A has two negative eigenvalues, -0.2 and -0.3. Figure 7.1 shows the corresponding vector field locally at $(0, 0)$, together with two trajectories starting from $(-4, 4)$ and $(4, -4)$, respectively. We see, locally, that the 'particles' of the vector field drift towards the origin, the unique sink of the vector field.

Example 7.7 A linear dynamical system with a saddle point

Consider now the system (7.8) with

$$A = \begin{pmatrix} 4 & 0 \\ 1 & -4 \end{pmatrix}.$$

Since A has a positive and a negative eigenvalues, namely, 4 and -4, the fixed point $(0, 0)$ is unstable. Figure 7.2 shows four trajectories starting from the following points (states)

$$(1, 4),\ (1, -4),\ (-1, 4) \text{ and } (-1, -4).$$

The origin is a saddle point; the 'particles' drift in some directions towards and in other directions away from the origin.

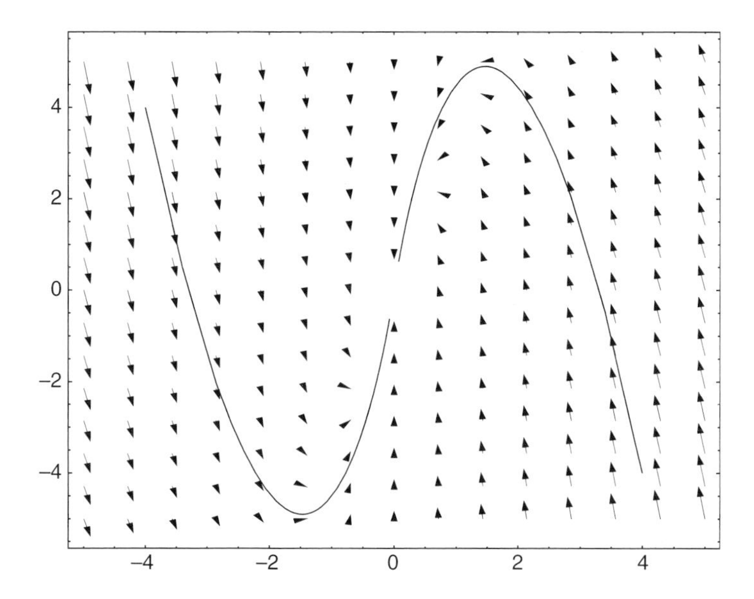

Fig. 7.1 A linear dynamical system with a sink at the origin.

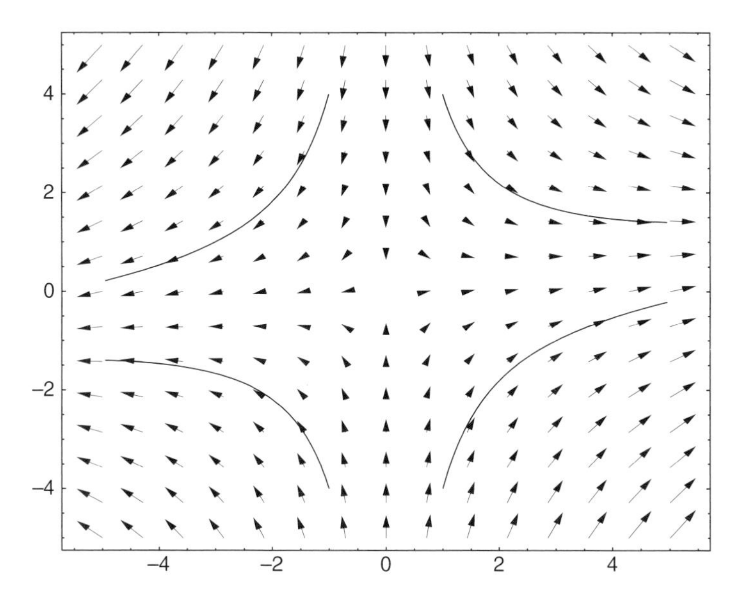

Fig. 7.2 A linear dynamical system with a saddle point at the origin.

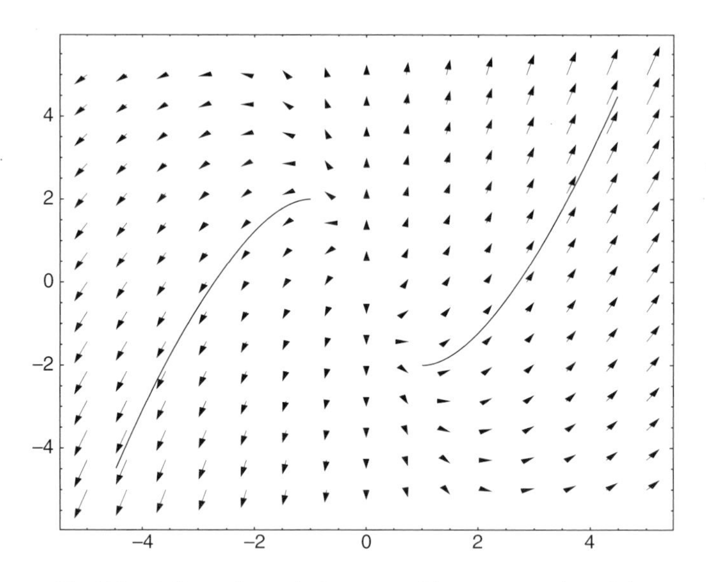

Fig. 7.3 A linear dynamical system with a source at the origin.

Example 7.8 A linear dynamical system with a source

Similar to the Example 7.7, the fixed point (0, 0) of the (7.8) with

$$A = \begin{pmatrix} 0.5 & 0 \\ 1 & 0.5 \end{pmatrix}$$

is also unstable since both of the eigenvalues of A, 0.5, 0.5, are positive. The origin in this case is a source. Figure 7.3 shows two trajectories starting from $(1, -2)$ and $(-1, 2)$, respectively.

Example 7.9 A linear dynamical system with a Liapunov stable fixed point

The linear system (7.8) with

$$A = \begin{pmatrix} 0 & 1 \\ -1 & 0 \end{pmatrix}$$

has a fixed point at the origin, which is Liapunov stable but not locally stable. This is because A has eigenvalues the pure imaginary numbers, $\pm\sqrt{-1}$. The trajectories are concentric circles (see Figure 7.4).

Example 7.10 Logistic regression with measurement errors

In Section 5.10.4 we considered a logistic regression model when measurement errors are present. We found that the estimating function (5.97)

$$g(y, \theta) = \sum_{i=1}^{n} (1,\ d_i)^{\mathrm{t}}(y_i - \mu_i^c) \tag{7.9}$$

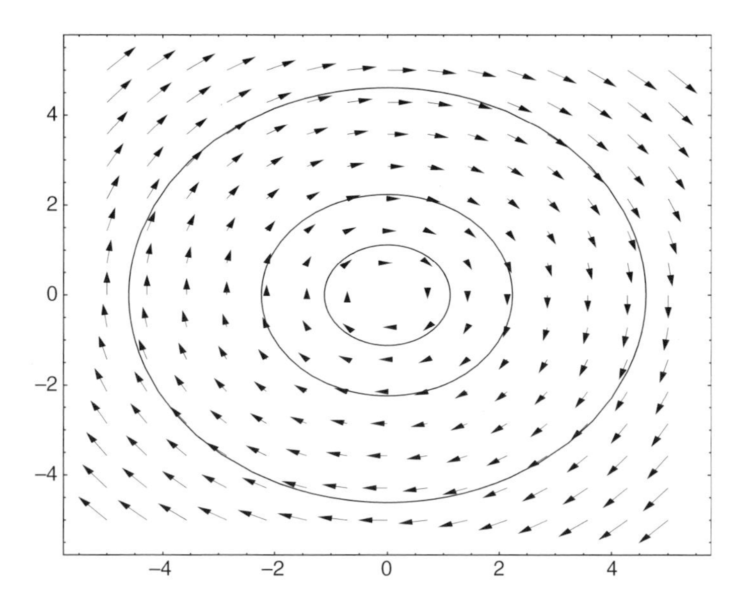

Fig. 7.4 A linear dynamical system with a Liapunov stable fixed point at the origin.

often gives multiple roots. We considered a bootstrap test for helping selecting a root as an estimator for the regression parameter θ. Here again we consider the case for $\theta = (\alpha, \beta)^t$.

Using the same model with $p = \dim \theta = 2$ as in Section 5.10.4, an artificial sample of size 100 was generated from a model with the true parameter value set to

$$(\alpha_0, \beta_0) = (-1.4, 1.4).$$

We found that the estimating function (7.9) had two roots for this particular data set. Figure 7.5 shows the associated dynamical estimating system linearised at the roots. The directions represented by the eigenvectors are also shown in the plots. In these plots the horizontal axis and the vertical axis represent the α and β values, respectively. While the first root (top plot), $\hat{\theta}_1 = (-1.05, 2.33)$, is a sink mimicking a maximum likelihood estimator, the second root (bottom plot), $\hat{\theta}_2 = (-0.96, 8.83)$, is a saddle point of the estimating system. We can show that the distribution of the bootstrap quadratic likelihood ratio at the sink of the vector field approximates the chi-squared distribution with two degrees of freedom very well. Thus the bootstrap test discussed in Section 5.10.4 indicates that this root is preferable to the saddle point shown in the bottom plot.

7.3 Stability of roots to estimating equations

How, then, should we determine the stability of roots to an estimating equation? Since estimating functions are in general non-linear, a systematic study of the

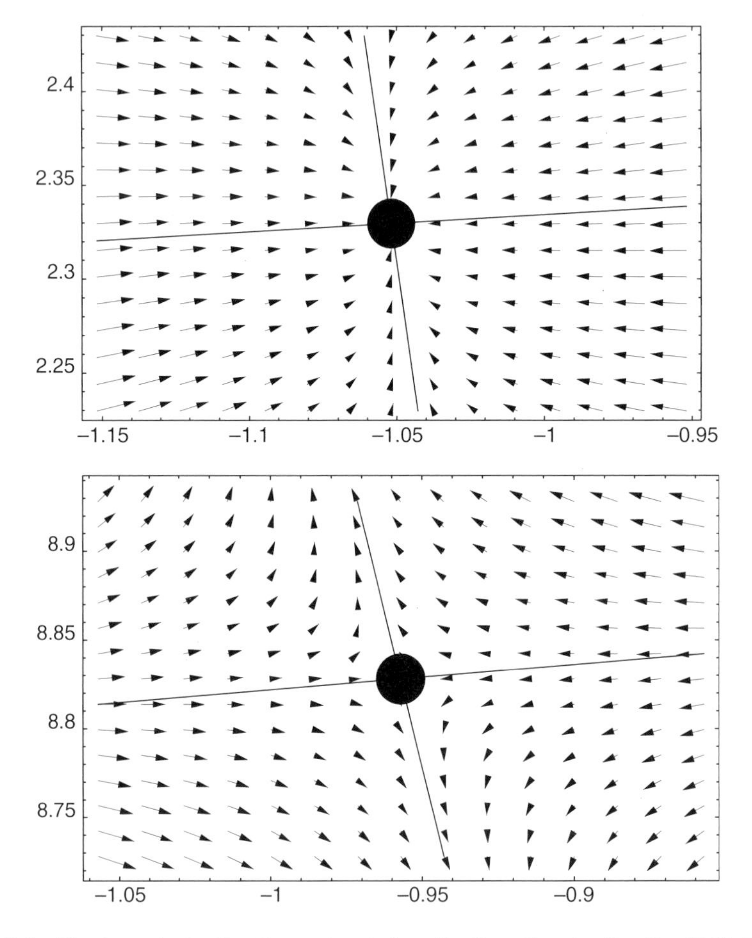

Fig. 7.5 The dynamical estimating system induced by the estimating function (7.9) shown locally at the two fixed points. Shown in these plots are also the eigenvectors at the respective roots.

global stability of such systems would be difficult. For this reason we shall focus on local stability of the roots, often making use of our knowledge on linear dynamical systems.

Let $g(\theta)$ be an estimating function and $\hat{\theta}$ be a root of it. A small perturbation of g at $\hat{\theta}$ usually will not greatly affect the *phase portrait*, the partition of the phase space Θ by the orbits of the vector field. Also, we note that no singular feature of a non-linear system will occur at *regular points*, states other than fixed points of the system. We therefore may study the stability of roots by linearising a dynamical estimating system at the roots of the corresponding estimating equation. The theory of linear dynamical systems introduced in the previous section may then be applied for this purpose.

An alternative, and more elegant, approach is to use the celebrated *Liapunov's method* (Liapunov (1947), a French translation). We shall state the Liapunov's Theorem of stability without proof. First, we introduce some basic concepts. We shall use the same notation $g(\theta)$, with dependency on data y suppressed, to denote the vector field induced by the same estimating function $g(\theta, y)$.

Let

$$h(\theta) : \Theta \to R$$

be a C^r function with $r \geq 1$. The derivative of $h(\theta)$ in the direction of the vector field g, or *directional derivative* of $h(\theta)$ with respect to g, is a C^{r-1} function $D_g h$ defined by

$$D_g h(\theta) = \sum_{j=1}^{p} g_j(\theta) \left\{ \frac{\partial}{\partial \theta_j} h(\theta) \right\} \tag{7.10}$$

where $g_j(\theta)$ is the jth component of $g(\theta)$. Since

$$\frac{\mathrm{d}}{\mathrm{d}t}\theta(t) = g(\theta(t))$$

is an autonomous system, it follows then

$$\begin{aligned} \frac{\mathrm{d}}{\mathrm{d}t} h(\theta(t)) &= \sum_{j=1}^{p} \left\{ \frac{\mathrm{d}}{\mathrm{d}t}\theta_j \right\} \left\{ \frac{\partial}{\partial \theta_j} h(\theta) \right\} \\ &= \sum_{j=1}^{p} g_j(\theta) \left\{ \frac{\partial}{\partial \theta_j} h(\theta) \right\} \\ &= D_g h(\theta). \end{aligned}$$

That is, the directional derivative (7.10) is the derivative of $h(\theta)$ *with respect to time t*.

Definition 7.11 *A scalar function $h(\theta)$ defined on the phase space Θ is called* positive definite *at $\hat{\theta} \in \Theta$, if*

1. $h(\theta) \in C^1$;
2. $h(\hat{\theta}) = 0$; *and*
3. $h(\theta) > 0$ *if* $\theta \neq \hat{\theta}$.

Positive semi-definiteness can be defined by replacing the third condition by $h(\theta) \geq 0$; negative (semi-) definiteness can be defined by changing the directions of the inequalities.

Definition 7.12 *A scalar function $h(\theta)$ is called a* Liapunov function *of a vector field $g(\theta)$ at $\hat{\theta}$, if*

1. *$h(\theta)$ is positive definite at $\hat{\theta}$; and*
2. *$D_g h(\theta)$ is negative semi-definite.*

In addition, we say that $h(\theta)$ is strongly Liapunov *if $D_g h(\theta)$ is negative definite.*

Theorem 7.13 (Liapunov's Theorem) *Let $g(\theta)$ be a vector field (an estimating function) and $\hat{\theta}$ an equilibrium state (a root of $g(\theta) = 0$). If there exists a Liapunov function $h(\theta)$ for $g(\theta)$ at $\hat{\theta}$, then $\hat{\theta}$ is Liapunov stable. The fixed point $\hat{\theta}$ will be stable if $h(\theta)$ is strongly Liapunov.*

The interested reader may consult Irwin (1980, p. 131) for a proof. Without resort to local linearisation, the elegance, as well as the difficulty, of Liapunov's method lies in the use of the Liapunov function. It is often non-trivial to find such a function. Another advantage of Liapunov's method is that it may be used to study the *global stability* of dynamical systems, where the method of local linearisation is irrelevant.

A useful, and often constructive, way of interpreting a Liapunov function $h(\theta)$ is to think of a system developing along the time so that $h(\theta)$ acts as an *energy function* dying asymptotically at the equilibrium state. A quadratic form

$$h(\theta) = (\theta - \hat{\theta})^{\mathrm{t}} \Sigma (\theta - \hat{\theta}) \tag{7.11}$$

thus is often useful in this respect, where Σ is a positive definite matrix. It is obvious that the scalar function $h(\theta)$ of (7.11) is positive definite for any $g(\theta)$ at $\hat{\theta}$. Whether (7.11) is a Liapunov function or not depends then on the choice of Σ.

Another complexity entering into our argument is that the vector field $g(\theta) = g(\theta, y)$ depends on the data y as well. We shall show, however, that with a mild restriction on the second moments of $g(\theta)$, the quadratic form (7.11) is a Liapunov function at a *consistent root* $\hat{\theta}$ of $g(\hat{\theta}) = 0$. That is, a consistent root $\hat{\theta}$ of a regular estimating equation $g(\theta) = 0$ will be stable with high probability when $n \to \infty$.

First, we shall consider the much simpler case, namely, the stability of a maximum likelihood estimator, a root to a regular likelihood system. Here the likelihood plays the role of a potential function. The Liapunov function in such a case is particularly simple to construct.

Definition 7.14 *A fixed point $\hat{\theta}$ is called* hyperbolic *if it is isolated. That is, $\hat{\theta}$ is hyperbolic if there exists a neighbourhood $\mathcal{N}$ of $\hat{\theta}$ so that $\hat{\theta}$ is the unique fixed point in $\mathcal{N}$.*

Proposition 7.15 *Let*

$$\ell_n(\theta) = \sum_{j=1}^{n} \log p(\theta; y_j)$$

be the log likelihood based on n independent observations. Let the maximum likelihood estimator $\hat{\theta}$ be in the interior of the parameter space. Assume that $\ell_n(\theta)$ is differentiable with respect to θ. Let $u_n(\theta) = \dot{\ell}_n(\theta)$ be the score function. In addition we suppose that $\hat{\theta}$ is hyperbolic.

Then $\hat{\theta}$ is a stable fixed point of the likelihood estimating system

$$\frac{\mathrm{d}}{\mathrm{d}t}\theta(t) = u_n(\theta). \tag{7.12}$$

Proof. Since $\hat{\theta}$ is hyperbolic, there exists a neighbourhood $\mathcal{N}_n$ of $\hat{\theta}$ so that $\hat{\theta}$ is the unique root of $u_n(\theta) = 0$. Consider the log likelihood ratio

$$\gamma_n(\theta) = \ell_n(\hat{\theta}) - \ell_n(\theta), \quad \theta \in \mathcal{N}_n \tag{7.13}$$

in this neighbourhood. Note that minus $\gamma_n(\theta)$ is the potential function of (7.12).

The log likelihood ratio $\gamma_n(\theta)$ is a strong Liapunov function of the likelihood estimating system (7.12) at $\hat{\theta}$. This is because $\gamma_n(\theta)$ is positive definite. To check this, we first note that the following properties:

1. $\gamma_n(\hat{\theta}) = 0$
2. $\gamma_n(\theta) > 0, \ \hat{\theta} \neq \theta \in \mathcal{N}_n$

hold for $\gamma_n(\theta)$. That the directional derivative is also negative definite can be seen as follows:

$$\begin{aligned}\frac{\mathrm{d}}{\mathrm{d}t}\gamma_n(\theta) &= -\{u_n(\theta)\}^{\mathrm{t}}\left\{\frac{\mathrm{d}}{\mathrm{d}t}\theta(t)\right\} \\ &= -||u_n(\theta)||^2 \\ &< 0 \quad \text{for } \theta \in \mathcal{N}_n \text{ and } \theta \neq \hat{\theta}.\end{aligned}$$

That $\hat{\theta}$ is stable follows from Theorem 7.13. This completes the proof.

The careful reader may have noticed that consistency of $\hat{\theta}$ is not required for its stability. In fact, the proof can be slightly modified to show that *each* local maximum of the likelihood function is stable. The assumption that the maximum likelihood estimator is a hyperbolic fixed point can be derived from the condition that the maximum likelihood estimator is a consistent estimator. See the proof of Proposition 7.16.

For a general non-conservative estimating function, since the roots do not correspond to local extrema of a potential function, the stability of a root must be proved using a Liapunov function of a different sort.

Proposition 7.16 *Let θ_0 be a true value of the parameter θ. Let $\hat{\theta}$ be a root to $g(\hat{\theta}) = 0$. Suppose that the following conditions hold:*

1. *The estimating function $g(\theta)$ is twice continuously differentiable in a neighbourhood $\mathcal{N}$ of θ_0*
2. *The second derivatives of $g(\theta)/n$ are uniformly bounded in probability in $\mathcal{N}$.*
3. *The root $\hat{\theta}$ is $\sqrt{n}$-consistent.*
4. *The covariance matrix Σ of $g(\theta_0)/\sqrt{n}$ is positive definite and is of constant nonsingular order.*
5. *The estimating function $g(\theta)$ is first-order information unbiased.*

Then the root $\hat{\theta}$ is a stable fixed point as n goes to infinity.

Proof. Again we shall use Liapunov's method, this time in an asymptotic fashion.

By Assumption 3, namely $\hat{\theta}$ is $\sqrt{n}$-consistent for θ_0, there exists a neighbourhood $\mathcal{N}_n \subset \mathcal{N}$ of θ_0 so that $\hat{\theta} \in \mathcal{N}_n$ and

$$\theta = \hat{\theta} + O_p(n^{-1/2}) \quad \text{for any } \theta \in \mathcal{N}_n. \tag{7.14}$$

We show that $\hat{\theta}$ is the only fixed point in $\mathcal{N}_n$. If there were two roots in $\mathcal{N}_n$, then $\dot{g}(\theta)$ must be singular at some point in $\mathcal{N}_n$ since $g(\theta)$ is a smooth function by Assumption 1. We show that this is impossible. To see this, we first note that by Assumption 5 we have

$$E_\theta\left[\dot{g}(\theta) + g(\theta)\{g(\theta)\}^{\mathrm{t}}\right] = O(n^{1/2}). \tag{7.15}$$

Using (7.15) and Assumption 4, we have that $\dot{g}(\theta_0)$ is of nonsingular order $O_p(n)$. By Assumption 2, the second-order derivatives of $g(\theta)$ in $\mathcal{N}_n$ are of order $O_p(n)$. A first-order Taylor expansion of $\dot{g}(\theta)$ at θ_0 in $\mathcal{N}_n$ then shows that the leading term is $\dot{g}(\theta_0)$ and is of nonsingular order $O_p(n)$. It follows therefore that $\dot{g}(\theta)$ is of nonsingular order $O(n)$.

Now consider the following quadratic form:

$$h(\theta) = \tfrac{1}{2}(\theta - \hat{\theta})^{\mathrm{t}}\Sigma^{-1}(\theta - \hat{\theta}) \tag{7.16}$$

in the neighbourhood $\mathcal{N}_n$. Obviously, $h(\theta)$ is differentiably in θ and satisfies the property that $h(\hat{\theta}) = 0$. Also, by Assumption 4, we have

$$h(\theta) > 0 \quad \text{for } \theta \neq \hat{\theta}.$$

By definition, therefore, the function $h(\theta)$ is positive definite at $\hat{\theta}$.

To show that $h(\theta)$ is a strong Liapunov function, we now show that the directional derivative $D_g h$ is negative definite at $\hat{\theta}$ when n goes to infinity. First, in the neighbourhood $\mathcal{N}_n$, by (7.14), Assumptions 1 and 2, we have

$$\begin{aligned} g(\theta) &= g(\hat{\theta}) + \dot{g}(\hat{\theta})(\theta - \hat{\theta}) + O_p(1) \\ &= \dot{g}(\hat{\theta})(\theta - \hat{\theta}) + O_p(1). \end{aligned}$$

The quantity $\dot{g}(\hat{\theta})$, in $\mathcal{N}_n$, can be expressed, using Assumptions 1 and 2, as follows:

$$\dot{g}(\hat{\theta}) = \dot{g}(\theta_0) + O_p(n^{1/2}).$$

Combing these expressions, we have, in $\mathcal{N}_n$, that

$$g(\theta) = \dot{g}(\theta_0)(\theta - \hat{\theta}) + O_p(1).$$

To show the negative definiteness of $D_g h$, as n goes to infinity, we compute, again in $\mathcal{N}_n$, the directional derivative as follows:

$$\begin{aligned} D_g h &= \{g(\theta)\}^{\mathrm{t}}\, \Sigma^{-1}(\theta - \hat{\theta}) \\ &= (\theta - \hat{\theta})^{\mathrm{t}} \left[\{\dot{g}(\theta_0)\}^{\mathrm{t}}\, \Sigma^{-1}\right](\theta - \hat{\theta}) + O_p(n^{-1/2}) \\ &= -n\|\theta - \hat{\theta}\|^2 + O_p(n^{-1/2}) \end{aligned} \tag{7.17}$$

where in passing from the second equality to (7.17) Assumptions 4 and 5 are used. It follows from (7.17), when n is large, that $h(\theta)$ is a strong Liapunov function for the vector field $g(\theta)$ at $\hat{\theta}$. By Theorem 7.13, $\hat{\theta}$ is a stable fixed point when n is large. This completes the proof. □

Proposition 7.16 essentially says that consistency of a root to an information unbiased estimating equation implies its local stability. The reverse of this is *not* true. For instance, every local maximum of the likelihood function is a locally stable fixed point of the likelihood estimating system.

7.4 A modified Newton's method

In Section 3.6 we introduced Newton's method. Using the notation of this chapter, Newton's method may be rewritten as

$$\theta(t+1) = \theta(t) + c(\theta(t))g(\theta(t)) \tag{7.18}$$

with the choice of c given by

$$c(\theta) = -\{\dot{g}(\theta)\}^{-1}. \tag{7.19}$$

The rate of convergence of the algorithm (7.18) will be maximised using c given by (7.19). From computational perspectives, in Section 3.6 we also discussed several modifications to Newton's method. Whittaker's method and Fisher's method of scoring, for instance, both concern the computation of (7.19). In Whittaker's method, it is suggested that one use a fixed value of (7.19) in later iterations. On the other hand, in the scoring method one replaces (7.19) by its expected value.

Newton's method was discussed further in Section 5.4, where we emphasised the advantages as well as the shortcomings of the method when an estimating equation has possibly more than one root. In Sections 5.5 and 5.6 we considered two alternative methods, each may be viewed as a modification of Newton's method, designed especially for solving an estimating equation when multiple roots are present. In this section we study the theoretical aspects of these algorithms from the viewpoint of dynamical systems. We begin with a brief review of Newton's method by regarding this method as a discrete time dynamical system. The modified iterative algorithms will be generalised along this line. The notations used in this section here differ slightly from those used in Chapter 3.

7.4.1 NEWTON'S METHOD REVISITED

Let $g(\theta) = g(\theta, y)$ be a scalar estimating function for θ. Suppose that $g(\theta)$ is differentiable with respect to θ. To find the roots of $g(\theta) = 0$ starting from an initial point $\theta(0)$, the celebrated *Newton–Raphson method* updates the estimate by the rule

$$\theta(t+1) = \theta(t) - \frac{g(\theta(t))}{\dot{g}(\theta(t))} \tag{7.20}$$

where $t = 0, 1, \ldots,$ and $\dot{g}(\theta(t)) \neq 0$. We use $\theta(T)$ as our final estimate of θ for a large value of T with regard to some stopping rule. In Section 5.4 we saw that in many cases a one-step iteration $\hat{\theta}(1)$ is a reasonable estimate if $\hat{\theta}(0)$ is suitably chosen.

To gain a different perspective of the algorithm (7.20), let us consider a function

$$w(\theta) = \theta - \frac{g(\theta)}{\dot{g}(\theta)} \tag{7.21}$$

where $\dot{g}(\theta) \neq 0$. The function (7.21) induces a *discrete time dynamical system*

$$\theta(t+1) = w(\theta(t)) \tag{7.22}$$

where θ is considered to be a function of t. The *fixed points* of the system (7.22) are defined by solutions to the following equation:

$$\theta(t+1) = \theta(t). \tag{7.23}$$

The set of points satisfying (7.23) is the same as the set of roots to $g(\theta) = 0$. In other words, the problem of solving the estimating equation $g(\theta) = 0$ can be transformed to the problem of finding fixed points of the dynamical system defined by (7.21) and (7.22). Since the iteration (7.22) is identical to the algorithm (7.20), we can therefore study properties of Newton's method by studying properties of the dynamical system induced by function (7.21).

Let $w^{(k)}$ be the kth iteration of w, i.e.,

$$w^{(k)}(\theta) = \underbrace{w \circ w \circ \cdots \circ w}_{k \text{ times}}(\theta).$$

Then the Newton–Raphson iteration at the T-th step can be written as

$$\theta(T) = w^{(T)}(\theta(0)).$$

We have seen that when the sequence

$$w^{(t)}(\theta(0)), \quad t = 1, 2, \ldots \tag{7.24}$$

converges, it converges to one of the fixed points of (7.22). That is, it must converge to a root of $g(\theta) = 0$. When there are multiple roots, does the sequence (7.24) converge to the *desired* root, or indeed to *any* root $\hat{\theta}$ of $g(\hat{\theta}) = 0$ as $t \to \infty$? The answer obviously depends on the choice of the initial starting point $\theta(0)$.

The sequence (7.24) will tend to $\hat{\theta}$ if $\theta(0)$ lies in the *basin of attraction* of $\hat{\theta}$, which is defined as follows:

$$A(\hat{\theta}) = \{\theta \in \Theta;\ w^{(t)}(\theta) \to \hat{\theta}, t \to \infty\}.$$

Clearly, $\hat{\theta} \in A(\hat{\theta})$. So the basin of attraction $A(\hat{\theta})$ is not empty. What else can be said about $A(\hat{\theta})$? The structure of $A(\hat{\theta})$ depends on the function $g(\theta)$ and the particular algorithm (7.21). If $A(\hat{\theta})$ is essentially empty, or no neighbourhood of $\hat{\theta}$ is contained in $A(\hat{\theta})$, then a point, however close to $\hat{\theta}$, when iterated under (7.20), may be sent away from the target value $\hat{\theta}$. Such a point is said to be a *repelling fixed point*. This will not happen if $\hat{\theta}$ is an *attracting fixed point*, a point satisfying the following condition:

$$w^{(t)}(\theta) \to \hat{\theta} \quad \text{as } t \to \infty$$

for θ in a neighbourhood of $\hat{\theta}$.

Newton's method has the property that *any* root $\hat{\theta}$ of *any* smooth function $g(\theta)$ is an attracting fixed point of the iterative algorithm (7.20). In other words, any root $\hat{\theta}$ of $g(\theta) = 0$ is guaranteed to be found by algorithm (7.20) provided that $\theta(0)$ is chosen to be close enough to $\hat{\theta}$.

To see why this is so, we study a general class of discrete time dynamical systems induced by the function

$$h(\theta) = \theta + \alpha(\theta)g(\theta) \tag{7.25}$$

where $\alpha(\theta)$ is a scalar function of θ and $\alpha(\theta) \neq 0$. The funcion $h(\theta)$ of (7.25) was used in Chapter 3 to form the iterative algorithm (3.15). The function $h(\theta)$ reduces to $w(\theta)$ of (7.21) if

$$\alpha(\theta) = -\frac{1}{\dot{g}(\theta)}.$$

A discrete time dynamical system

$$\theta(t+1) = h(\theta(t)) \tag{7.26}$$

associated with (7.25) for any $\alpha(\theta)$ enjoys the property that fixed points of (7.26) are roots to the equation $g(\theta) = 0$. Thus we may investigate the properties of Newton's method through properties of (7.26) by varying the choice of $\alpha(\theta)$.

Recall that a fixed point b of a discrete dynamical system

$$\theta(t+1) = r(\theta(t))$$

induced by a function $r(\theta)$ is attracting if the derivative $\dot{r}(\theta)$ at b has absolute value less than 1. That is,

$$|\dot{r}(b)| < 1.$$

Let $g(\hat{\theta}) = 0$. Then we have

$$\dot{h}(\hat{\theta}) = 1 + \alpha(\hat{\theta})\dot{g}(\hat{\theta}).$$

Thus, for $h(\theta)$ of (7.25) to satisfy the condition $|\dot{h}(\hat{\theta})| < 1$, we require $\alpha(\theta)$ to satisfy the following condition:

$$-2 < \alpha(\hat{\theta})\dot{g}(\hat{\theta}) < 0. \tag{7.27}$$

Any root $\hat{\theta}$ of $g(\hat{\theta}) = 0$ will be an attracting fixed point of the associated dynamical system using $h(\theta)$ of (7.25) if $\alpha(\theta)$ satisfies (7.27).

If $\alpha(\theta) = -1/\dot{g}(\theta)$, then we see that every fixed point of (7.22) is attracting because

$$\alpha(\hat{\theta})\dot{g}(\hat{\theta}) = -1 \in (-2, 0).$$

It is a curious fact that the value of $\alpha(\hat{\theta})\dot{g}(\hat{\theta})$ for Newton's method is the middle point of the permissible range $(-2, 0)$.

Properties of Newton's method in higher dimensions can be studied along the same line. Let $g(\theta)$ be a p-dimensional estimating function for a p-dimensional parameter θ for $p \geq 1$. In this case we define a family of dynamical systems as follows:

$$\begin{aligned}\theta(t+1) &= h(\theta(t)) \\ &= \theta(t) + \alpha(\theta)g(\theta(t))\end{aligned} \tag{7.28}$$

where $\alpha(\theta)$ is a $p \times p$ matrix with $\det\{\alpha(\theta)\} \neq 0$. The fixed points of (7.28) are the roots to $g(\theta) = 0$, and vice versa. The theory of dynamical systems shows that a fixed point $\hat{\theta}$ of (7.28) is attracting if any (complex) eigenvalue λ of $h(\hat{\theta})$ has absolute value less than unity, i.e., $|\lambda| < 1$. Note that

$$\dot{h}(\hat{\theta}) = \mathrm{I} + \alpha(\hat{\theta})\dot{g}(\hat{\theta})$$

where I is the $p \times p$ identity matrix. So if we let

$$\alpha(\theta) = -\left\{\dot{g}(\theta)\right\}^{-1}$$

to be the value for Newton's method, then $\dot{h}(\hat{\theta})$ becomes a zero matrix having only zero eigenvalues for *any* root $\hat{\theta}$. We conclude therefore that Newton's method in the multi-dimensional case

$$\theta(t+1) = \theta(t) - \{\dot{g}(\theta(t))\}^{-1} g(\theta(t))$$

again has the property that every fixed point is attracting.

The above view of Newton's method helps to motivate a number of iterative algorithms useful for the purpose of choosing an appropriate root of an estimating equation. In Sections 5.5 and 5.6 we introduced two such algorithms. In the rest of this section we shall see how these algorithms may be derived by considering appropriate dynamical systems. Convergence properties of these algorithms will also be proved here. See Sections 5.5 and 5.6 for further exploration of these methods and numerical examples.

7.4.2 A METHOD WITH CONTINUITY CORRECTION

In Section 5.5 we considered a modification of Newton's method so that convergence to roots of an estimating equation mimicking local minima or saddle points of a log likelihood function will be avoided by the modified algorithm. We now give a derivation of the algorithm and give a proof of the convergence properties of the algorithm. We shall discuss the scalar case first and then generalise the results to the multivariate case.

Consider a dynamical estimating system associated with an estimating function $g(\theta, y)$. Suppose that we wish to compute the trajectories of this system. The orbits are usually not available in closed forms due to non-linearity of g in θ. An approximate form may however be obtained by considering a first-order approximation of the system at a point θ_0

$$\begin{aligned}\frac{\mathrm{d}}{\mathrm{d}t}\theta(t) &= g(\theta_0) + \dot{g}(\theta_0)\{\theta(t) - \theta_0\} \\ &= \dot{g}(\theta_0)\theta(t) + \{g(\theta_0) - \dot{g}(\theta_0)\theta_0\}. \qquad (7.29)\end{aligned}$$

For (7.29) to be of use, we assume that θ_0 is a non-degenerate point, i.e., $\dot{g}(\theta_0) \neq 0$. By solving (7.29) with the initial condition that

$$\theta(0) = \theta_0$$

we find the unique trajectory of (7.29) passing through θ_0, which is explicitly given by

$$\theta(t) = \left\{\theta_0 - \frac{g(\theta_0)}{\dot{g}(\theta_0)}\right\} + \frac{g(\theta_0)}{\dot{g}(\theta_0)} \exp\{\dot{g}(\theta_0)t\}. \qquad (7.30)$$

Suppose that $\dot{g}(\theta_0)$ is a negative number. So as $t \to \infty$ we have that

$$\dot{g}(\theta_0)t \quad \to \quad -\infty$$

which implies that the second term on the right-hand side of (7.30) vanishes. So we arrive at the following formula when $t \to \infty$:

$$\theta(\infty) = \theta_0 - \frac{g(\theta_0)}{\dot{g}(\theta_0)} \tag{7.31}$$

the right-hand side being identical with the one-step iteration using Newton's method. The formula (7.30) may be regarded as a continuous version of Newton's method.

More formally, we may define the continuous version of Newton's method as follows

$$\theta(t+1) = \theta(t) - \Big[1 - \exp\{\gamma \dot{g}(\theta(t))\}\Big] \frac{g(\theta(t))}{\dot{g}(\theta(t))} \tag{7.32}$$

where and γ is a positive constant functionally independent of the data. The algorithm (7.32) was initially considered in Section 5.5.2; see equation (5.34). Since $\dot{g}\{\theta(t)\}$ is usually $O_p(n)$, n being the sample size, iterative algorithm (7.32) differs from the discrete Newton's method, when it converges, essentially for the first few steps.

Now we restate Proposition 5.9 using the notation of this chapter and give a proof of it.

Proposition 7.17 *Let $g(\theta)$ be a smooth one-dimensional estimating function for θ and $\hat{\theta}$ be a root of $g(\hat{\theta}) = 0$. Let γ be a positive constant. Define the following function:*

$$w(\theta) = \theta - \Big[1 - \exp\{\gamma \dot{g}(\theta)\}\Big] \frac{g(\theta)}{\dot{g}(\theta)} \tag{7.33}$$

for which $\dot{g}(\theta) \neq 0$.

Then the iterative algorithm $\theta(t+1) = w(\theta(t))$ has the following properties for any estimating function $g(\theta)$:

1. *If $\dot{g}(\hat{\theta}) < 0$, then $\hat{\theta}$ is an attracting fixed point.*
2. *If $\dot{g}(\hat{\theta}) > 0$, then $\hat{\theta}$ is a repelling fixed point.*

Proof. We first note that the dynamical system

$$\theta(t+1) = w(\theta(t))$$

with $w(\theta)$ given by (7.33) has the property that each root $\hat{\theta}$ of $g(\hat{\theta}) = 0$ is a fixed point of the dynamical system, and *vice versa*.

To investigate the stability of each fixed point $\hat{\theta}$ of the system, we differentiate $w(\theta)$ with respect to θ and evaluate at $\hat{\theta}$ to give

$$\dot{w}(\hat{\theta}) = \exp\{\gamma \dot{g}(\hat{\theta})\}.$$

By the *local linearisation method*, a fixed point $\hat{\theta}$ is an attracting fixed point if and only if

$$|\dot{w}(\hat{\theta})| = \exp\{\gamma \dot{g}(\hat{\theta})\} < 1$$

which is equivalent to

$$\dot{g}(\hat{\theta}) < 0$$

since γ is positive. This proves part 1 of the proposition.

On the other hand, a fixed point $\hat{\theta}$ is a repelling fixed point if and only if

$$|\dot{w}(\hat{\theta})| = \exp\{\gamma \dot{g}(\hat{\theta})\} > 1$$

which is equivalent to

$$\dot{g}(\hat{\theta}) > 0$$

which proves part 2 of the proposition. □

We have seen that the condition $\dot{g}(\hat{\theta}) < 0$ is satisfied by a consistent root under mild moment conditions. So a desirable root is guaranteed to be found by the algorithm (7.32). At the same time, the algorithm has the merit over Newton's method in the presence of multiple roots in that convergence to undesirable roots with $\dot{g}(\hat{\theta}) > 0$ will be avoided. The algorithm, for instance, will never converge to local minima of a log likelihood function. As discussed in Section 5.5.2, practically when the sample size is large, one may simply choose $\gamma = 1$ so to use the following version of the algorithm:

$$\theta(t+1) = \theta(t) - \Big[1 - \exp\{\dot{g}(\theta(t))\}\Big] \frac{g(\theta(t))}{\dot{g}(\theta(t))}$$

which is the algorithm proposed in (5.39).

Now we show that the iterative algorithm (5.48) in multivariate cases discussed in Chapter 5 has similar convergence properties.

Proposition 7.18 *Let $g(\theta)$ be a p-dimensional estimating function. Define a discrete time dynamical system as follows:*

$$\begin{aligned} \theta(t+1) &= w\{\theta(t)\} \\ w(\theta) &= \theta - \Big[\mathrm{I} - \exp\{\dot{g}(\theta)\}\Big]\{\dot{g}(\theta)\}^{-1} g(\theta) \end{aligned} \tag{7.34}$$

on the set $\{\theta : \det|\dot{g}(\theta)| \neq 0, \theta \in \Theta\}$. Let $\hat{\theta}$ be a root of g. Let λ_j, $j = 1, \ldots, p$, be the complex eigenvalues of $\dot{g}(\hat{\theta})$.

Then the iterative algorithm defined by (7.34) has the following properties.

1. *If the real parts $\Re(\lambda_j)$ for any j are negative, then there is an ϵ-neighborhood $\mathcal{N}_\epsilon(\hat{\theta})$ of $\hat{\theta}$ such that*

$$\lim_{t\to\infty} \theta(t) = \hat{\theta}, \quad \textit{for any } \theta(0) \in \mathcal{N}_\epsilon(\hat{\theta}).$$

2. *If there exists at least one* λ_j *such that* $\Re(\lambda_j) > 0$, *then* $\theta(t)$ *typically diverges as* t *tends to infinity.*

Proof. The proof is similar to that in the scalar case. The algorithm (7.34) defines a p-dimensional discrete time dynamical system, each fixed point $\hat{\theta}$ being the root of $g(\hat{\theta}) = 0$ and *vice versa.*

A fixed point $\hat{\theta}$ of (7.34) will be locally stable if and only if the absolute value of each eigenvalue η_j of $\dot{w}(\hat{\theta})$ is less than one, and $\hat{\theta}$ is repelling if and only if $|\eta_j| > 1$ for at least one η_j. Using the fact that each fixed point $\hat{\theta}$ of (7.34) is a root of $g(\hat{\theta}) = 0$, we have

$$\dot{w}(\hat{\theta}) = \exp\left\{\dot{g}(\hat{\theta})\right\}. \tag{7.35}$$

Let ξ_j be the eigenvector corresponding to each eigenvalue λ_j. Suppose that the matrix

$$\Gamma = (\xi_1, \ldots, \xi_p)$$

with ξ_j as its j-th column vector is invertible. Then we have

$$\begin{aligned}\dot{w}(\hat{\theta}) &= \exp\left\{\Gamma \mathrm{diag}\{\lambda_1, \ldots, \lambda_p\}\Gamma^{-1}\right\}\\ &= \Gamma \mathrm{diag}\{e^{\lambda_1}, \ldots, e^{\lambda_p}\}\Gamma^{-1}.\end{aligned}$$

Since each eigenvalue of the matrix $B^{-1}AB$ is also an eigenvalue of matrix A, we then conclude that

$$\eta_j = \exp \lambda_j, \quad j = 1, \ldots, p$$

by appropriately reordering these values.

Finally, we have that $\hat{\theta}$ is stable if and only if

$$|\eta_j| = \left|\exp\left\{\Re(\lambda_j) + \Im(\lambda_j)i\right\}\right| = \exp \Re(\lambda_j) < 1$$

or, equivalently,

$$\Re(\lambda_j) < 0 \quad \text{for } j = 1, \ldots, p.$$

Similarly, $\hat{\theta}$ is repelling if there exists η_j such that $|\eta_j| > 1$ or $\Re(\lambda_j) > 0$. This completes the proof. □

7.4.3 A MODIFICATION BASED ON THE INFORMATION IDENTITY

The continuous version of Newton's method discussed in the previous section is based on the fact that a consistent root $\hat{\theta}$ of $g(\hat{\theta}) = 0$ is an asymptotically stable fixed point of the system $(\mathrm{d}/\mathrm{d}t)\theta(t) = g(\theta)$. Undesirable convergence to sources and saddle points are therefore avoided by this method. However, convergence to sinks other than the consistent root remains a possibility. To remedy this problem, we now discuss another method based on the generalised discrete-type Newton's method. This algorithm was first introduced in Section 5.6.

We shall focus on the scalar case. A brief discussion on the multivariate case will be given later. Consider a class of discrete time dynamical systems induced by the functions

$$w(\theta) = \theta + \alpha(\theta)g(\theta) \tag{7.36}$$

indexed by a unction $\alpha(\theta) \neq 0$. Each system in the class (7.36) has the property that the set of the fixed points of

$$\theta(t+1) = w(\theta(t))$$

using $w(\theta)$ of (7.36) equals the set of zeros of $g(\theta)$. We have seen previously that such a root $\hat{\theta}$ is stable if and only if $\alpha(\theta)$ satisfies the condition

$$-2 < \alpha(\hat{\theta})\dot{g}(\hat{\theta}) < 0 \tag{7.37}$$

where $\hat{\theta}$ is a zero of g. The condition (7.37) is key to designing a new algorithm. An ideal function $\alpha(\theta)$ is one so that (7.37) will be satisfied only by a consistent root $\hat{\theta}$. To explore the condition (7.37), we make the assumption that $g(\theta)$ is information unbiased to the first order. That is,

$$E_\theta \{g(\theta)\}^2 = -E_\theta \{\dot{g}(\theta)\} \left\{1 + O(n^{-1/2})\right\}.$$

This condition is automatically satisfied by all information unbiased estimating functions. As a direct consequence of this condition, if $\hat{\theta}$ is consistent and we may apply the law of large numbers to $\dot{g}(\hat{\theta})$, then we have

$$\begin{aligned}\lim_{n\to\infty} \frac{n^{-1}\dot{g}(\hat{\theta})}{n^{-1}E_{\theta_0}\{g(\theta)\}^2} &= \frac{E_{\theta_0}\{\dot{g}(\theta_0)\}}{E_{\theta_0}\{g(\theta_0)\}^2} \\ &= -1.\end{aligned}$$

Therefore, if we let

$$\alpha(\theta) = \frac{1}{v(\theta)}$$

where $v(\theta)$ is a consistent estimate of the variance of $g(\theta)$, then the modified Newton's method

$$\theta(t+1) = \theta(t) + \frac{g(\theta(t))}{v(\theta(t))} \tag{7.38}$$

will have the properties stated in Proposition 5.12.

Now suppose that $g(\theta)$ is p-dimensional. Let $\alpha(\theta)$ be a $p \times p$ non-singular matrix and define as in the scalar case $w(\theta) = \theta + \alpha(\theta)g(\theta)$. Since $\det\{\alpha(\theta)\} \neq 0$, the set of fixed points of the system $\theta(t+1) = w\{\theta(t)\}$ and the set of roots of $g(\theta) = 0$ will coincide. At each root $\hat{\theta}$, we then have

$$\dot{w}(\theta) = \mathrm{I} + \alpha(\hat{\theta})\dot{g}(\hat{\theta}), \tag{7.39}$$

where I is the $p \times p$ identity matrix. Let λ_j be the (complex) eigenvalues of $\alpha(\hat{\theta})\dot{g}(\hat{\theta})$. The eigenvalues of $\dot{w}(\hat{\theta})$ are then given by $1 + \lambda_j$. A root $\hat{\theta}$ is stable if and only if the absolute value of $1 + \lambda_j$ is less than 1, i.e.

$$|1 + \lambda_j| < 1, \quad j = 1, \ldots, p.$$

The above argument then suffices to establish the results stated in Proposition 5.14.

7.5 Complex estimating functions and Julia sets

In Sections 4.3 and 5.8 we considered estimation of the correlation coefficient for a bivariate normal distribution. The score function, ignoring a multiplier factor, can be written

$$P(\rho) = \rho(1 - \rho^2) + (1 + \rho^2)\frac{\sum xy}{n} - \rho\left[\frac{\sum(x^2 + y^2)}{n}\right] = 0.$$

This estimating equation can be solved explicitly and we have found that the formula, which asymptotically gives the correct answer, can have a complex-valued solution. More generally, if the estimating function is a polynomial of degree $d = d(n)$ in the parameter, then there exist exactly d complex roots (counting multiplicities). By considering the complex roots of such an equation, we can learn about the number and distribution of the real roots. Moreover, if the estimating function yields a minimal sufficient partitioning, as is the case for the score function, then the complex roots may themselves be regarded as a minimal sufficient statistic. There are also cases where a real starting point for an iterative algorithm will lead to an inconsistent root or not converge at all, but iterating from a slightly perturbed complex number may lead to the consistent root.

In this section, we shall consider estimating functions defined for complex-valued parameters by analytic continuation into the complex plane.

7.5.1 JULIA SETS

For iterative algorithms defined for complex numbers *Julia sets* are of basic interest. We now describe the concept of Julia sets through a classical example. Consider a family of functions defined on the complex plane

$$w_c(z) = z^2 + c$$

where c is a complex number. Functions f_c define a family of discrete time dynamical systems indexed by c. For each c, $w_c(z)$ partitions the complex plane $\mathbf{C}$ into two disjoint regions

$$\mathbf{C} = B_c \cup U_c$$

where

$$B_c = \left\{z : |w_c^k(z)| \not\to \infty \text{ as } k \to \infty\right\}$$

$$U_c = \left\{z : |w_c^k(z)| \to \infty \text{ as } k \to \infty\right\}.$$

That is, B_c consists of all values z for which the iterates $w_c^k(z)$ stay bounded, and U_c the complementary set of B_c, i.e., the set of z for which $w_c^k(z)$ explodes.

The boundary, J_c, between B_c and U_c is called the *Julia set* of the function w_c, and B_c the *filled-in Julia set* of w_c. Now we have defined the (filled-in) Julia set for the specific family $w_c(z)$. Obviously the definition applies to any discrete time dynamical system. For example, we may study Julia sets for the dynamical systems corresponding to the Newton–Raphson method or the modified algorithms discussed in the previous section.

Of interest to us, in our present context, is the ability of an algorithm to converge to the consistent root of an estimating equation in question. The starting points of the algorithm may lie in the real or the extended complex parameter space. This property is summarised by the concept of *basins of attraction* discussed earlier. It turns out that the concepts of Julia set and basin of attraction are closely related to each other. Therefore, visualisation of the filled-in Julia set can provide considerable information about the convergence property of a concerned algorithm.

Now we introduce a popular algorithm called the *escape-time algorithm*, again through the example of $w_c(z) = z^2 + c$, for drawing the filled-in Julia sets.

Let $w_c(z) = z^2 + c$. For each $z \in \mathbf{C}$, z will belong to U_c if and only if $|w_c^k(z)| \to \infty$ as $k \to \infty$. So to judge whether $z \in U_c$, an approximate algorithm is first to choose a sufficiently large integer k, then to see if

$$|w_c^k(z)| > M$$

for a large $M > 0$. For this particular function, we let

$$M = \max\{2, |c|\}.$$

This is because, if

$$|z| > \max\{2, |c|\}$$

then we will have

$$|z| \geq 2 + \epsilon$$

for some $\epsilon > 0$. It is also easy to see that

$$\begin{aligned}|w_c(z)| &\geq |z^2| - |c| \\ &\geq |z^2| - |z| \\ &\geq |z|(1 + \epsilon).\end{aligned}$$

It now follows that

$$|w_c^k(z)| \geq |z|(1 + \epsilon)^k \to \infty \quad \text{as } k \to \infty.$$

For each $z \in U_c$, there exists a smallest integer T_z such that

$$|w_c^k(z)| > M = \max\{2, |c|\}.$$

This integer T_z is called the *escape-time* of the point z. The escape-time of any point in B_c is infinite.

Finally, we have to determine the iteration number k. Obviously, the larger k is, the more accurate the result will be. For the specific function $w_c(z)$, $k = 20$ is enough to reveal the main features of the Julia sets. Summing up, we may draw the filled-in Julia set, for a fixed k and M, by plotting the points z if

$$|w_c^k(z)| \leq M.$$

The complementary region, U_c, may be coloured using the information of the escape-time. Figure 7.6 shows the filled-in Julia set using the function $w_c(z)$ with $c = -0.85 + 0.18\sqrt{-1}$. See Figure 7.7 for a close-up of Figure 7.6.

7.5.2 THE CORRELATION COEFFICIENT

Now we consider the example of estimating the correlation coefficient ρ using a set of independent observations

$$(x_i, y_i), \quad i = 1, \ldots, n$$

from a bivariate normal distribution having means

$$\mu_x = \mu_y = 0$$

and variances

$$\sigma_x^2 = \sigma_y^2 = 1$$

and an unknown correlation coefficient ρ. The score function for ρ takes the form

$$u(\rho) = \frac{n}{(1-\rho^2)^2} P(\rho) \tag{7.40}$$

Fig. 7.6 The filled-in Julia set B_c, where $c = -0.85 + 0.18\sqrt{-1}$.

Fig. 7.7 A close-up of Figure7.6.

where $P(\rho)$ is a polynomial given by

$$P(\rho) = \rho(1 - \rho^2) + (1 + \rho^2)S_1 - \rho S_2. \tag{7.41}$$

In (7.41), S_1 and S_2 are defined by

$$S_1 = \frac{1}{n}\sum_{i=1}^{n} x_i y_i \quad \text{and} \quad S_2 = \frac{1}{n}\sum_{i=1}^{n}(x_i^2 + y_i^2)$$

respectively. Note that both $u(\rho)$ and $P(\rho)$ depend on the data through (S_1, S_2), a complete sufficient statistic for ρ.

It is customary to solve for ρ the cubic equation $P(\rho) = 0$, which can have as many as three real roots in the interval $(-1, 1)$. If three roots are present, then these will correspond to two relative maxima and one relative minimum of the likelihood. Although estimating functions (7.40) and (7.41) have the same roots, they display different convergence properties for some iterative methods such as Newton's method. This is because an iterative algorithm associated with the discrete dynamical system based on the following function:

$$\theta + \alpha(\theta)g(\theta)$$

is not invariant under scale transformation of $g(\theta)$. In fact, an estimating function $g(\theta)$ induces a class of functions

$$\mathcal{L}_g = \{a(\theta)g(\theta) \colon a(\theta) \neq 0\}$$

each member in the class having the same roots. It is of interest to choose a function in $\mathcal{L}_g$ so that good convergence properties can be obtained for finding the Godambe efficient root.

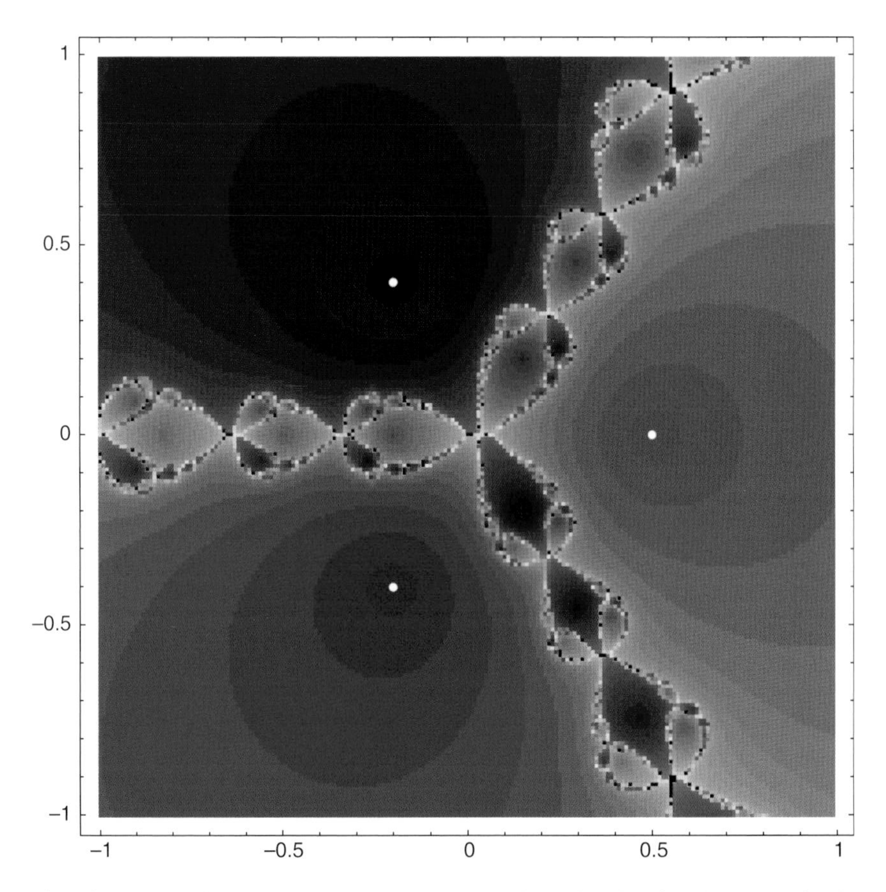

Fig. 7.8 Basins of attraction of Newton's method using estimating function (7.41) for the correlation coefficient with $S_1 = 0.1$, $S_2 = 1.0$.

Figures 7.8 and 7.9 show basins of attraction of roots to estimating functions for ρ based on both the cubic equation (7.41) and likelihood equation (7.40). We have used the Newton–Raphson method for these plots using the values $S_1 = 0.1$, $S_2 = 1.0$. There is one real root, 0.5, and two conjugate complex roots, $-0.2 \pm 0.4\sqrt{-1}$. The roots are plotted as white points in the figures. The real root is the maximum likelihood estimate, basins of attraction to which are coloured red in both Figures 7.8 and 7.9. In both figures we have coded a point that converges to none of the root as a black point. By comparing Figures 7.8 and 7.9 we see that Newton's method has a much wider basin of attraction to the real root for the cubic equation than for the likelihood equation.

The complete sufficient statistics was changed to $S_1 = 0.0001$, $S_2 = 0.99$ in Figures 7.10 and 7.11. There are three distinct real roots, −0.09, −0.01, 0.10, of the estimating equations in this case. Of these roots, the first and the third are

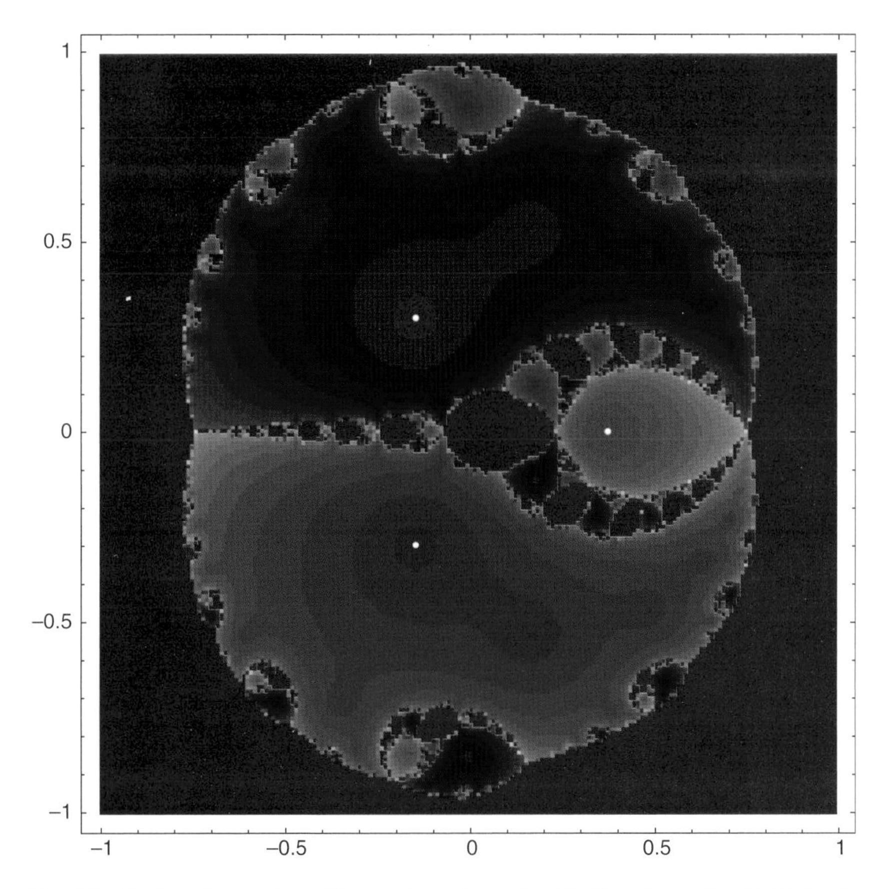

Fig. 7.9 Basins of attraction of Newton's method using likelihood equation (7.40) for the correlation coefficient with $S_1 = 0.1$, $S_2 = 1.0$.

local maxima, and the second root is a local minimum. The third root, 0.10, is the maximum likelihood estimate. These three roots are plotted as white points in the figures. Figure 7.10 plots the basins of attraction of Newton's method for the cubic equation (7.41), and Figure 7.11 plots the basins of attraction of Newton's method for the likelihood equation (7.40).

The coding conventions of the points in these two plots are the same as in Figures 7.8 and 7.10. For example, points coloured red form the basins of attraction of the maximum likelihood estimate 0.01 in both Figure 7.10 and Figure 7.11. For the second artificial data set, again the cubic equation displays better convergence property than the likelihood equation. Namely, the basin of attraction of the maximum likelihood estimate using Newton's method for solving the cubic equation is much wider than the corresponding one when the likelihood equation is used instead.

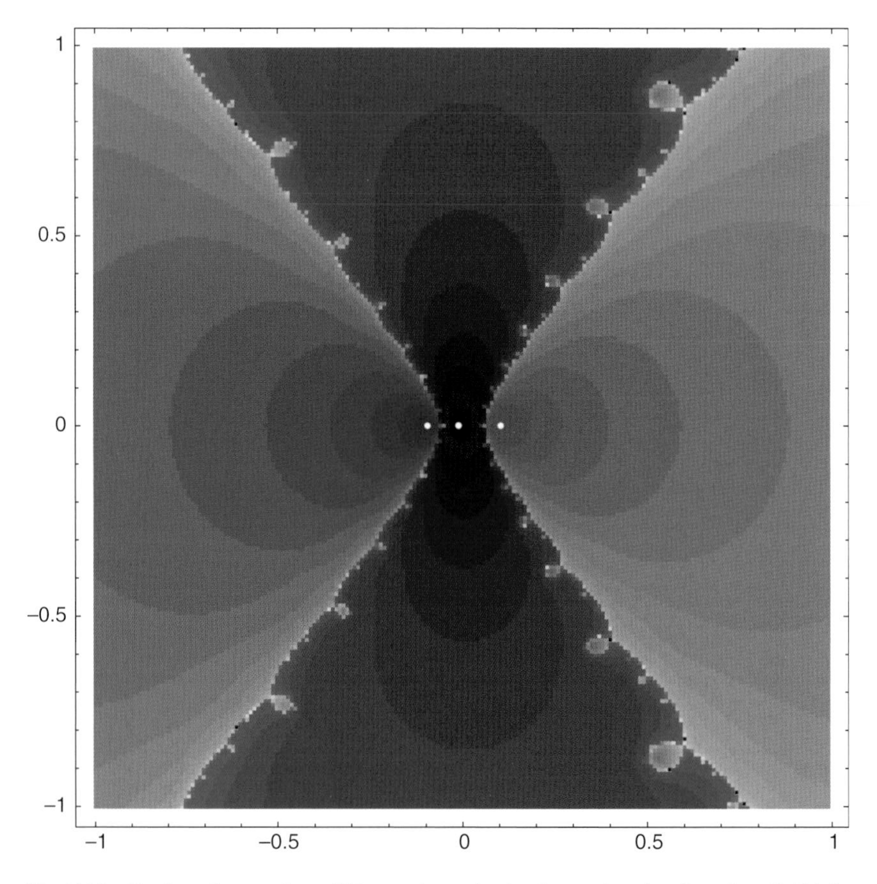

Fig. 7.10 Basins of attraction of Newton's method using estimating function (7.41) for the correlation coefficient with $S_1 = 0.0001$, $S_2 = 0.99$.

The basins of attraction for each of the above examples divide the complex plane into three regions, each point on the boundary bordering all three basins. The basins of attraction of an algorithm display some fractal geometrical structure, exploration of which is beyond the scope of the present book.

The above plots were obtained using a *converge-time algorithm*, about which we now describe briefly. Basins of attraction are similar to Julia sets. The filled-in Julia sets of a system $w(z)$ contain all the points z such that iterates $w^{(k)}(z)$ stay bounded. On the other hand, the basin of attraction of a root $\hat{z}$ to $g(z) = 0$ consists of all points z such that

$$\lim_{k \to \infty} w^{(k)}(z) = \hat{z}$$

where

$$w(z) = z - \alpha(z)g(z)$$

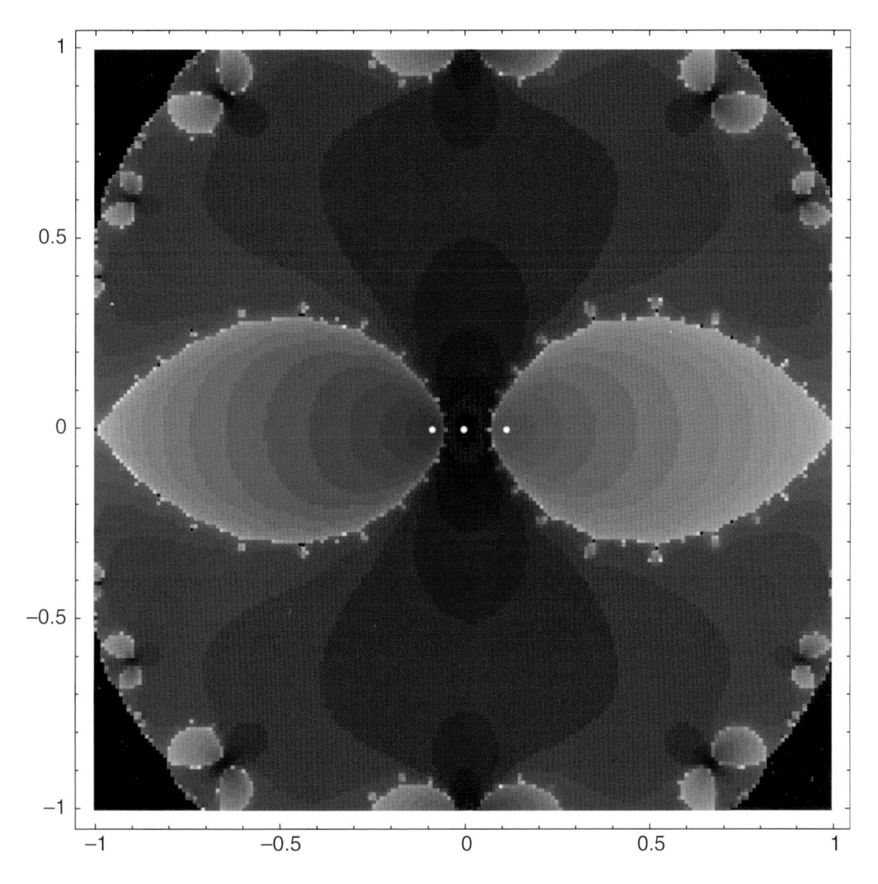

Fig. 7.11 Basins of attraction of Newton's method using likelihood equation (7.40) for the correlation coefficient with $S_1 = 0.0001$, $S_2 = 0.99$.

for a suitably chosen $\alpha(z)$, a function characterising the underlying iterative algorithm. For Newton's method, for instance, we choose $\alpha(z) = 1/\dot{g}(z)$. For a given precision $\epsilon > 0$, we define the *converge-time* of z to be the smallest integer k_z so that

$$|w^{(k_z)}(z) - \hat{z}| \leq \epsilon.$$

So we can distinguish each basin of attraction corresponding to each root by using different colours, and points within the same basin by using their respective converge-times. In these plots a point that converges slowly is cloured light.

The basins of attraction in the complex plane give detailed information on the convergence properties of an algorithm. A particularly interesting feature in these plots is that an algorithm can converge to the desirable root for many *complex starting points*, which are far away in Euclidean distance from the consistent root.

8
Bayesian estimating functions

8.1 Introduction

In the previous chapters, we found that the score function has maximal Godambe efficiency in a general class of unbiased estimating functions for a parametric model. Since the score function is the gradient of the log-likelihood, maximising the efficiency criterion can be regarded as a method for deriving the techniques of likelihood inference. As the Godambe efficiency of an estimating function is defined by the expectations of functionals of the estimating function, the approach to likelihood developed in Chapter 2 is *frequentist* in nature.

The concept of likelihood is also important in Bayesian inference. The fundamental difference from the classical frequentist approach is that in Bayesian inference we make decisions by combining information from two sources: the likelihood given the parameter and the information about the parameter prior to making observations. The first source of information is the familiar one in the frequentist framework. We have called this the *likelihood* function, $p(y|\theta)$, which is the conditional density function (or probability function in the discrete case) of an observable random variable Y given the value of an interest parameter θ. The second source of information is often summarised in the form of a *prior distribution*, $p(\theta)$, when the parameter θ is assumed to be a random variable.

Our notation in this chapter will differ slightly from the notation used so far. For example, the likelihood function $p(y|\theta)$ was denoted by $f(\theta; y)$ or $p(\theta; y)$ in previous chapters.

By regarding the parameter as a random variable, we derive, using *Bayes' Theorem*, the *posterior distribution* of θ given data y

$$p(\theta|y) = \frac{p(\theta)p(y|\theta)}{\int p(\theta)p(y|\theta)\,\mathrm{d}\theta} \tag{8.1}$$

by combining the prior distribution $p(\theta)$ and the likelihood $p(y|\theta)$. In (8.1), the numerator is the *joint* density of the data and the parameter, and the denominator is the *marginal* density of the data. The Bayesian approach in inference postulates that the posterior distribution, $p(\theta|y)$, summarises all the relevant information for making inference about θ.

The Bayesian school of inference, originated in Bayes (1763), has a history longer than the frequentist approach to inference. It is now a well-established

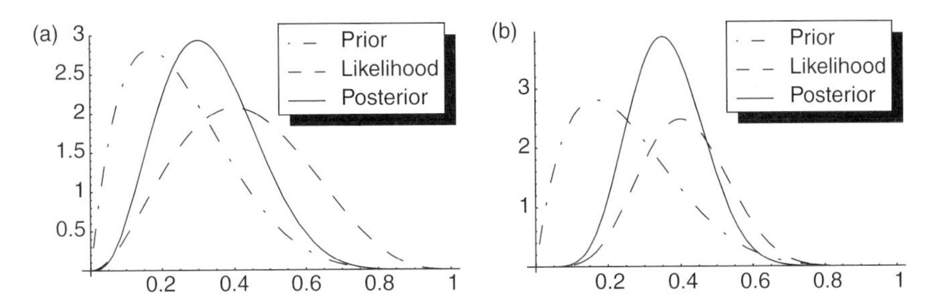

Fig. 8.1 Two binomial likelihoods with the same Beta prior.

alternative to the classical inference (O'Hagan, 1994). Recent advances in computational techniques, notably the *Markov chain Monte Carlo* methods (see e.g. Gamerman, 1997), make this approach particularly popular among practitioners.

Example 8.1 Binomial-Beta model

From (8.1) we see that the posterior density $p(\theta|y)$ is proportional to the product of the prior density and the likelihood. So $p(\theta|y)$ will be low if either the prior or the likelihood is low. The Posterior density will be high at values of θ at which both the prior and the likelihood are reasonably high. So the Bayes' Theorem balances the two sources of information from the prior and the likelihood. This can be clearly seen from Figure 8.1, which compares the prior Beta distribution $\text{Be}(p, q)$ with density function

$$p(\theta) = \frac{\Gamma(p+q)}{\Gamma(p)\Gamma(q)}\theta^{p-1}(1-\theta)^{q-1}, \quad 0 \leq \theta \leq 1$$

with the binomial likelihood

$$p(y|\theta) = \binom{n}{y}\theta^{y}(1-\theta)^{n-y}, \quad y = 0, 1, \ldots, n$$

and the posterior Beta distribution $\text{Be}(p+y, q+n-y-1)$. In Figure 8.1, we let $(p, q) = (2, 6)$ and set $(y, n) = (2, 5)$ in (a), and $(y, n) = (6, 15)$ in (b). For better visual effects, we multiplied the binomial likelihoods by a factor free of θ.

8.2 Bayes consistency

Under mild regularity conditions, estimation of the parameter based on the posterior distribution given by (8.1) is *consistent*. This means that as the sample size goes to infinity, the posterior distribution of θ will converge stochastically to the true value of the parameter. The regularity conditions needed to ensure such consistency requires that the prior distribution assigns a positive probability to every neighbourhood of the true value of the parameter. *Bayes consistency* essentially

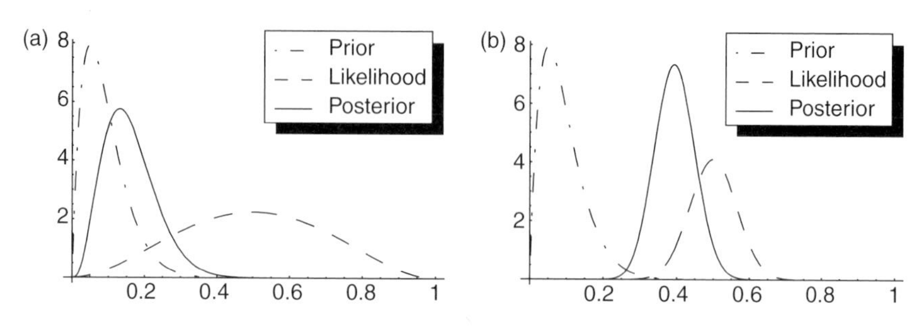

Fig. 8.2 Bayes consistency.

states that data will eventually become overwhelming so that prior information is asymptotically irrelevant for inference.

Example 8.2 Binomial-Beta model (continued)

The principle that the data eventually overwhelm the prior as the sample size goes to infinity can be illustrated by the binomial example once again. Figure 8.2, which is similar to Figure 8.1, compares the Beta prior distribution, the binomial likelihood and the Beta posterior distribution, with $(y, n) = (2, 4)$ in (a) and $(y, n) = (30, 60)$ in (b). The posterior distribution is pulled over to the likelihood in (b) due to the large sample size.

In the frequentist framework, we are often interested in estimating θ with a point estimate $\hat{\theta}$ found as the root to an estimating equation. While the Bayesian can interpret the posterior distribution of the parameter as, in some sense, the inferential conclusion of the analysis, it is also of interest in Bayesian inference to summarise the information about the parameter using a point estimate. As we shall consider below, the estimating equations of classical frequentist methods have their analogs in Bayesian estimating equations. However, like their classical counterparts, Bayesian estimating equations can have multiple solutions. So it is important to consider the consistency of any point estimate derived using such an equation. The frequentist concept of consistency can be modified as follows to the notion of Bayes consistency.

Definition 8.3 *An estimate $\hat{\theta}_n$ is said to be Bayes consistent if $\hat{\theta}_n - \theta$ converges in probability to zero as $n \to \infty$, where probabilities are calculated using the posterior distribution of θ given y. That is,*

$$P\{|\hat{\theta}_n - \theta| < \epsilon|y\} > 1 - \eta, \quad n > N \tag{8.2}$$

holds for some N given any $\epsilon, \eta > 0$.

Unlike the frequentist definition of consistency, where the parameter θ is fixed and the estimate $\hat{\theta}$ is a random variable, the roles are reversed here. The estimate $\hat{\theta}$ is a function of the data y. Since the posterior distribution is calculated conditionally on y, the point estimate is a constant in the posterior distribution. On the other hand, the posterior distribution is on the parameter space, and the parameter θ is regarded as a random variable in the definition of Bayes consistency.

8.3 A Bayesian approach to estimating functions

Recall that an estimating function $g(\theta, y)$ is a function,

$$g\colon \Theta \times \mathcal{Y} \to \mathbb{R}^k$$

where Θ and $\mathcal{Y}$ are the parameter space and the sample space, respectively. In Bayesian framework, Θ plays the role as a sample space for prior probability measures $p(\theta)$.

Definition 8.4 *A Bayesian estimating function $g(\theta, Y)$ is a function, $g\colon \Theta \times \mathcal{Y} \to \mathbb{R}^k$; where $\mathcal{Y}$ is the sample space of an observable random variable Y and Θ the sample space of an unobservable random variable θ.*

It is evident from their respective definitions that classical and Bayesian estimating functions are rather general real or vector-valued functions of the data and the parameter. The difference between a Bayesian estimating function and a classical one is not to be found in the formal properties of the function, but in the context in which the estimating function is derived and used.

Among the estimating functions of this definition are the estimating functions that were discussed in the previous chapters. However, to this class of estimating functions we can add many others, such as the *prior score*

$$u(\theta) = \frac{\partial}{\partial \theta} \log p(\theta) \tag{8.3}$$

or the *posterior score*

$$\begin{aligned} u(\theta|y) &= \frac{\partial}{\partial \theta} \log p(\theta|y) \\ &= \frac{\partial}{\partial \theta} \log p(\theta) + \frac{\partial}{\partial \theta} \log p(y|\theta). \end{aligned} \tag{8.4}$$

Equation (8.4) follows from the fact that

$$p(\theta|y) \propto p(\theta)p(y|\theta)$$

where the constant of proportionality does not involve θ.

In the previous chapters, the concept of the unbiasedness of an estimating function $g(\theta, Y)$, using the notation of this chapter, corresponded to the equation

$$E\{g(\theta, Y)|\theta\} = 0$$

and we saw that within certain class of unbiased estimating functions, the score function is Godambe-optimal. We shall have a similar result about the posterior score.

First, we discuss the basic concepts of unbiasedness and information unbiasedness of Bayesian estimating functions $g(\theta, y)$. To motivate our discussion, we consider the following simple example.

Example 8.5 Gaussian–Gaussian model

Suppose that we have normal likelihood and normal prior

$$Y \sim N(\theta, \sigma^2) \quad \text{and} \quad \theta \sim N(m, \tau^2)$$

where σ^2, m and τ are known. The posterior distribution $p(\theta|y)$ is again normal

$$(\theta|y) \sim N(\mu, v^2) \tag{8.5}$$

where μ and v are given by

$$\mu = \frac{\tau^2}{\tau^2 + \sigma^2} y + \frac{\sigma^2}{\tau^2 + \sigma^2} m, \quad v^2 = \frac{\sigma^2 \tau^2}{\tau^2 + \sigma^2}.$$

So the posterior score is

$$\begin{aligned} u(\theta|y) &= \frac{1}{v^2}(\mu - \theta) \\ &= \frac{1}{\sigma^2}(y - \theta) + \frac{1}{\tau^2}(m - \theta). \end{aligned} \tag{8.6}$$

By (8.5) and (8.6), and the fact that

$$E(Y) = E\{E(Y|\theta)\} = m$$

we have

$$E\{u(\theta|Y)\} = 0. \tag{8.7}$$

However, it is also clear that

$$E\{u(\theta|Y)|Y\} = 0. \tag{8.8}$$

So $u(\theta|y)$ is an estimating equation that is unbiased in two senses according to (8.7) and (8.8). It is also natural to ask whether it is unbiased according to the classical frequentist definition of Chapter 2. Classical unbiasedness would require

$$E[u(\theta|Y)|\theta] = 0.$$

If we calculate this expectation, we find instead that

$$E\{u(\theta|Y)|\theta\} = \frac{1}{\tau^2}(m - \theta)$$

which does not vanish, except in the limiting case when $\tau^2 \to \infty$. The limiting form of this prior distribution is the improper uniform prior on the entire line of $\mathbb{R}$. For such an improper prior, the posterior density reduces to the likelihood function.

The above example shows that the classical concept of unbiasedness of estimating functions needs to be modified in the Bayesian context. Equations (8.7) and (8.8) give rise two possible definitions of unbiasedness for a Bayesian estimating function. Using the relation between expectation with respect to a joint density and that with respect to a conditional expectation, we can see that (8.8) implies (8.7) in general. So (8.8) is the stronger condition of the two.

Definition 8.6 *A Bayesian estimating function $g(\theta, Y)$ is said to be conditionally unbiased, if for all θ*

$$E\{g(\theta, Y)|Y\} = 0 \tag{8.9}$$

holds with probability one.

The function $g(\theta, Y)$ is said to be average unbiased if

$$E\{g(\theta, Y)\} = 0. \tag{8.10}$$

Now we establish a condition for (8.9) and (8.10) to hold for a posterior score in more general models. Suppose that θ is a real-valued parameter. Let $p(y|\theta)$, $p(\theta)$ and $p(\theta|y)$ be the likelihood, prior distribution and the posterior distribution, respectively. Denote the respective logarithms by $\ell(y|\theta)$, $\ell(\theta)$ and $\ell(\theta|y)$. Again, we shall denote the derivatives of these logarithms with respect to θ by $u(y|\theta)$, $u(\theta)$ and $u(\theta|y)$, respectively. We assume that the prior distribution $p(\theta)$ is supported on $(a\ b)$, where both a and b may be infinite. By (8.1), the posterior log likelihood and posterior score can be written as

$$\ell(\theta|y) = \ell(y|\theta) + \ell(\theta) - c(y) \tag{8.11}$$

$$u(\theta|y) = u(y|\theta) + u(\theta). \tag{8.12}$$

Since the classical score function $u(Y|\theta)$ is unbiased with respect to the model density, we have

$$E\{u(Y|\theta)|\theta\} = 0.$$

And therefore we have

$$E\{u(Y|\theta)\} = E[E\{u(Y|\theta)|\theta\}] = 0.$$

Let $E\{u(\theta)\} = A$ where

$$A = \lim_{\theta \to b^-} p(\theta) - \lim_{\theta \to a^+} p(\theta). \tag{8.13}$$

It follows that

$$E\{u(\theta|y)\} = 0$$

if (8.13) vanishes.

Although the posterior score function is not unbiased in the classical sense, it shares with the classical score the property of being average unbiased. Under standard regularity conditions, it is also conditionally unbiased, for reasons that are similar to the classical case,

$$\begin{aligned} E\left\{\frac{\partial}{\partial\theta}\log p(\theta|Y)|Y\right\} &= \int \frac{\partial/\partial\theta p(\theta|y)}{p(\theta|y)} p(\theta|y)\, \mathrm{d}\theta \\ &= \int \frac{\partial}{\partial\theta} p(\theta|y)\, \mathrm{d}\theta \\ &= \frac{\partial}{\partial\theta}\int p(\theta|y)\, \mathrm{d}\theta \\ &= \frac{\partial}{\partial\theta}(1) \\ &= 0. \end{aligned}$$

Summarising the above discussion, we have the following result.

Proposition 8.7 *Let $u(\theta|Y)$ be any posterior score function. Then we have*

1. *under standard regularity, $u(\theta|Y)$ is conditionally unbiased, and*
2. *$u(\theta|Y)$ is average unbiased provided*

$$\lim_{\theta\to b^-} p(\theta) - \lim_{\theta\to a^+} p(\theta) = 0$$

where $p(\theta)$ is the prior density function with support $(a\ b)$.

In the development of Bayesian estimating functions, we frequently need to combine non-Bayesian estimating functions with prior estimating functions. For such combined estimating functions we have the following.

Proposition 8.8 *Let $g(\theta, Y)$ be an estimating function that is unbiased in the classical sense, so that $E\{g(\theta, Y)|\theta\} = 0$. Let $g(\theta)$ be a prior estimating function satisfying $E\{g(\theta)\} = 0$. Then the following Bayesian estimating function:*

$$g(\theta|Y) = g(\theta, Y) + g(\theta) \tag{8.14}$$

is average unbiased.

In the same way that classical unbiasedness implies average unbiasedness, the property of conditional unbiasedness implies average unbiasedness. Thus average unbiasedness is the weakest, or least restrictive of the three properties.

Now suppose that an estimating function $g(\theta, Y)$ is given. Consider the estimating functions defined by

$$h_1(\theta, Y) = a(\theta)g(\theta, Y), \quad h_2(\theta, Y) = b(Y)g(\theta, Y)$$

where $a(\cdot)$ and $b(\cdot)$ are functions of θ and Y, respectively. If $g(\theta, Y)$ is conditionally unbiased, then so is $h_2(\theta, Y)$. But $h_1(\theta, Y)$ is in general not conditionally unbiased. In other words, conditional unbiasedness is not invariant under scaling transformation by a function containing the parameter. When $g(\theta, Y)$ is average unbiased, it turns out that neither $h_1(\theta, Y)$, nor $h_2(\theta, Y)$ is in general average unbiased.

Now we turn to discuss the concept of information unbiasedness. This concept, as defined in equation (2.4), plays an important role in the theory of classical estimating functions. It plays a similar important role in the development of Bayesian estimating functions. To begin with, we first note the following property of density functions.

Lemma 8.9 *Let $p(y)$ be a density function, $y \in (a, b) \subset \mathbb{R}$. Suppose that $p(y)$ is differentiable with respect to y. Let E denote the expectation with respect to $p(y)$. Further, let*

$$R = \lim_{y \to b^-} \frac{\partial}{\partial y} p(y) - \lim_{y \to a^+} \frac{\partial}{\partial y} p(y).$$

Then the following identity holds

$$E\left\{\frac{\partial}{\partial y} \log p(y)\right\}^2 = -E\left\{\frac{\partial^2}{\partial y^2} \log p(y)\right\} + R. \tag{8.15}$$

The proof is straightforward, and is therefore omitted.

Now let us compare (8.15) with information unbiasedness as in (2.4). As a direct consequence of Lemma 8.9, we have the following property for posterior score functions.

Proposition 8.10 *Let $u(\theta|y) = (\partial/\partial\theta) \log p(\theta|y)$ be a posterior score function. Suppose that the posterior density $p(\theta|y)$ is defined on an open interval (a, b). Let*

$$R = \lim_{\theta \to b^-} \frac{\partial}{\partial \theta} p(\theta|y) - \lim_{\theta \to a^+} \frac{\partial}{\partial \theta} p(\theta|y). \tag{8.16}$$

Then

$$E\{u^2(\theta|Y)|Y\} = -E\{\dot{u}(\theta|Y)|Y\} + R.$$

Definition 8.11 *A Bayesian estimating function $g(\theta, y)$ is said conditionally information unbiased, if the following holds:*

$$E\{g^2(\theta, Y)|Y\} = -E\{\dot{g}(\theta, Y)|Y\}. \tag{8.17}$$

By Proposition 8.10, we immediately have the following result.

Proposition 8.12 *A posterior score function* $u(\theta|y) = (\partial/\partial\theta)\log p(\theta|y)$ *is conditionally information unbiased if*

$$\lim_{\theta\to b^-}\frac{\partial}{\partial\theta}p(\theta|y) - \lim_{\theta\to a^+}\frac{\partial}{\partial\theta}p(\theta|y) = 0. \tag{8.18}$$

Condition (8.18) is satisfied by most posterior densities. As can be directly verified, (8.18) is satisfied by the posterior scores in both Examples 8.1 and 8.5. The following result gives a general method to achieve conditional information unbiasedness starting from any sufficiently regular estimating function.

Proposition 8.13 *Let* $g(\theta, Y)$ *be a conditionally unbiased estimating function, which is differentiable in* θ. *Then*

$$h(\theta, Y) = -E\{\dot{g}(\theta, Y)|Y\}[E\{g^2(\theta, Y)|Y\}]^{-1}g(\theta, Y)$$

is conditionally information unbiased.

Definition 8.14 *Similarly to (8.17), we say that* $g(\theta, y)$ *is average information unbiased, if*

$$E\{g^2(\theta, Y)\} = -E\{\dot{g}(\theta, Y)\}. \tag{8.19}$$

Again, under mild conditions, posterior scores are average information unbiased. This follows immediately from the fact that posterior scores are conditionally information unbiased. So we have the next result.

Proposition 8.15 *Let* $p(\theta)$ *be a prior density function with support* (a, b). *Let* $u(\theta|Y)$ *be the posterior score. Then* $u(\theta|Y)$ *is average information unbiased, if the following condition is met*

$$\lim_{\theta\to b^-}\frac{\partial}{\partial\theta}p(\theta) - \lim_{\theta\to a^+}\frac{\partial}{\partial\theta}p(\theta) = 0.$$

Among estimating functions that are conditionally unbiased, it is natural to look for examples of functions which satisfy some optimality criterion for Bayesian point estimation. As we shall see, the theory of optimal Bayesian estimating functions is in some sense simpler than the classical theory, as the expectations which are conditional on the data involve integrations over the parameter alone.

Let $\mathcal{G}$ be the class of all estimating functions $g(\theta, Y)$ which are conditionally unbiased and conditionally square integrable, so that

$$E\{g^2(\theta, Y)|Y\} < \infty.$$

If g is also differentiable with respect to θ, then it is convenient to standardise g in a manner that is analogous to the classical case, by defining

$$g_s(\theta, Y) = \frac{g(\theta, Y)}{E\{\dot{g}(\theta, Y)|Y\}}. \tag{8.20}$$

Note that if g is conditionally unbiased, then so is g_s. In the definition of optimality that follows, it will be convenient to appeal to the standardised form of each estimating function. Therefore we shall impose the additional restriction that all estimating functions in $\mathcal{G}$ are differentiable with respect to θ, and that

$$E\{\dot{g}^2(\theta, Y)|Y\} > 0$$

with probability one. This will ensure the existence of the standardised form of every estimating function in $\mathcal{G}$.

Definition 8.16 *Among standardised functions, it is natural to call a function $g^*(\theta, Y)$ conditionally optimal if it has minimum variance. That is, if for all $g \in \mathcal{G}$,*

$$E[\{g_s^*(\theta, Y)\}^2|Y] \leq E[\{g_s(\theta, Y)\}^2|Y] \tag{8.21}$$

holds with probability one.

The following proposition, which is analogous to the classical result, is due to Ghosh (1993).

Proposition 8.17 *Suppose that the posterior density $p(\theta|y)$ satisfies the regularity assumptions below, namely that $p(\theta|y)$ has support set (a, b), and*

1. *$p(\theta|y) \to 0$ as $\theta \to a$ and $\theta \to b$;*
2. *$p(\theta|y) \log p(\theta|y) \to 0$ as $\theta \to a$ and $\theta \to b$; and*
3. *for all $g \in \mathcal{G}$,*

$$g(\theta, y)p(\theta|y) \to 0$$

as $\theta \to a$ and $\theta \to b$.

Then, under these conditions, the posterior score function $u(\theta|y)$ is conditionally optimal.

The roots of the classical score function can be associated with the stationary points of the likelihood. Similarly, the roots of the posterior score can be associated with the stationary points of the posterior density. So if the solution to the equation $u(\hat{\theta}|y) = 0$ is unique, then under mild conditions, the root $\hat{\theta}$ corresponds to the *generalised maximum likelihood estimator*, which is a Bayesian variant of the maximum likelihood estimator.

8.4 Bayes linear and semi-parametric estimation

Subjective Bayesian inference starts with the adoption of a subjective prior distribution, $p(\theta)$, which expresses one's personal knowledge prior to making observations. It was Savage (1954) who axiomatically argued on favour of the Bayesian methodologies using the concept of subjective probability advocated earlier by de Finetti. In reality, however, it is often difficult, if not impossible, to express one's knowledge about the parameter of interest in the form of a distribution. A practical way of overcoming this difficulty is to specify a few functionals of the prior distribution and then to fit one of the familiar probability laws. For example, to fit a prior distribution for the parameter θ of a binomial model, one may choose a prior Beta distribution, $\text{Be}(p, q)$, and estimate the two parameters p, q using assumptions on the first two moments about $\text{Be}(p, q)$. Replacing such summary information by a prior distribution could give rise to a *sensitivity* issue, that is, a small change of the form of the prior may greatly affects the inference for the parameter. When this is the case, more *robust* methods, which are insensitive to the choice of prior distributions, will be of importance.

The so-called linear Bayes or *Bayes linear* methods (Whittle, 1958; Stone, 1963; Hartigan, 1969; Goldstein, 1975*a*; Smouse, 1984, etc.) for estimation are devised in response to this need of robustness. Bayes linear methods only make use of prior information, be it subjective or otherwise, in the form of a few moment assumptions. As a consequence, the standard Bayesian approach for inference based on posterior distributions must be altered, since we are no longer able to use Bayes' Theorem to combine the prior information with the likelihood.

Now consider a scalar parameter θ and suppose that a location summary for θ is desired. The Bayes linear method starts by considering estimates *linear* in Z

$$d(Y) = a + b^{\mathrm{t}} Z \tag{8.22}$$

where $Z = z(Y)$ is a known column vector function of the data Y. The coefficient a and the column vector b are to be determined by minimising the *unconditional quadratic risk*

$$R = E[\{d(Y) - \theta\}^2] \tag{8.23}$$

where the expectation is taken with respect to the joint distribution of Y and θ.

From (8.22) and (8.23), after some algebraic manipulation, the optimal estimate can be shown to be

$$d(Y) = E(\theta) + \{\text{Cov}(Z, \theta)\}^{\mathrm{t}} \text{Var}^{-1}(Z)\{Z - E(Z)\} \tag{8.24}$$

where $\text{Cov}(Z, \theta)$ is the column vector of covariances between the entries of Z and θ, and $\text{Var}(Z)$ is the covariance matrix for the vector Z. All expectations are calculated with respect to the joint distribution of Z and θ. The estimate $d(y)$ of (8.24) is called the *Bayes linear estimate* of θ. Note that to define a Bayes linear estimate, one needs only to specify the first two joint moments of θ and $z(Y)$. So estimate (8.24) is derived under more robust assumptions than an estimate derived from a posterior distribution.

Since the unconditional quadratic risk (8.23) may also be rewritten as

$$R = E[\{d(Y) - E(\theta|Y)\}^2] + E\{\text{Var}(\theta|Y)\} \tag{8.25}$$

it is clear that the posterior mean $E(\theta|y)$ is the optimal estimate under a quadratic loss function. The Bayes linear estimator (8.24) is therefore the best approximation to the posterior mean within the class of linear functions of $z(y)$. Hence if the posterior mean is exactly linear, then it equals the Bayes linear estimate.

In certain special cases, the linearity of the posterior mean uniquely specifies the class of prior distributions. Diaconis and Ylvisaker (1979, 1985) show that, under certain conditions, if y is exponentially distributed as

$$p(y|\eta) = \exp\left\{\sum_{j=1}^{k} \eta_j A_j(y) - \psi(\eta)\right\}$$

then the linearity of the posterior mean $E(\theta_j|y)$ in $z_j(y)$ implies that the prior distribution is necessary a member of the natural conjugate family; where

$$\theta_j = \frac{\partial \psi(\eta)}{\partial \eta_j} = E\{A_j(y)|\eta\} \quad \text{and} \quad z_j(y) = \frac{1}{n}\sum_{i=1}^{n} A_j(y_i).$$

We shall now turn to the Bayesian formulation of semi-parametric models and inference. Suppose that, conditionally on θ, $Y_1, Y_2, \ldots, Y_n$ are independently distributed with means and variances

$$E(Y_j|\theta) = \mu_j(\theta), \quad \text{Var}(Y_j|\theta) = \sigma_j^2(\theta).$$

As usual, we let Y denote the column vector of $Y_1, \ldots, Y_n$. As above, we shall seek to obtain the model robustness of a semi-parametric framework with Bayesian assumptions. The first and second conditional moments of each Y_j will be assumed to be known; the remaining aspects of these distributions will be unspecified. Similarly, the prior distribution of θ will also be unspecified except for its first two moments, namely,

$$E(\theta) = \mu, \quad \text{Var}(\theta) = \sigma^2.$$

A natural class of estimating functions for this model is the collection of all conditionally unbiased functions of the form

$$g(\theta, Y) = b(\theta)(\theta - \mu) + \sum_{j=1}^{n} a_j(\theta)\{Y_j - \mu_j(\theta)\} \tag{8.26}$$

where the coefficient functions $b(\theta)$ and $a_j(\theta)$, $j = 1, \ldots, n$, are arbitrary.

Under the optimality condition specified in Definition 8.16, the conditionally optimal estimating function within this class is obtained by setting

$$a_j^*(\theta) = -\frac{\dot{\mu}_j(\theta)}{\sigma_j^2(\theta)} \quad \text{and} \quad b^*(\theta) = \frac{1}{\sigma^2}.$$

The proof of this result is similar to Godambe and Thompson (1989), and is therefore omitted.

8.5 An application to credibility estimation in actuarial science

While Bayesian assumptions are commonly adopted for philosophical reasons, there are applications in which prior distributions have frequentist interpretations. Examples in actuarial science are cases in point. Suppose that in successive years an individual makes successive claims on an insurance policy. Let $Y^t = (Y_1, Y_2, \ldots, Y_n)$ denote the claim expenses for the first n years. Assuming that the risk factors associated with successive claims remain constant, then the claims Y_j to the insurance company can be regarded as a sequence of identically distributed random variables. A basic actuarial problem is to try to predict the size of the future claim expense Y_{n+1} in the $(n+1)$-st year from the individual.

Individual policy holders are themselves grouped according to certain characteristics, such as age, location and other factors. While the risk factors for an individual claimant may be partially determined by the group characteristics, there is usually heterogeneity within each group. So it makes sense to apply a Bayesian model to the data on claim expenses for an individual. So, for each claimant, there is a random unknown parameter vector θ which controls the distribution of the claims process $Y_1, \ldots, Y_n$. The prior distribution on θ represents the random risk factors associated with population heterogeneity. Characteristics of the population or a predetermined subgroup of the population are then controlled by the hyper-parameters on the prior distribution.

By a *credibility estimator*, we shall mean a statistic whose purpose is to estimate the value of the future claim Y_{n+1}. Such credibility estimators are functions $d(Y)$ of the claims process, where the coefficients can be chosen to minimise an expected squared error of the form

$$E[\{Y_{n+1} - d(Y)\}^2].$$

The solution to this problem has a formal similarity to the problem of Bayes linear estimation discussed in Section 8.4, except that the parameter has been replaced by the $(n+1)$-st observation.

The problem of prediction can be related to the problem of estimation. Suppose that, conditionally on θ, the claims $Y_1, Y_2, \ldots,$ are independent. In this case, we can parameterise the model so that

$$\eta = E(Y_{n+1}|\theta).$$

Then

$$E[\{Y_{n+1} - d(Y)\}^2] = E\{\mathrm{Var}(Y_{n+1}|\theta)\} + E[\{\eta - d(Y)\}^2].$$

So expected squared error is minimised by that choice of $d(Y)$ which is the Bayes linear estimate for η.

Example 8.18 Bühlmann model

To illustrate the application of these methods to actuarial science, we shall consider the *Bühlmann model*, which is perhaps the simplest and most straightforward model for such claims processes. Let us assume that, conditionally on θ, the claims Y_j of a given policy holder are independent and identically distributed, with mean $\mu(\theta)$ and variance $\sigma^2(\theta)$.

The Bayesian semi-parametric approach needs only the specification of the mean and variance for $\mu(\theta)$ and the specification of mean for $\sigma^2(\theta)$. Now assume that the following quantities

$$E\{\mu(\theta)\} = \mu_0, \quad \mathrm{Var}\{\mu(\theta)\} = \sigma_0^2, \quad E\{\sigma^2(\theta)\} = \nu \tag{8.27}$$

are known. From (8.27) the unconditional moment structure of the Y_j can be computed. We obtain

$$\begin{aligned} E(Y_j) &= E\{\mu(\theta)\} = \mu_0, \\ \mathrm{Var}(Y_j) &= \mathrm{Var}\{\mu(\theta)\} + E\{\sigma^2(\theta)\} = \sigma_0^2 + \nu \end{aligned}$$

and

$$\mathrm{Cov}(Y_j, Y_k) = \mathrm{Var}\{\mu(\theta)\} = \sigma_0^2.$$

With this structure, the conditionally optimal estimating function for $\mu(\theta)$ becomes

$$\frac{\hat{\mu} - \mu_0}{\sigma_0^2} - \sum_{j=1}^{n} \frac{Y_j - \hat{\mu}}{\hat{\sigma}^2} = 0$$

where

$$\hat{\mu} = \mu(\hat{\theta}) \quad \text{and} \quad \hat{\sigma}^2 = \sigma^2(\hat{\theta}).$$

In practice, it is common to replace $\hat{\sigma}^2$ in this equation by $\nu = E\{\sigma^2(\theta)\}$, for reasons of algebraic simplicity. With this substitution, the equation can be solved explicitly. We can see that

$$\hat{\mu} = \frac{n\sigma_0^2}{n\sigma_0^2 + \nu}\bar{Y} + \frac{\nu}{n\sigma_0^2 + \nu}\mu_0 \tag{8.28}$$

which is a shrinkage estimator.

Note that the estimator (8.28) may be rewritten as

$$\hat{\mu} = \frac{\sigma_0^2}{\sigma_0^2 + \nu/n}\bar{Y} + \frac{\nu/n}{\sigma_0^2 + \nu/n}\mu_0.$$

From this form we see that the optimal Bayesian semi-parametric estimator for the mean, $\hat{\mu}$, formally coincides with the posterior mean derived from the full Bayesian Gaussian–Gaussian model,

$$N(\bar{Y}, \nu/n) \quad \text{and} \quad N(\mu_0, \sigma_0^2).$$

The advantage of the Bayesian semi-parametric approach is the distribution-robustness: in deriving the estimator (8.28) neither prior knowledge in the form of a distribution nor a specific parametric model is assumed.

8.6 Bibliographical notes

The Bayesian philosophy for inference can be traced back to Thomas Bayes' (1763) original writing, which was published two years after his death. Bayes' Theorem, as generalised by Laplace (1814), which combines prior knowledge with information contained in the data, forms the starting point for Bayesian inference.

The Bayes linear methods have been developed independently by a number of authors. More technical and philosophical aspects of the methods can be found in Goldstein (1975*b*, 1976, 1979, 1981, 1985, 1986, 1990) and Goldstein and O'Hagan (1996). That linearity of posterior mean implies the form (8.24) was proved by Ericson (1969). We also note that Bayes linear estimate (8.24) may be obtained without referring to any loss function, but we have to make instead a somewhat obscure assumption that $\theta - \{\text{Cov}(z, \theta)\}^{\text{t}}\{\text{Var}(z)\}^{-1}z$ and z are independent. See Harrison (1996) for details.

The derivation of the posterior score as the optimal estimating function is due to Ferreira (1982*b*) and Ghosh (1993). In this chapter, we have followed Ghosh's definitions and development. Ferreira defined the optimality of estimating functions using the joint distribution of the parameter and the data, in contrast to Ghosh, whose optimality criterion was based upon the posterior distribution. These ideas were extended to semi-parametric models by Li (1995) and Godambe (1999). The discussion of the actuarial applications described in this chapter is based upon Li (1995).

Bibliography

Amemiya, T. (1973). Regression analysis when the dependent variable is truncated normal. *Econometrica* **41**, 997–1016.

Bahadur, R. R. (1958). Examples of inconsistency of maximum likelihood estimates. *Sankhyā* **20**, 207–210.

Barbeau, E. J. (1989). *Polynomials*. Springer-Verlag, New York.

Barndorff-Nielsen, O. E. (1983). On a formula for the distribution of the maximum likelihood estimator. *Biometrika* **70**, 343–365.

Barndorff-Nielsen, O. E. (1995). Quasi profile and directed likelihoods from estimating functions. *Ann. Inst. Statist. Math.* **47**, 461–464.

Barnett, V. D. (1966). Evaluation of the maximum likelihood estimator where the likelihood equation has multiple roots. *Biometrika* **53**, 151–165.

Bartlett, M. S. (1936). The information available in small samples. *Proc. Cambridge Philos. Soc.* **32**, 560–566.

Basford, K. E. and McLachlan, G. J. (1985). Likelihood estimation with normal mixture models. *Appl. Statist.* **34**, 282–289.

Bayes, Rev. T. (1763). An essay toward solving a problem in the doctrine of chances, *Philos. Trans. R. Soc.* **53**, 370–418; Reprinted in *Biometrika*, 1958, **45**, 293–315.

Berger, J.O. and Wolpert, R.L. (1988). *The Likelihood Principle* (2nd). Inst. of Math. Statistics: Hayward.

Bergström, H. (1952). On some expansions of stable distributions. *Arkiv for Matematik*, vol. 2 no. 18, 375–378.

Bickel, P. J., Klaassen, P., Ritov, C.A.J. and Wellner, J. (1993). *Efficient and Adaptive Estimation for Semiparametric Models*. Baltimore: Johns Hopkins University Press.

Billingsley, P. (1995). *Probability and Measure*(3rd ed.). Wiley: New York.

Birnbaum, A. (1977). The Neyman-Pearson theory as decision theory, and as inference theory; with a criticism of the Lindley-Savage argument for Bayesian theory. *Synthése*, **36**, 19–49.

Bissel, A. F. (1972). A negative binomial model with varying element sizes. *Biometrika* **59**, 435–441.

Boyles, R. A. (1983). On the convergence of the EM algorithm. *J. Roy. Statist. Soc. Ser. B* **45**, 47–50.

Broyden, C. G. (1965). A class of methods for solving nonlinear simultaneous equations. *Math. Comput.* **19**, 577–593.

Burridge, J. (1981). A note on maximum likelihood estimation for regression models using grouped data. *J. Roy. Statist. Soc. Ser. B* **43**, 41–45.

Buse, A. (1982) The likelihood ratio, Wald, and Lagrange multiplier test: an expository note. *Am. Statist.*, **36**, 153–157.

Carroll, R. J., Ruppert, D., and Stefanski, L. A. (1995), *Measurement Error in Nonlinear Models*. London: Chapman and Hall.

Chandrasekar, B. (1983). Contributions to the theory of unbiased statistical estimation functions. Ph.D. thesis, Univ. of Poona, Pune-7, India.

Chandrasekar, B. and Kale, B. K. (1984). Unbiased statistical estimation functions in presence of nuisance parameter. *J. Statist. Plan. Inf.* **9**, 45–54.

Chaubey, Y. P. and Gabor, G. (1981). Another look at Fisher's solution to the problem of the weighted mean. *Comm. Statist. A* **10**, 1225–1237.

Copas, J. B. (1972). The likelihood surface in the linear functional relationship problem. *J. R. Statist. Soc. B* **34**, 374–278.

Copas, J. B. (1975). On the unimodality of the likelihood for the Cauchy distribution. *Biometrika* **62**, 701–704.

Cox, D. R. and Hinkley, D. V. (1974). *Theoretical Statistics*. Chapman & Hall, London.

Cramér, H. (1946). *Mathematical Methods of Statistics*. Princeton University Press, Princeton.

Cressie, N. A. (1993). *Statistics for Spatial Data*. Wiley, New York.

Crowder, M. (1986). On consistency and inconsistency of estimating equations. *Econ. Theory* **2**, 305–330.

Daniels, H. E. (1960). The asymptotic efficiency of a maximum likelihood estimator. *Proc. 4th Berkeley Symp. Math. Statist. and Prob.* **1**, 151–163.

Daniels, H. E. (1983). Saddlepoint approximations for estimating equations. *Biometrika* **70**, 89–96.

Darling, R. W. R. (1994). *Differential Forms and Connections*. Cambridge University Press, Cambridge.

Davison, A. C. and Hinkley, D. V. (1997). *Bootstrap Methods and Their Application*, Cambridge University Press: Cambridge.

Dempster, A. P., Laird, N. M. and Rubin, D. B. (1977). Maximum likelihood from incomplete data via the EM algorithm. *J. Roy. Statist. Soc. Ser. B* **39**, 1–22.

Dennis, J. E. (1996). *Numerical methods for unconstrained optimization and nonlinear equations*. Society for Industrial and Applied Mathematics, Philadelphia.

Diaconis, P. and Ylvisaker, D. (1979). Conjugate priors for exponential families. *Ann. Statist.* **7**, 269 –281.

Diaconis, P. and Ylvisaker, D. (1985). Quantifying prior opinion. In: *Bayesian Statistics 2*, J. M. Bernardo et al. (ed.), 133–156. North-Holland: Amsterdam.

Diggle, P., Liang, K.-Y. and Zeger, S.L. (1994). *Analysis of Longitudinal Data*. Oxford University Press: Oxford.

DuMouchel, W. H. (1973). On the asymptotic normality of the maximum-likelihood estimate when sampling from a stable distribution. *Ann. Statist.* **1**, 948–957.

DuMouchel, W. H. (1975). Stable distributions in statistical inference 2: information from stably distributed samples. *J. Amer. Statist. Assoc.* **70**, 386–393.

Durbin, J. (1960). Estimation of parameters in time series regression models. *J. Roy. Statist. Soc. Ser. B* **22**, 139–53.

Edwards, A.W.F. (1972). *Likelihood*. Cambridge University Press: Cambridge.

Edwards, A.W.F. (1974). The history of likelihood. *Int. Statist. Rev.* **42**, 9–15.

Efron, B. and Hinkley, D. (1978). Assessing the accuracy of the maximum likelihood estimator: Observed vs. expected Fisher information. *Biometrika*. **65**, 457–481.

Efron, B. and Tibshirani, R. J. (1993). *An Introduction to the Bootstrap*. Chapman and Hall: New York.

Eguchi, S. (1983). Second order efficiency of minimum contrast estimators in a curved exponential family. *Ann. Statist.* **11**, 793–803.

Engle, R. F. (1981) Wald, likelihood ratio, and Lagrange multiplier tests in econometrics. In *Handbook of Econometrics* (eds. Griliches, Z. and Intriligator, M.), Amsterdam: North Holland.

Fahrmeir, L. and Tutz, G. (1994) *Multivariate Statistical Modelling Based on Generalized Linear Models*. New York: Springer.

Fama, E. and Roll, R. (1968). Parameter estimates for symmetric stable distributions, *J. Amer. Statist. Assoc.* **63**, 817–836.

Feller, W. (1968). *An Introduction to Probability Theory and Its Application, Vol. 1 (3rd ed.)*. Wiley: New York.

Ferguson, T. S. (1958). A method of generating best asymptotically normal estimates with application to the estimation of bacterial densities. *Annals of Mathematical Statistics* **29**, 1046–1062.

Ferguson, T. S. (1982). An inconsistent maximum likelihood estimate. *J. Amer. Statist. Assoc.* **77**, 831–834.

Ferreira, P. E. (1982a). Multiparametric estimating equations. *Ann. Statist. Math.* **34A**, 423–431.

Ericson, W. A. (1969). A note on the posterior mean of a population mean. *J. R. Statist. Soc.* B., **31**, 195–233.

Ferreira, P. E. (1982b). Estimating equations in the presence of prior knowledge, *Biometrika*, **69**, 667–669.

Feuerverger, A. and McDunnough, P. (1981). On the efficiency of empirical characteristic function procedures. *J. Roy. Statist. Soc. Ser. B* **43**, 20–27.

Finch, S. J., Mendell, N. R. and Thode, H. C. (1989). Probability measures of adequacy of a numerical search for a global maximum. *J. Amer. Statist. Assoc.* **84**, 1020–1023.

Fisher, R. A. (1922). On the mathematical foundations of theoretical statistics. *Phil. Trans. Roy. Soc. London, Ser. A* **222**, 309–368.

Fisher, R. A. (1925). Theory of statistical estimation. *Proc. Camb. Phil. Soc.* **22**, 700–725.

Foxman, B., Marsh J. V., Gillespie, B., Rubin, K. N., Koopman, J. S. and Spear, S. (1997). Condom Use and First Time Urinary Tract Infection. *Epidemioly*, **8**, 637–641.

Fuller, W. A. (1987). *Measurement Error Models*. New York: Wiley.

Gamerman, D. (1997). *Markov Chain Monte Carlo : Stochastic Simulation for Bayesian Inference*. Chapman & Hall, London.

Gan, L. and Jiang, J. (1999). A test for global maximum. *J. Amer Statist. Assoc.* **94**, 847–854.

Ghosh, M. (1993). On a Bayesian analog of the theory of estimating function. *C. G. Khatri Memorial Volume Gujarat Statistical Review*, **17A**, 47–52.

Gilmore, R. (1981). *Catastrophe Theory for Scientists and Engineers*. Dover, New York.

Godambe, V. P. (1960). An optimum property of regular maximum likelihood estimation. *Ann. Math. Statist.* **31**, 1208–1212.

Godambe, V. P. (1991). Orthogonality of estimating functions and nuisance parameters. *Biometrika* **78**, 143–151.

Godambe, V. P. (1999). Linear Bayes and optimal estimation. *Ann. Inst. Statist. Math.* **51**(2), 201–215.

Godambe, V.P. and Thompson, M.E. (1989). An extension of quasi-likelihood estimation (with discussion). *J. Statist. Plan. Inference* **22**, 137–172.

Goldstein, M. (1975a). Approximate Bayes solutions to some nonparametric problems. *Ann. Statist.* **3**, 512–517.

Goldstein, M. (1975b). A note on some Bayes non-parametric estimates. *Ann. Statist.* **3**, 736–740.

Goldstein, M. (1976). Bayesian analysis of regression problems. *Biometrika*, **63**, 51–58.

Goldstein, M. (1979). The variance modified linear Bayes estimator. *J. Roy. Statist. Soc. Ser. B*, **41**, 96–100.

Goldstein, M. (1981). Revising previsions: a geometric interpretation. *J. Roy. Statist. Soc. Ser. B.* **58**, 105–130.

Goldstein, M. (1985). Temporal coherence. In: *Bayesian Statistics 2*, J.M. Bernardo et al. (ed.), 231–248. North-Holland: Amsterdam.

Goldstein, M. (1986). Exchangeable belief structures. *J. Amer. Statist. Ass.* **81**, 971–976.

Goldstein, M. (1990). Influence and belief adjustment. In: *Inference Diagrams 3*. J.M. Bernardo et al. (ed.), 189–209. Oxford University Pres: Oxford.

Goldstein, M. and O'Hagan, A. (1996). Bayes linear sufficiency and systems of expert posterior assessments. *J. Roy. Statist. Soc. Ser. B.* **58**, 301–316.

Good, I. J. (1953). The population frequencies of species and the estimation of population parameters. *Biometrika* **40**, 237–264.

Greene, W. (1990). Multiple roots of the Tobit log-likelihood. *J. Econometrics* **46**, 365–380.

Habbema, J.D.F., Hermans, J. and van den Broek, K. (1974). A stepwise discriminant analysis program using density estimation. In *Compstat 1974, Proceedings in Computational Statistics*, pp. 101–110. Physica Verlag, Wien.

Haberman, S.J. (1974). *The Analysis of Frequency Data*. Chicago: University of Chicago Press.

Haberman, S.J. (1977). Maximum likelihood estimates in exponential response models. *Ann. Statist.* **5**, 815–41.

Hall, P. (1992). *The Bootstrap and Edgeworth Expansion*. New York: Springer.

Hanfelt, J.J. and Liang, K.-Y. (1995). Approximate likelihood ratios for general estimating functions. *Biometrika* **82**, 461–477.

Hanfelt, J.J. and Liang, K.-Y. (1997). Approximate likelihood for generalized linear errors-in-variables models. *J. R. Statist. Soc. B* **59**, 627–637.

Hansen, L. P. (1982). Large sample properties of generalized method of moments estimators. *Econometrica*, **50**, 1029–1054.

Harrison, P.J. (1996). Weak probability. *Research Report*, Department of Statistics, University of Warwick.

Hartigan, J.A. (1969). Linear Bayes methods. *J. Roy. Statist. Soc. Ser. B.* **31**, 446–454.

Henrici, P. (1958). The quotient-difference algorithm. *Nat. Bur. Standards Appl. Math. Series*, **49**, 23–46.

Henrici, P. (1964). *Elements of Numerical Analysis*. Wiley, New York.

Heyde, C. C. (1997). *Quasi-Likelihood And Its Application*. Springer-Verlag, New York.

Heyde, C. C. and Morton, R. (1998). Multiple roots in general estimating equations. *Biometrika* **85**, 954–959.

Hu, F. and Kalbfleisch, J. D. (2000). The estimating function bootstrap (with discussions). *Can. J. Statist.* **28**, 449–672.

Hutton, J. E. and Nelson, P. I. (1986). Quasi-likelihood estimation for semimartingales. *Stochastic Process. Appl.* **22**, 245–257.

Huzurbazar, V. S. (1948). The likelihood equation, consistency and the maxima of the likelihood function. *Ann. Eugen.* **14**, 185–200.

Irwin, M. C. (1980). *Smooth Dynamical Systems*. Academic Press, London.

Isaacson, E. and Keller, H. B. (1966). *Analysis of Numerical Methods*. Wiley, New York.

Iwata, S. (1993). A note on multiple roots of the Tobit log likelihood. *J. Econometrics*, **56**, 441–445.

Jensen, J. L. and Wood, A. T. A. (1998). Large deviation and other results for minimum contrast estimators. *Ann. Inst. Statist. Math.* **50**, 673–695.

Johnson, N. L. and Kotz, S. (1970) *Continuous Univariate Distributions* **2**, New York: Wiley.

Kalbfleisch, J. D. and Sprott, D. A. (1970). Applications of likelihood methods to models involving large numbers of parameters (with discussion). *J. Roy. Statist. Soc. Ser. B* **32**, 175–208.

Kale, B. K. (1962) An extension of Cramer-Rao inequality for statistical estimation functions. *Skand. Aktur.* **45**, 60–89.

Kale, B. K. (1985) *Theory of unbiased statistical estimation function*. Lecture Notes, Depart. of Statistics, Iowa State University.

Kass, R. E. and Vos, P. W. (1997) *Geometrical Foundations of Asymptotic Inference* New York: Wiley.

Kaufmann, H. (1988). On existence and uniqueness of maximum likelihood estimates in quantal and ordinal response models. *Metrika*, **35**, 291–313.

Kendall, M. G. (1951). Regression, structure and functional relationship-I. *Biometrika* **38**, 11–25.

Koutrouvelis, I. A. (1980). Regression type estimation of the parameters of stable laws. *J. Amer. Statist. Assoc.* **75**, 918–928.

Kraft, C. H. and LeCam, L. M. (1956). A remark on the roots of the maximum likelihood equation. *Ann. Math. Statist.* **27**, 1174–1177.

Kullback, S. (1978). *Information Theory and Statistics*. Gloucester, Mass: Peter Smith.

Lange, K. (1999). *Numerical analysis for statisticians*. Springer, New York.

Laplace, P. S. (1814). *A Philosophical Essay on Probabilities*, unabridged and unaltered reprint of Truscott and Emory translation (1951), Dover Publications: New York.

Le Cam, L. (1969). *Théorie Asymptotique de la Décision Statistique*. Montréal: Les Presses de l'Université de Montréal.

Le Cam, L. (1979). *Maximum Likelihood: An Introduction*. Lecture Notes in Statistics No. 18. University of Maryland, College Park, Md.

Le Cam, L. (1990a). On the standard asymptotic confidence ellipsoids of Wald, *Int. Statist. Rev.* **58**, 129–152.

Le Cam, L. (1990b). Maximum likelihood: an introduction. *Int. Statist. Rev.* **58**, 153–171.

Lehmann, E. L. (1983). *Theory of Point Estimation*. Wiley, New York.

Lehmann, E. L. and Casella, G. (1998). *Theory of Point Estimation* (2nd ed). Springer.

Li, B. (1993). A deviance function for the quasi-likelihood method. *Biometrika* **80**, 741–753.

Li, B. and McCullagh, P. (1994). Potential functions and conservative estimating functions. *Ann. Statist.* **22**, 340–356.

Li, X. (1995). *An Estimating Function Approach To Credibility Theory.* Ph. D. Thesis, Dept. of Statistics and Actuarial Science, University of Waterloo.

Liang, K.-Y. and Zeger, S. L. (1986). Longitudinal data analysis using generalized linear models. *Biometrika* **73**, 13–22.

Liapunov, A. M. (1947). Problème général de la stabilité du mouvement. *Ann. Math. Study*, **17**, Princeton University Press, Princeton.

Lindley, D.V. ad Phillips, L.D. (1976). Inference for a Bernoulli process (a Bayesian view). *Amer. Statist.*, **30**, 112–119.

Lindsay, B. G. (1980). Nuisance parameters, mixture models, and the efficiency of partial likelihood estimators. *Phil. Trans. R. Soc. Lond. A* **296**, 639–665.

Lindsay, B. G. (1982). Conditional score functions: some optimality results. *Biometrika* **69**, 503–512.

Lindsay, B. and Yi, B. (1996). On second-order optimality of the observed Fisher information. Technical Report No. 95–2, Center for Likelihood Studies. Pennsylvania State University.

Liu, R. Y. (1990). On a notion of data depth based upon random simplices. *Ann. Statist.* **18**, 405–414.

Lubischew, A. (1962). On the use of discriminant functions in taxonomy. *Biometrics* **18**, 455–477.

Mandelbrot, B. (1963). The variation of certain speculative prices. *J. Business* **36**, 394–419.

Markatou, M., Basu, A. and Lindsay, B. G. (1998). Weighted likelihood equations with bootstrap root search. *J. Amer. Statist. Assoc.* **93**, 740–750.

McCullagh, P. (1991). Quasi-likelihood and estimating functions. In *Statistical Theory and Modelling, in honour of Sir David Cox.* Edited by D. V. Hinkley, N. Reid, E. J. Snell. Chapman & Hall, London, 267–286.

McCullagh, P. and Nelder, J. A. (1989). *Generalized Linear Models*, 2nd ed. Chapman & Hall, London.

McCullagh, P. and Tibshirani, R. (1990). A simple method for adjustment of profile likelihoods. *J. R. Statist. Soc. B.* **52**, 325–344.

McCulloch, J. H. (1998a). Linear regression with stable disturbances. In Adler, R, Feldman, R. and Taqqu, M., eds. *A Practical Guide to Heavy Tails*. Birkäuser, Boston, 359–376.

McCulloch, J. H. (1998b). Numerical approximation of the symmetric stable distribution and density. In Adler, R., Feldman, R. and Taqqu, M., eds. *A Practical Guide to Heavy Tails*. Birkäuser, Boston, 489–499.

McLeish, D. L. and Small, C. G. (1988). *The Theory and Applications of Statistical Inference Functions.* Springer Lecture Notes in Statistics 44, Springer-Verlag, New York.

McLeish, D. L. and Small, C. G. (1992). A projected likelihood function for semiparametric models. *Biometrika.* **79**, 93–102.

Nelder, J. A. and Mead, R. (1965). A simplex method for function minimization. *Computer J.* **7**, 308–313.

Neyman, J. (1949). Contribution to the theory of the χ^2 test. *Berkeley Symposium on Mathematical Statistics and Probability*, University of California Press.

Neyman, J. and Scott, E. L. (1948). Consistent estimates based on partially consistent observations. *Econometrica* **16**, 1–32.

Nolan, J. P. (1997). Numerical computation of stable densities and distribution functions. *Commun. Stat.: Stochastic Models* **13**, 759–774.

O'Hagan, A. (1994). *Kendall's Advanced Theory of Statistics Volume 2b: Bayesian Inference*. Edward Arnold: London.

Olsen, R. (1978). Note on the uniqueness of the maximum likelihood estimator of the Tobit model. *Econometrica* **46**, 1211–1215.

Orme, C. (1990). On the uniqueness of the maximum likelihood estimator in truncated regression models. *Economic Reviews* **8**, 217–222.

Paulson, A. S., Holcomb, E. W. and Leitch, R. (1975). The estimation of the parameters of the stable laws. *Biometrika* **62**, 163–170.

Pearson, K. (1894). Contributions to the mathematical theory of evolution. *Phil. Trans. Roy. Soc. London A* **185**, 71–110.

Perlman, M. D. (1983). The limiting behavior of multiple roots of the likelihood equation. In *Recent Advances in Statistics: Papers in Honor of Herman Chernoff on his Sixtieth Birthday*, Edited by M. Rizvi, J. Rustagi, D. Siegmund. Academic Press, New York, 339–370.

Pratt, J. W. (1981). Concavity of the log likelihood. *J. Amer. Statist. Assoc.* **76**, 103–106.

Qu, A., Lindsay, B. G. and Li, B. (2000). Improving generalised estimating equations using quadratic inference functions. *Biometrika* **87**, 823–836.

Protter, P. (1990). *Stochastic Integration and Differential Equations*, Springer Verlag, New York.

Rao, C. R. (1947). Large sample tests of statistical hypotheses concerning several parameters with applications to problems of estimation. *Proc. Camb. Phil. Soc.* **44**, 50–57.

Rao, C. R. (1973). *Linear Statistical Inference and Its Applications* 2nd ed. Wiley, New York.

Reeds, J. A. (1985). Asymptotic number of roots of Cauchy location likelihood equations. *Ann. Statist.* **13**, 775–784.

Robbins, H. (1968). Estimating the total probability of the unobserved outcomes of an experiment. *Ann. Math. Statist.* **39**, 256–257.

Savage, L. L. (1954). *The Foundation of Statistics*. John Wiley & Sons: London.

Schafer, D. W. (1987) Covariate measurement error in generalized linear models. *Biometrika*, **74**, 385–391.

Scheinerman, E. R. (1995). *Invitation to Dynamical Systems*. Prentice Hall.

Severini, T. A. (1998). Likelihood functions for inference in the presence of a nuisance parameters. *Biometrika* **85**, 507–522.

Shao, J. and Tu, D. (1995). *The Jackknife and Bootstrap*. New York: Springer.

Silvapulle, M.J. (1981). On the existence of maximum likelihood estimates for the binomial response models. *J. Roy. Statist. Soc. Ser. B.* **43**, 310–313.

Singh, A. C. and Mantel, H. J. (1998). Minimum chi-square estimating function and the problem of choosing among multiple roots. SMRD Technical paper. Statistics, Canada

Skovgaard, I. M. (1990). On the density of minimum contrast estimators. *Ann. Statist.* **18**, 779–789.

Small, C. G. and McLeish, D. L. (1988). *The theory and applications of statistical inference functions*. Lecture Notes in Statistics, **44**, Springer-Verlag, New York.

Small, C. G. and McLeish, D. L. (1994). *Hilbert Space Methods in Probability and Statistical Inference*. Wiley, New York.

Small, C.G., Wang, J. and Yang, Z. (2000). Eliminating multiple root problems in estimation (with discussions). *Statist. Sci.*, **15**, 313–341.

Small, C. G. and Yang, Z. (1999). Multiple roots of estimating functions. *Can. J. Statist.* **27**, 585–598.

Smouse, E.P. (1984). A note on Bayesian least squares inference for finite population models. *J. Amer. Statist. Ass.*, **79**, 390–392.

Solari, M. (1969). The "maximum likelihood solution" of the problem of estimating a linear functional relationship. *J. R. Statist. Soc. B* **31**, 372–375.

Starr, N. (1979). Linear estimation of the probability of discovering a new species. *Ann. Statist.* **7**, 644–652.

Stefanski, L. A. and Carroll, R. J. (1985) Covariate measurement error in logistic regression. *Ann. Statist.*, **13**, 1335–1351.

Stefanski, L. A. and Carroll, R. J. (1987). Conditional scores and optimal scores for generalized linear measurement-error models. *Biometrika* **74**, 703–716.

Stone, M. (1963). Robustness of non-ideal decision procedures. *J. Amer. Statist. Ass.*, **58**, 480–486.

Stuart, A. (1958). Note 129: Iterative solutions of likelihood equations. *Biometrics*, **14**, 128–130.

Stuart, A. and Ord, J. K. (1991). *Kendall's Advanced Theory of Statistics, Vol.2, Classical Inference and Relationship.* Edward Arnold, London.

Tzavelas, G. (1998). A note on the uniqueness of the quasi-likelihood estimator. *Statist. Prob. Letters* **38**, 125–130.

Wald, A. (1949). Note on the consistency of the maximum likelihood estimate. *Ann. Math. Statist.* **20**, 595–601.

Wang, C.Y. and Pepe, M. S. (2000). Expected estimating equations to accommodate covariate measurement error, *J. Roy. Statist. Soc. Ser. B*, **62**, 509–524.

Wang, J. (1999). Nonconservative estimating functions and approximate quasi-likelihoods. *Ann. Inst. Statist. Math.* **51**, 603–619.

Wedderburn, R. W. M. (1974). Quasi-likelihood functions, generalized linear models, and Gauss-Newton method. *Biometrika* **61**, 439–447.

Wedderburn, R. W. M. (1976). On the existence and uniqueness of the maximum likelihood estimates for generalized linear models. *Biometrika* **63**, 27–32.

White, H. (1982). Maximum likelihood estimation of misspecified models. *Econometrica*, **50**, 1–25.

Whittle, P. (1958). On the smoothing of probability density functions. *J. Roy. Statist. Soc. Ser. B*, **20**, 334–343.

Wilks, S. S. (1938). Shortest average confidence intervals from large samples . *Ann. Math. Statist.* **9**, 166–75.

Wolfowitz, J. (1957). The minimum distance method. *Ann. Math. Statist.* **28**, 75–88.

Yanagimoto, T. and Yamamoto, E. (1991). The role of unbiasedness in estimating equations. In *Estimating Functions*, Ed. Godambe, V. P., pp. 89–101. Clarendon, Oxford.

Index